Economy & Environment

VOLUME 10

The titles published in this series are listed at the end of this volume.

Environmental Options: Accounting for Sustainability

Environmental Options: Accounting for Sustainability

by

Kimio Uno
Keio University at Shonan Fujisawa,
Fujisawa, Japan

KLUWER ACADEMIC PUBLISHERS
DORDRECHT / BOSTON / LONDON

A C.I.P. Catalogue record for this book is available from the Library of Congress.

ISBN 0-7923-3513-9

Published by Kluwer Academic Publishers,
P.O. Box 17, 3300 AA Dordrecht, The Netherlands.

Kluwer Academic Publishers incorporates
the publishing programmes of
D. Reidel, Martinus Nijhoff, Dr W. Junk and MTP Press.

Sold and distributed in the U.S.A. and Canada
by Kluwer Academic Publishers,
101 Philip Drive, Norwell, MA 02061, U.S.A.

In all other countries, sold and distributed
by Kluwer Academic Publishers Group,
P.O. Box 322, 3300 AH Dordrecht, The Netherlands.

Printed on acid-free paper

Printed in the Netherlands

To Yoshiko

Contents

Preface

The concept of sustainability was the consensus reached at the United Nations Conference on Environment and Development held in June 1992. The operational framework to put this concept into policy action is yet to be devised.

This study aims at constructing a systematic statistical framework concerning environment, technology, economy, and society, and carrying out a series of analyses regarding the impact of human activities on environment. Environment (energy consumption, resource use, air and water pollution, land use), technology (conservation of energy and resources, recycling, pollution prevention), economy (industrial structure, input structure, transportation, final demand), and society (employment, income and consumption, diffusion of consumer durables including motor vehicles, lifestyle, value) are all intertwined. It is therefore necessary to develop a statistical system encompassing these areas in order to operationalize the concept of "sustainable growth". The scope of the work includes social and technological variables and is wider than the framework suggested by the UN manual on an integrated environmental economic accounting. In fact, I believe that a wider scope is needed to avoid the pitfall of posing the issue as an alternative between environment and economy. We will have to seek technological progress and changes in lifestyle in order to make our society an environmentally sustainable one.

My approach to the environmental issue has been to seek a common point of reference where information is the basis of individual choice which collectively makes up social choice. This is particularly important in environmental spheres because of externalities involved and intergenerational equity problem in an attempt to achieve sustainability. For this purpose, it is important to provide a picture not only at the aggregate level but also at disaggregated, individual industry level and for various social groups. Equally important, the framework has to allow for international linkages. Environmental issue has a global span whereas our policy machinery is traditionally developed with national constituency.

For this purpose, I have based the analysis on a multisectoral framework with industrial details. The approach maintains the criteria which I have spelled out above. It also maintains consistency with my previous work dealing with sources of growth, service orientation of the economy, and technological change.

There are many attempts dealing with environmental issue, and a question may be asked what my contributions are.

First, this volume is among the early attempts to examine the feasibility of the System for Integrated Environment and Economic Accounting (SEEA) proposed by the United Nations. Whereas the UN's framework is focused on integrating environment and economy, mine is broader in including social and technological factors. It is hoped that the scheme I am proposing will serve as a common yardstick in formulating environmental policies in developed as well as less developed countries and set standards for international comparisons. It should be noted that the development of the System of National Accounts itself points to the inclusion of environmental spheres as an integral part. In the 1993 revision of the SNA, environment is included as a satellite account. The framework adopted here also constitutes an important expansion of input-output analysis in environment, material flow, and lifestyle while preserving the theoretical consistency.

Second, pertaining to the policy application of the accounts, there have been attempts to estimate "green GNP". As is widely recognized, conventional macroeconomic performance indicators such as GNP, does measure positive output but fails to take account of negative side effects. "Green GNP" can be conceived of as an effort to arrive at some kind of economy-wide cost-benefit analysis of growth, where cost in this case is represented by environmental degradation. Along this line of thinking, the present study derives measures of economic wellbeing.

Third, the volume introduces Japanese experience in environment-related fields. Japan's experience is unique in that it compresses industrial expansion and urbanization in a short time span of 40 years or so which is fully captured by various statistics, providing a valuable case study in a global context where many societies are about to take off in their industrialization drive. Japan has accumulated a pioneering experience in energy conservation, pollution prevention, and other related fields, while witnessing a radical change in values and lifestyle of the people. In this sense, the Japanese experience goes beyond a case study of a specific country. This study intends to construct a statistical framework pertinent to these aspects and to carry out quantitative analysis of policy scenario leading to sustainable growth.

The book consists of five parts. Part 1 spells out the policy issues to be covered by the statistical framework which is intended to operationalize the concept of sustainability. Part 2 deals with energy. Following a discussion on policy options concerning energy supply, energy use in industry, household, and transportation is examined. Part 3 takes up pollution prevention investment and how it relates to treatment capacity. A need for a shift from produce-consume-and-forget regime to produce-consume-and-recycle regime is suggested. Research and development in pollution prevention and energy fields are then discussed. Part 4 focuses on the changes in land use and international linkages through trade in resources. Part 5 includes a controversial chapter which examines the relation between the quality of life and envionment. It is important to note that the concept of econmic wellbeing defined in this chapter is derived, for the most part, from the System of National Accounts. The final chapter examines a comprehensive statistical framework encompassing environment, economy, technology, and society. This chapter is closely related to the Integrated Environmental and Economic Accounting proposed by the United Nations and satellite accounts on environment contained in the Revised System of National Accounts also by the United Nations. In fact, various topics in previous chapters fit into the comprehensive framework proposed in this chapter.

Finally, I want to mention the assistance of my supporting staff, in particular Keiko Yamahira and Terumitsu Koshimizu.

Part of this research was supported by Grant-in-Aid for Scientific Research entitled "Diffusion of Advanced Technology and Sustainability of the Global Community" (grant number 04210126). This was a project conducted during the 1990-1993 period as one of scientific research on priority areas entitled "Perspectives of Advanced Technology Society" organized by the Research Center for Advanced Science Technology, University of Tokyo under the sponsorship of the Ministry of Education, Science, and Culture. In 1993, this project was supported by Grant-in-Aid for Scientific Research entitled "Analysis of Impacts of Human Activity based on an Integrated Statistical System on Environment, Technology, Economy, and Society" (grant number 05278119) under the priority area "The Man-Earth System" organized by the Institute of Industrial Science, University of Tokyo. The final phase of the project was supported by Keio University Fukuzawa Memorial Fund for Education and Research.

Shonan Fujisawa Campus

May 8, 1994

Kimio Uno

List of Tables

List of Figures

Abbreviations

ATR: Advanced thermal converter reactor
bbl: Barrels
BOD: Biological oxygen demand
BWR: Boiling water reactor
C2O: Nitrous oxide
Cd: Cadmium
CFC: Chlorofluorocarbons
CH4: Methane
CHP: Combined heat and power
CNG: Compressed natural gas
CO: Carbon monoxide
CO2: Carbon dioxide
COD: Chemical oxygen demand
Cr: Chromium
dB: Decibel
dBA: Decibel A-weighted
DID: Densely inhabited districts
EC European Community
FAO: Food and Agriculture Organization of the United Nations
FBD: Fluidised bed combustion
FBR: Fast breeder reactor
Fe: Iron
FGD: Flue gas desulphurization
g: Grams (1g = 0.0353oz.)

GDP:	Gross domestic product
GJ:	Giga joule
ha:	Hectares (1ha = 1km-2)
HC:	Hydrocarbons
Hg:	Mercury
HM:	Tonnage of heavy metal contained (nuclear spent fuel)
IAEA:	International Atomic Energy Agency
IEA:	International Energy Agency
IMF:	International Monetary Fund
IPCC:	The Intergovernmental Panel on Climate Change
kcal:	Kilocalories
kg:	Kilograms (1kg = 1000g = 2.2046lb.)
kl:	Kiloliters (1kl = 6.29bbl.)
km:	Kilometers (1km =0.6214ml.0)
km2:	Square kilometers (1km2 = 100ha = 0.3861mi.2)
kwh:	Kilo watt hours
l:	Liter
LEQ:	Equivalent sound level
LNG:	Liquified natural gas
m2:	Square meters
m3:	Cubic meters
mg:	Milligrammes (1mg = 10-3g)
MJ:	Mega joule
Mt:	Million metric tons
MTOE:	Million tons of oil equivalent
Nm3:	Normal cubic meters
NOX:	Nitrogen oxides
NO3:	Nitrates
ODA:	Official development assistance
OECD:	Organisation for Economic Cooperation and Development
PCBs:	Polychlorinated biphenyls
PET:	Polyethylene terephtalate resin
pH:	Hydrogen power

POM:	Polycyclic organic matter
ppbv:	1/billion in volume
ppmv:	1/million in volume
PPP:	Polluter pays principle
pptv:	1/trillion in volume
PWR:	Pressurized water reactor
SCR:	Selective catalytic reduction
SOX:	Sulphur oxides
st:	Short tons
t:	Metric tons
TFC:	Total final consumption
TPER:	Total primary energy requirements
TOE:	Ton of oil equivalent
Tons:	Metric tons
UNCED:	United Nations Conference on Environment and Development
UNEP:	United Nations Environment Program
WHO:	World Health Organization

Table A. Conversion tables

a. Conversion to kcal (kilo calories)

Energy source	Unit	Heat value (kcal)
Electricity	kWh	(efficiency 38.1%) 2250
Petroleum		
Crude oil	l	9250
Benzine	l	8400
Naphtha	l	8000
Kerosine	l	8900
Jet fuel	l	8700
Light oil	l	9200
Heavy oil A	l	9300
Heavy oil B	l	9600
Heavy oil C	l	9800
LPG	kg	12000
Refinery gas	m3	9400
Other petroleum products	l	10100
Natural gas	m3	9800
LNG	kg	13000
Furnace gas	m3	800
Coke oven gas	m3	4800
City gas	m3	10000
Lignite	kg	4100
Charcoal	kg	7000
Coal		
Domestic	kg	5600 - 5800
Imported coking coal	kg	7600
Imported general	kg	6200
Cokes	kg	7200

b. Volume

kl	bbl	1000 gal
1	6.29	0.264
0.159	1	0.042
3.79	23.8	1

c. Heat value

MJ	kwh	kcal	kl	TOE	BTU
1	0.278	239	0.0258	0.0239	948
3.60	1	860	0.0930	0.0860	3412
0.00419	0.00116	1	$1.08 * 10^7$	10^{-7}	3.97
$3.87 * 10^4$	$1.08 * 10^4$	$9.25 * 10^6$	1	0.925	$3.67 * 10^7$
$4.19 * 10^4$	$1.16 * 10^4$	10^7	1.08	1	$3.97 * 10^7$
0.00155	$2.93 * 10^{-4}$	0.252	$2.72 * 10^{-5}$	$2.52 * 10^{-5}$	1

Note: Converted as 9250kcal per litre, 1 TOE = 10^7cal.

Table B. The yen exchange rate

(unit: yen per currency unit)

Currency Year	US$	German mark	French franc	British pound	ECU
1970	358.1	98.20	64.78	857.8	—
1971	349.3	100.07	63.03	850.4	—
1972	303.2	95.08	60.04	758.5	—
1973	271.7	101.66	60.95	666.3	—
1974	292.1	112.87	60.67	683.2	—
1975	296.8	120.63	69.24	659.4	—
1976	296.6	117.77	62.05	535.6	—
1977	268.5	115.63	54.65	468.7	—
1978	210.4	104.77	46.63	403.9	268.1
1979	219.1	119.56	51.51	464.9	293.0
1980	226.7	124.74	53.66	527.5	315.4
1981	220.5	97.58	40.58	447.2	246.5
1982	249.1	102.65	37.90	436.0	244.4
1983	237.5	93.02	31.16	360.3	211.7
1984	237.5	83.46	27.18	317.4	187.4
1985	238.5	81.03	26.55	309.2	181.8
1986	168.5	77.61	24.33	247.2	165.4
1987	144.6	80.47	24.06	237.1	167.0
1988	128.2	72.97	21.51	228.3	151.7
1989	138.0	73.38	21.62	226.2	152.1
1990	144.8	89.61	26.59	258.4	184.3
1991	134.7	81.18	23.88	238.4	167.1

Source: IMF, International Financial Statistics.

Part I

What is at Issue

Chapter 1

Global Environment and Governance

1.1 AGENDA21 and beyond

The role of the North is indisputable in overcoming the global environmental problem. Industrial development since the 18th century has occurred solely in the North, albeit at varying paces. Rapid industrialization is taking place in the South, but this is a fairly recent phenomenon. Thus, the North is largely responsible for the vast increase in CO_2 emission which resulted, first, from steam engine and iron smelting based on coal and, second, from electric power generation, metallurgy, and automobile transportation. If these remain the key technologies in the global society, environmentally sound development is indeed untenable. But the North also has at its disposal new technology which imposes less of a burden on the global resource base and environment: information technology based on micro-electronics, new materials, and biotechnology are examples. These technologies have emerged in the late 20th century, and they will change the development pattern in the 21st century. It is still true that at a certain stage of economic development, the demand for energy and steel tends to expand very rapidly. But in this field as well, the North has already produced resource-saving technology.

The global community is plagued by wide income differentials, and we have to create a policy scenario which enables us to narrow this gap while

Table 1-1 The principal greenhouse gases

	CO_2	CH_4	CFC-11	CFC-12	N_2O
Concentration:	ppmv	ppmv	pptv	pptv	ppbv
pre-industrial	280	0.79	0	0	280
present	353	1.72	280	484	310
Radiative forcing per molecule (CO_2 as 1)	1	21	12400	15800	206
Lifetime in atmosphere (year)	50-200	10	65	130	150
Global warming potential relative to CO_2					
20 years	1	63	4500	7100	270
100 years	1	21	3500	7300	290
500 years	1	9	1500	4500	190
Contribution to total radiative forcing, 1980-1990 (%)	55	15	all CFCs 24		6
Current increase (percent/year)	0.5	0.9	all CFCs 4		0.25

Note: The warming effect of an emission of 1 kg of each gas relative to CO_2 based on the present day atmosphere.
ppmv=1/million in volume; ppbv=1/billion in volume;
pptv=1/trillion in volume.

Source: Data from IPCC Working Group 1.

safeguarding the environment at the regional and global levels.

Discrepancies have widened between convention in macroeconomic accounting and what is relevant to today's society, which is beleaguered by environmental disruption and resource exhaustion. In fact, the whole realm of environmental issue spans such diverse aspects as:

a. contamination of air, water, and soil;
b. disposal of industrial and other wastes;
c. resource depletion;
d. global warming;
e. destruction of the ozone layer in the atmosphere;
f. soil erosion;
g. deforestation;

h. endangered species;
i. urban congestion;
j. change in land use; and
k. destruction of scenic beauty, among others.

It is widely recognized that these phenomena are related to economic activities in production, transportation, and consumption. In many cases, economic development invites deterioration of environmental quality in one sense or another. The environmental impact of human activities usually goes unrecorded, whereas an increased level of current production is often used as a measure of socio-economic performance. An accounting system which keeps track of the positive side of economic development while neglecting its negative side is not particularly desirable as it leads to biases in our policy formulation. In order to describe each of these aspects sufficiently, we would need a much wider framework than one which can be adequately dealt with in a single book. I therefore have confined myself to narrower topics, namely (a)(b)(i) and (j), and examined the methodology of establishing linkages between the macroeconomic statistical framework and these increasingly important aspects. Contributing factors to (d) also come within this scope, although global implications cannot be drawn until similar data are made available for the entire world.

It is recognized today that the composition of the earth's atmosphere is a key factor in determining its surface temperature and climate. Solar radiation penetrates the atmosphere and warms the earth, while invisible infrared radiation is emitted by the earth and cools it down. There is increasing concern about the greenhouse effect, in which this radiation is trapped in the atmosphere by trace gases, causing global warming.

Although water vapor is the most important natural greenhouse gas in the atmosphere, man-made additions are believed to have triggered the warming process. These include carbon dioxide (CO_2), methane (CH_4), nitrous oxide (N_2O), and the halocarbons such as chlorofluorocarbons (CFC)s. CO_2 is now seen as contributing 55% to the total radiative force as shown in Table 1-1. CO_2 levels began to rise in the 19th century with the Industrial Revolution, which was realized through greater use of fossil fuels. Methane levels have also been rising since the beginning of the 19th century. There is a feedback mechanism: atmospheric warming brought about by man-made gases increases water evaporation, thus allowing the atmosphere to contain more vapor, and leading to further warming.

According to an IPCC (the Intergovernmental Panel on Climate Change estimate, the earth's average temperature has risen 0.3 to 0.6 degrees centigrade during the past 100 years, and as a result, the sea level has risen by as much as 10 to 20 cm. It is still being vigorously debated whether this is solely attributable to global warming. However, consensus has been reached in the global community to adopt a 'no-regret' scenario. Rather than wait for scientific understanding of all the factors which may be contributing to these complex phenomena, we had better start adjusting our mode of production and transportation and way of life, which may be resulting in irreversible damage.

In order to avoid global warming, we need to control fossil fuel burning. It would not be practical, however, to put the issue as a choice between growth or no growth, especially when serious income differentials persist in the world.

"Sustainable development is development that meets the needs of the present without compromising the ability of future generations to meet their own needs. It contains within it two key concepts:
the concept of 'needs', in particular the essential needs of the world's poor, to which overriding priority should be given; and
the idea of limitations imposed by the state of technology and social organization on the environment's ability to meet present and future needs.
Thus the goals of economic and social development must be defined in terms of sustainability in all countries — developed or developing, market-oriented or centrally planned. Interpretations will vary, but must share certain general features and must flow from a consensus on the basic concept of sustainable development and on a broad strategic framework for achieving it." [The World Commission on Environment and Development, p.43.]

The concept of sustainable development became a central theme at the Earth Summit which took place in Rio de Janeiro in 1992. The world community has accepted the concept as a matter of fact, although policy measures leading to sustainability are far from clearly defined.

The sustainability concept includes a North-South issue, as has been pointed out by the Brundtland report. The South accuses the North, in its industrialization drive since the Industrial Revolution , of using up natural resources and the assimilative capacity of nature, which are the common assets available to human beings. The North accuses the South of destroying the global environment and exhausting natural resources while preoccupied with short-term economic gains. Centrally planned economies have collapsed, and the East-West confrontation of the Cold War era has been reduced to something resembling the North-South problem.

From a policy point of view, it is imperative to come up with an accurate understanding of the various issues plaguing the world community, such as demographic trend, food production, energy, industrialization, urbanization and lifestyle, poverty and unemployment, capital formation, research and development, international trade and investment, technology transfer, pollution, and global environment, taken individually and, more importantly, as an integral whole. No one can stand outside this vital issue in a world which has become increasingly interdependent.

It is also important to realize that sustainability entails a potential conflict of interest between the present generation and the generations to come. In this respect, both North and South undoubtedly belong to the present generation. They may jointly take advantage of the future generations for their own interest. Future generations will be left with the aftermath of whatever the present generation does today for the sake of betterment of their well-being.

Resource exhaustion in an example. Natural assets (stock) are being transformed into income (flow), part of which is channeled into the formation of tangible capital (stock). On the surface, one type of asset is transformed into another, maintaining the society's portfolio intact. It is also true that endowment of natural resources does not generate income or employment, whereas industrial capital will. But, in the long run, tangible capital stock will wear out or become technologically obsolete. In other words, the formation of tangible capital stock is not necessarily conducive to sustained growth. It must be supplemented by the formation of intangible assets (technology) and human assets (health, education, skill).

Another example is nuclear energy. Because of the concern today about global warming, a reduction of greenhouse gas emission is being sought. To this end, CO_2 emission has to be contained by restraining fossil fuel burning. Hence, it is concluded, we must increase our dependency on nuclear energy. This will enable both North and South to pursue their immediate policy goals. The other side of the coin is that future generations will inherit nuclear wastes which will need careful management for a period which exceeds recorded human history. "... The radioisotope plutonium-239, for example, with a half-life of 24,400 years, is dangerous for a quarter of a million years, or 12,000 human generations. And as it decays it becomes uranium-235, its radioactive 'daughter,' which has a half-life of its own of 710,000 years" [Lenssen, 1992, p.50]. The Nuclear Energy Agency observed that "for all reactor types considered the volume of decommissioning waste is about the same order of magnitude as the volume of wastes from their operations during a 25-year period." [Nuclear Energy Agency, 1986, p.34] Although it is true that the amount of decommissioning wastes generated will be small over the next decade, their share of the total wastes will become significant in the future. There are numerous other examples such as the production and consumption of CFCs and other chemicals which generate ozone holes, erosion of soil due to deforestation, etc.

The above observations suggest that the environmental issue poses a real challenge to the governance of the global community in terms of policy agents and policy tools. Is a national government, which is chosen through democratic process but in which future generations are not properly represented, ready to respond to this challenge? Is a national government, which is responsible largely for domestic issues, ready to respond to global concerns? Is the economists' standard tool of cost-benefit analysis, discounting the future income and cost to the present value, still appropriate for this particular challenge? Do we not tend to discount the future rather rapidly? Given the extreme income differentials between the North and the South, how much should we rely on market solutions? Are the market prices still viable indicators of resource allocation where various externalities are involved in international trade of environment resources?

It has been pointed out that the government, caught between the micro-standpoint of businesses and pressure groups (small and medium businesses, agriculture, etc.), has become increasingly subject to political constraints. With the liberalization of trade and capital transfer, the influence any one government can exercise on an economy has become limited. Now, the global environmental issue seems to constitute yet another field where the effectiveness of our policy institution is put under strain.

1.2 Technology and society in an environmental accounting

Compared with the challenge in front of us, the scope of this book is very modest. But I believe this step is indispensable in describing the environmental issue in operational terms. As it stands today, there is no such framework. It is probably useful to distinguish an accounting aspect of the framework from its use. The accounting framework describes the state of the environment for a particular point in time (environment-related activities within a year, for example). The dynamic change of environment and related activities can be described by time-series comparison of such a framework. One example of indicators which can be derived from the accounting framework would be what is called "Green GNP." The relationship among various variables in the framework can be explained by theory and empirically described by econometric "models." Policy prescription can be arrived at by examining alternative scenarios. The purpose of this book is to suggest one such framework.
The relationship among various factors can be formulated as follows:[1]

$$\frac{Z}{A} = \frac{N}{A} \cdot \frac{Y}{N} \cdot \frac{E}{Y} \cdot \frac{Z}{E}$$

where A: geographical area,
E: resources (e.g. energy)
N: population,
Y: income (product), and
Z: pollutants.

Thus, environmental limit is definitionally expressed as the multiple of (1) population density, (2) per capita income, (3) resource intensity, and (4) efficiency of resource use.

The follow-up of the Earth Summit at the global, national, and regional levels has to be based on a common framework. The framework should allow aggregation/disaggregation among various aspects of the issue. It should provide a dynamic picture in which the activities today can exert an influence on the future generations. In discussing sustainability, it is essential to have a parallel framework dealing with both monetary and physical magnitudes.

Figure 1-1 Scope of the SEEDS (Social, Economic, and Environmental Data Set)

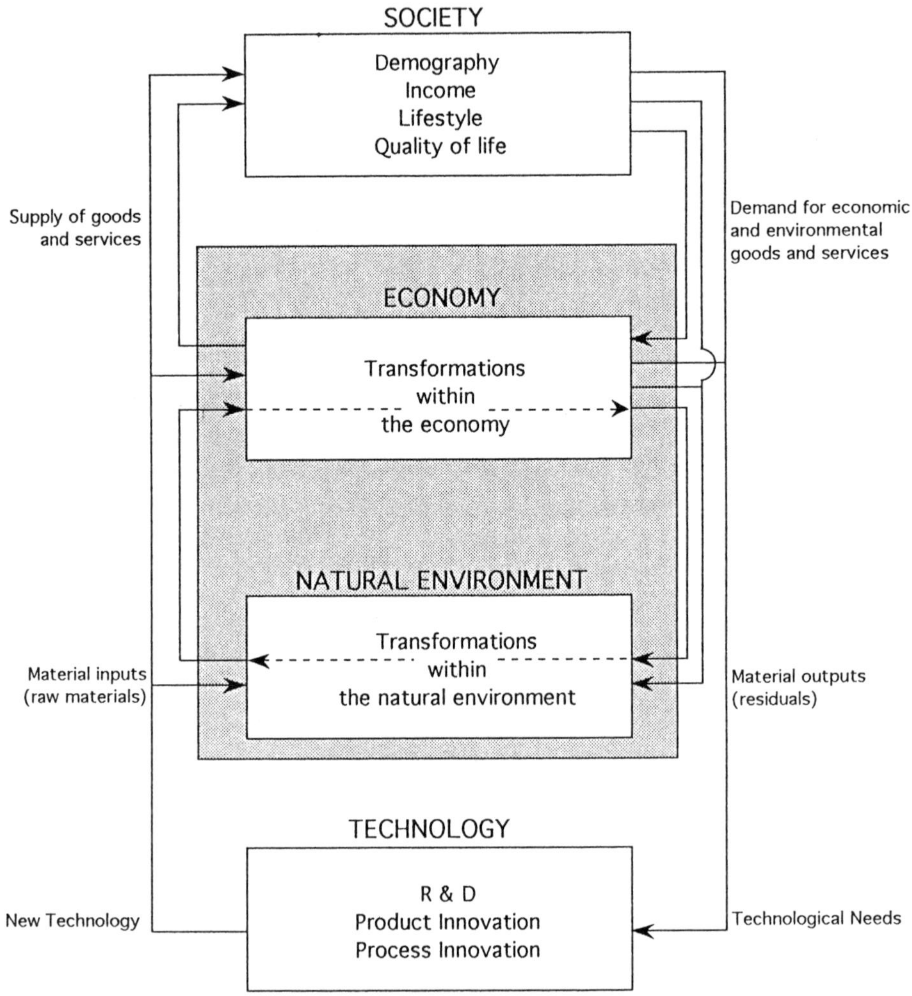

Note: The shaded area shows the scope of the UN's Satellite System for Integrated Environmental and Economic Accounting (SEEA). The figure has been adopted from the United Nations [1992] and modified by the present author.

One such framework is being proposed by the United Nations. Figure 1-1 presents the scope of the proposed scheme. It consists of establishing linkages between the economy and the natural environment, which is represented by the shaded area in the diagram. The framework, which is called a System of Integrated Environmental and Economic Accounts (SEEA), came out in 1992 in the form of a draft manual, and individual countries are being asked to try it out based on national experiences.[UN Department of Economic and Social Development, 1992]

"Materials/energy balances provide detailed information on the material inputs of an economy delivered by the natural environment, of their transformation and use in economic processes (extraction, conversion, manufacturing, consumption) and their return to the natural environment as residuals (wastes etc.). Their accounting concepts are firmly founded on the first law of thermodynamics stating that matter (mass/energy) is neither created nor destroyed by any physical process. Economic activities can be described as throughputs of materials/energy which do not change mass but which increase unavailable energy (second law of thermodynamics, law of entropy)."[UN Department of Economic and Social Development, 1992, p.115]

A detailed list of the items included in the UN's SEEA is provided in Table 1-2 together with suggested priorities for implementation in developed and developing countries.

The framework consists of three parts:

(1) Opening stock which describes the state of man-made assets (such as plant and equipment, social facilities, consumer durables , and housing) as well as environmental assets (stock of natural assets such as minerals and forests, the quality of environment including air, water, and soil, and land use).
(2) Current activities which include investment and consumption in the conventional sense and the use of environmental resources in various respects.
(3) Closing stock with identical scope as (1) above, with adjustment for losses due to natural disasters, new discoveries, and price fluctuations.

The account covers both monetary and physical spheres as far as possible. It is translated into matrix form, individual entries of which can be further

Table 1-2 The scope of the SEEA

		Physical acc.		Monetary acc.	
	Environmental issues	Developed country	Developing country	Developed country	Developing country
1.	Use of natural assets (except discharge of residuals):				
	Depletion of assets				
1.1	Biological	++	+++	++	+++
1.2	Subsoil assets	++	+++	++	+++
1.3	Water	+	+++	++	+++
	Degradation of land and landscape				
1.4	Restructuring (urbanization, changes in land use)	+++	+++	++	+
1.5	Agricultural use (soil erosion)	+	+++	+	+++
1.6	Recreational use	++	++	++	++
2.	Product flow analysis	+++	+	+	+
3.	Discharge of residuals				
3.1	Waste and land contamination	+++	+	++	++
3.2	Waste-water	+++	++	++	++
3.3	Air pollution	+++	++	++	++
4.	Actual environmental costs				
4.1	Environmental protection activities			+++	++
4.2	Damage costs			++	+

Note: +++: high priority, ++: medium priority, +: low priority.

Source: United Nations [1992], p. 56f.

disaggregated as the need develops. Environmental and economic changes over time can be described by a time series of such accounts.

In order to better suit the spirit of UNCED AGENDA 21, it is proposed to expand the SEEA in several ways. The framework here is wider in its scope than the ones presented above in explicitly focusing on (1) the social or human aspect and (2) the technology factor in addition to the interaction between (3) the economy and (4) the natural environment.

The human aspect is an important factor in the global environmental crisis, and may also prove important in finding a solution to the problem. This group of variables includes demography, income, lifestyle, and quality of life, among others.

Technology relates to the way in which the economy, environment, and society interact. We should note that technology has been changing over time, sometimes aggravating the environmental impacts of human activities and sometimes alleviating them. It is increasingly important to keep a watchful eye on the direction of technological change . The guiding principle has to be 'no regret' for future generations.

Focus on the human or social aspect has been stressed in AGENDA 21. It has been made clear that the primary objective of the effort should be placed on protecting human life and opening up human options. The proposed framework already spells out the household sector more explicitly than conventional economic accounts, both in terms of monetary and physical measures.

In order to describe the state of human welfare more explicitly, however, we should add another account with regard to the services of housing, consumer durables, and social capital stock (schools, medical facilities, transportation, etc.). Figure 1-1 makes this point. The SEEA is a framework focusing on the interaction between the economy and the natural environment. In contrast, the framework which is proposed here has two additional dimensions, social aspect (demography, income, lifestyle, quality of life, etc.) and technological aspect (research and development, process innovation, product innovation, pollution abatement, energy conservation, technology transfer, etc.).

Environmental damage such as degradation of air, water, and soil, exhaustion of resources, emission of greenhouse gases, etc., will have to be deducted from the total value added. Household production will have to be added in. Such a framework has been tested in some countries including the USA and

Japan. Private consumption in the conventional sense will have to be categorized by its purpose, such as education, health, energy, transportation, waste disposal, recreation, etc., and collective consumption will also have to be subdivided in order to provide an overall picture of resource allocation pertinent to the quality of life. Changing "lifestyle" is an important facet of social structure which can be captured in this framework in operational terms. As a matter of fact, this aspect is already included in the current version of the SNA, but its implementation is lagging in most countries. If this is done, then we can start discussing the state of the environment stock, the educational stock and health stock of a nation together with the more conventional housing, consumer durables, and productive capital stock.

The technology aspect is also explicitly included in the framework. It is very important in discussing the environment because existing practice varies drastically among nations, suggesting an enormous potential for improvement based on existing technology alone. New technologies can help in drastically reducing the input of energy and output of pollutants per unit of production, while not curtailing the production potentials and quality of life. This is pertinent to production technology, transportation, pollution prevention, energy transformation from primary to secondary and to final use, among others. Since the accounting framework is close to that of input-output tables as far as its production sphere is concerned, the input structure of a specific industry comes close to describing the production technology of the sector in question. This is also where monetary variables can be compared directly with physical variables.

Policy tools have to be embedded in the framework. It becomes necessary to provide appropriate incentives which reflect the correct value of the environment in order to influence individual decision-making. This may take the form of the carbon tax, subsidies, or anything else deemed appropriate from the theoretical point of view. The proposed framework, just like the existing input-output framework, is capable of incorporating such policy tools because it distinguishes indirect tax and subsidies for individual sectors, in addition to inputs of capital and labor in the production processes. Since the framework is consistent with existing economic accounts, one can derive a fiscal balance (revenues vs. expenditures), trade balance (exports vs. imports), and saving-investment balance. The sectoral demand-supply gap can also be calculated.

This will help identify feasible policy packages and unrealistic ones.

Lastly, the global aspect is encompassed in the framework. This can be achieved by linking country accounts by trade matrices pertinent to environmental issues. Examples are forest products, fossil oil and gas, and mineral resources. Such matrices are readily available from the UNSO. Country accounts will have to distinguish the use of environmental resources from domestic and foreign sources. Economic activities in one country may result in the depletion of environmental resources in others through international trade. Exporting of resources, on the other hand, may enhance income levels in the current period while exhausting future income. In the same vain, the international flow of financial resources in the forms of official development assistance (ODA) and direct foreign investment may be captured and integrated among the sources of financial resources available to a particular country.

1.3 Mapping AGENDA21 in the proposed framework

Various issues spelled out in the UNCED AGENDA 21 will have to be fitted into the proposed framework, which is composed of integrated environmental-economic accounts, trade matrices, and quality of life accounts. Depending on the issue, more detailed accounts may be needed to provide vivid pictures, but this can be treated as part of the matrix being disaggregated further and would not invalidate the integrity of the conceptual framework. Some missing links will have to be filled by statistical data and/or scientific research.

Data pertinent to individual countries will have to be prepared together with the ones on global resource endowment, etc. There is a large body of published data, although the sources are widely scattered. In some cases data are not readily available, and a temporary solution would be to fit comparable ones from neighboring countries or from industrial averages, etc.

Policy analysis can then be conducted regarding the sustainability of the global environmental resource base and the sustainability of the quality of life, two major issues spelled out in AGENDA 21. This can be done for a single country, for a group of countries (Asia, Africa, Latin America, Middle-East; developed vs. developing), or for the global community as a whole.

Some major items taken up in the UNCED AGENDA 21 are listed below together with their possible treatment in the proposed framework. They constitute the most important research topics as well as being urgent policy issues.

(1) *The definition and measurement of sustainable development* : The sustainability concept is examined from two perspectives in the proposed framework: the sustainability of the global environmental resources and the sustainability of quality of life. For the former, existing environmental stock or assets are explicitly recognized both in physical and monetary terms. Socioeconomic activities use up environmental assets in various forms, such as energy consumption, degradation of air and water, and transformation of forests and grass land into urban-type uses. Thus, the framework treats man-made assets and natural assets in a consistent manner.

For the latter, household assets (stock of consumer durables and housing) and activities (both marketed and nonmarketed) are included. The services rendered by social capital stock are also recorded. Damages attributable to environmental degradation are measured based on actual damage cost or on imputation and included in the framework. Thus, rather than looking at the level of current consumption alone, the framework describes a multifaceted "quality of life". The methodology has its precedence in such approaches as measures of economic welfare, social indicators, and total consumption of the population.

(2) *Integrating the environment and development in decision-making*: One of the serious drawbacks of conventional measurement of macroeconomic performance is that it does not recognize the fact that the growth in the current period is sometimes realized through depletion and/or degradation of environmental assets . Such "growth" is not sustainable. The proposed framework avoids pitfalls such as this by explicitly taking into account reproducible tangible wealth (structure and durable equipment), natural resource wealth (the stock of mineral, forest, water, climate assets), human wealth (the stock of skills, knowledge, and health), and reproducible intangible wealth (the stock of technology). Resource allocation such as government expenditures and personal consumption is reclassified according to purpose (education, health care, and research are examples), making some of the items "investment" in a wider sense.

From the point of view of the environment, activities which have been neglected in the conventional economic accounting need to be recognized explicitly. Take, for instance, the household sector which typically accounts for 1/3 of total energy consumption and 1/2 of transportation. For this reason alone, housing conditions, the stock of automobiles owned by the household sector (which in a conventional framework appears as a part of current consumption only), and other durable goods warrant a more appropriate treatment. This, incidentally, makes the account more consistent, contributing to more socially sound decision-making.

(3) *North-South interaction and sustainable development*: The proposed framework is first developed for individual countries which are then linked internationally through trade matrices. The trade matrices reveal importers and exporters of different commodities including, for example, wood products, energy, and metals. Thus, for instance, if the North consumes natural resources which are balanced by imports, the repercussion can be traced across national boundaries. On the side of the exporters, this will generate production and create employment, while the stock of environmental resources will be depleted. There may be other unintended consequences, such as soil erosion, flooding, and damage to biodiversity.

International linkages of the framework will also enable examination of the consequences of economic growth. While this will certainly improve the income levels of the population involved, it will undoubtedly be accompanied by environmental consequences, unless efficiency is increased at the same time. With an explicit inclusion of production technology in the proposed framework (i.e., represented by the input structure similar to the input-output tables), the effect of shifting to a more efficient technology can be demonstrated in a consistent manner. The case for technology transfer can be made more persuasive.

(4) *Poverty, demographic dynamics, and sustainability*: The framework pertaining to the quality of life can be appropriately disaggregated by income bracket, age bracket, or geographic area (urban-rural, for example), according to the purpose of the analysis. The proposed framework focuses on production spheres, although it is extended to include wealth of various kinds, as we have seen above. The production sphere is disaggregated to industrial sectors and can be linked with industry-occupation matrices, establishing linkages with employment and income generation.

On the other hand, the supply of labor can be determined in a demographic model which explains the population pyramid and labor force participation based on various socioeconomic variables. For example, it is generally agreed that a higher income is conducive to a lower fertility rate. It is proposed that government expenditures and private consumption should be classified by purpose such as health care and education, the benefit of which accrues differently according to age bracket. Thus, the framework provides a basis for analysis in tracing economic growth, demographic dynamics, and life styles with repercussions on the environment.

(5) *The development of sustainable human settlements*: The use of land for various purposes are recorded in the proposed framework together with the transformation for different purposes. The cases in question include forests, farm land, and reclaimed land which are turned into residential area, industrial sites, roads, and other social facilities.

Land and trees are treated in the conventional economic framework as forms of assets and are measured in monetary terms. For sustainable human settlement, land and natural environment have to be assessed from the point of view of habitat and have to be measured in physical terms (land area, for example). Quality of life is treated in items (1) and (2) above. Human settlement is the ultimate goal of our socioeconomic activities and will be the central theme of the proposed framework.

(6) *Environment, water, and health*: Air, water, and soil are affected by various pollutants, harming human health. Efforts have begun to remove such pollutants by retrofitting plant and equipment or by shifting to new production technology, which can be accommodated in the proposed framework by including capital formation and capital stock (which is the treatment capacity when measured in physical terms) for such purposes. In the case of industrial and household wastes, recycling and/or incineration is a possibility. Use of water can also be treated in the proposed framework.

However, health is a complicated problem to be dealt with within an aggregated framework such as the one presented here.

(7) *Biodiversity , biotechnology, and sustainable agriculture*: This topic is outside the scope of the proposed framework. The only possibility is to examine production technology in agriculture, especially with regard to its inputs such as fertilizers, pesticides, agricultural machinery, etc., and their

biological impact will have to be dealt with separately.

(8) *New directions in international environmental governance*: With explicit recognition of environmental assets , the proposed scheme will serve as a common frame of reference. The accounts for individual countries will have various environmental assets as well as their consumption for domestic socioeconomic purposes or for export. The framework will thus reveal a sort of trade-off between current income (via production) and future income . International linkages are revealed by trade matrices which record the value and physical quantity of transactions. The price may or may not be "fair" from an environmental perspective, although it may have been established in the market. In other words, the market mechanism is not well suited to accommodate the externalities which characterize environmental spheres. Theoretical prices can be fitted into the proposed framework by replacing the current input structure by a hypothetical one in order to examine interindustry repercussions.

(9) *Industrial metabolism* : The driving engine of our economies has shifted from agriculture to manufacturing and then to services. Among services, production of knowledge itself has become of ultimate importance. Also gaining importance are the quality-of-life-related activities such as recreation, education, and health care.

It is widely recognized that such socioeconomic change carries with it various environmental implications. If scarce factors tend to dominate the rules of the game, then the industrial North has shifted from the age of capital and energy dominance to an information-oriented one, while the less developed world is passing through the industrialization phase or just entering it. If industrialization is the key to higher income levels, as has been witnessed through historical experience, it is only natural that the less developed South would aim at industrialization.

The only problem is that, if the South succeeds in achieving the level of affluence (and wastefulness) of the North, our planet Earth cannot support it. Given the tremendous income differentials which prevail today, we have to agree that zero growth is not an answer. The fact remains, however, that continuation of the present course of development only aggravates the problem. A transfer of income from the North to the South is only part of the answer. The ultimate solution would have to come through transforming the societies in the North into environmentally compatible ones.

It is ironic to see that most affluent economies and most technologically advanced ones have yet to grasp these hard facts. It seems important to start sending correct signals to producers and consumers, ones which reflect the true environmental values accompanying various socioeconomic activities. At the same time, we should not resort to some kind of rigid bureaucratic guidance. It seems important to utilize the market signals which have proved to be so powerful in guiding individual motives. The prices, however, would have to be set to reflect environmentally sound values.

Based on the proposed framework, various policy schemes can be tried, including carbon tax and subsidies to pollution prevention activities. This can be done by inserting hypothetical values in the value-added sphere of the framework, which include subsidies and indirect taxes in addition to returns on labor and capital.

(10) *Energy and sustainable development*: Some of the energy resources are recorded in the form of subsoil environmental assets, which enter the economic sphere through mining activities and are either consumed domestically or exported. Primary energy sources including fossil fuels, hydro-, nuclear, and nonconventional energy (solar, wind, etc.) are either used directly for various socioeconomic purposes or converted to secondary energy. The entire flow of energy conversion and consumption (industry with sectoral breakdown, transportation by different modes, and various household uses) can be traced within the proposed framework.

Different energy sources have different environmental implications. It is well known that the burning of fossil fuels will result in an output of CO_2, possibly leading to global warming. Nuclear energy is free from the emission of greenhouse gases but entails a huge cost in processing and disposing of radioactive wastes at the backend. What is more, the environmental impact of radioactive waste is still under serious discussion, to say the least.

One last question is whether to try to cope with the ever growing demand for energy by increasing the supply capacity or to restrain the demand in the first place. New technologies are emerging which are more energy efficient. The cases in question range from production technology to transportation and household appliances.

Also, a case can be made for changing lifestyles as reflected in consumption patterns, time-use patterns, and diffusion of various household durable goods including automobiles.

(11) *Critical environmental zones and fragile ecosystems for sustainable development*: This item is outside the scope of the proposed framework. The only possibility is to distinguish critical environmental zones within the land use statistics. One such example is recording the transformation of land from farming, forest, and water surface into various urban (residential, industrial, public facilities, leisure facilities, among others) and rural-type (agriculture and forestry) uses. Another possibility would be to utilize Landsat data over time and trace a dynamic process involving urbanization /industrialization and land use. The analysis in this case can entail the spatial dimension of the shift in land use.

1.4 Virtue of the framework

The proposed framework, when fully implemented, will have the following characteristics:

(1) It provides a consistent picture linking economy, environment, people, and technology.

(2) Supplementary information, such as industry-occupation matrices, will further provide leads to measure employment impact upon shifting to an alternative growth path.

(3) The time series of such a framework will provide a dynamic picture. It can describe current activities and stock in economic and environmental spheres.

(4) It establishes linkages between monetary variables and physical variables.

(5) One can also derive quality-of-life from the identical framework, without disturbing the integrity of the accounts. This can be further disaggregated by region (urban-rural), income brackets, age brackets, etc.

(6) More specialized in-depth study in various fields can be fitted into a comprehensive framewrok.

(7) It incorporates technological change . It will enable us to examine, for instance, the effect of transferring more efficient or environmentally sound technology into an economy. The cost of such technology transfer can also be identified.

(8) It is policy oriented. It is able to describe the impact of policy incentives such as carbon tax on an industrial sector level, on a national level, as well as on a global level. It will also enable us to distinguish resource allocation by social purpose such as pollution abatement, energy conservation, education, health, etc.

(9) It has been developed with an intensive exchange of views within academic circles in economic and social spheres, and theoretical/statistical integrity and feasibility are guaranteed.

(10) The framework is already being implemented in some countries.

In the following chapters, we shall examine various aspects of interaction among the natural environment, economy, quality of life, and technology.

The final chapter contains an integration of the various aspects into a consistent framework. Based on the pieces of information examined in the previous chapters, a reconciliation table is derived for two points in time, namely 1970 and 1990. Existing statistical data are not intended to be a part of such an integrated system, and there remain considerable inconsistencies regarding conceptual definition, coverage of the data, and so on. However, even with these limitations, the prototype fulfills at least two purposes.

One is to demonstrate the feasibility of the proposed integrated framework. In this regard, the prototype will be of instrumental value in showing where the gaps exist with regard to definition, data availability, etc.

Second, the prototype tables based on empirical data are of considerable interest in shedding light on the use of the environment as a resource base and sink, resource allocation aimed at enhancing human aspiration and environmental maintenance, economic and environmental aspects of quality of life , and the roles of technology in moderating the potentially acute collision between economic development and environment.

It is hoped that, with the completion of a similar statistical information system among various members of the international society, it will become possible to establish linkages among them through the international flow of goods, direct investment, and technology so that sustainability of the global community as a whole can be analyzed.

FOOTNOTES

[1] Similar formulation is due to Commoner [1970] who expressed the environmental pollution in terms of three factors including population, per capita production, and pollution factors per unit product.

Part II

Energy and Environment

Chapter 2

Environmental Options

2.1 The background

This chapter looks into the policy options available to us that will maintain the quality of the earth's environment intact while permitting economic benefits to grow. We believe growth is needed, especially in those regions where productive capacity is kept low due to insufficient development or serious disruption from economic transition.

Environmental disruption can thus be envisaged as excessive use of environmental capital and is attributable to various causes. We have seen in Chapter 1 that the relation among them can be formulated as follows:

(1) Population density : Environmental pollution occurs where human habitation is dense. In sparsely populated areas, or when the global population was small, the output of gases, wastes, etc. was absorbed by Mother Nature.

(2) Environmental pollution accompanied industrialization. Modernization and improvement in the quality of life were made possible through increased industrial production and higher incomes. This process, however, was accompanied by an increased 'production' of unwanted gases, wastes, and polluted water.

(3) Pollution also depends on the resource intensity of economic activities. In other words, the magnitude of hazardous outputs depends on the industrial structure and pattern of household consumption.

Figure 2-1 Trends of World population

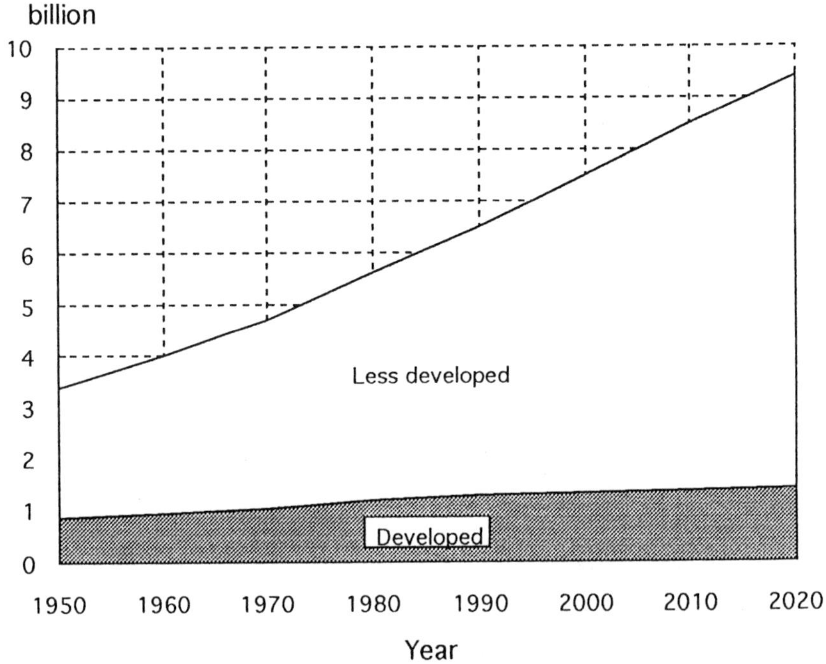

Source: United Nations estimate (revised 1990), medium case.

(4) Pollution prevention and energy conservation also count. Polluting gases, wastes, and water can be removed by proper provision of pollution prevention facilities. A shift to nonpollution processes instead of the conventional mode of production also helps.

The world population today stands at 5.4 billion (estimate for mid-1991), of which 4.1 billion (77%) reside in less developed countries and 1.2 billion (23%) in developed countries. It should be noted that, of the annual population growth amounting to about 100 million, 95% occurs in less developed areas. In the year 2000, the world population is projected to reach 6.3 billion, of which 5.0 billion will be found in less developed areas and the remaining 1.3 in developed areas. By 2025, it will have increased to 8.5 billion, of which 7.2 billion will be in less developed areas. This amounts to 84% of the world total (Figure 2-1). In many less developed countries, the rate of population growth exceeds 2% per annum, enough to double the population size every 35 years. It is imperative to restrain the growth of population. According to past experience, however, turning points in population growth only occurred as a result of industrial growth, urbanization, higher income, and higher literacy.

As for per capita income, which is the second term in the above formulation, we notice enormous differentials among countries (Table 2-1). No one can deny the imperative need to improve the economic conditions of the major part of the world community. If the catching up does take place, the impact of human activity will certainly exceed what the global environment can accommodate. This is the basic dilemma with which we are confronted.

This brings us to the examination of the third and fourth terms in the above formulation. There are considerable differentials among the countries of the world regarding energy efficiency and emissions of greenhouse gases (GHGs). Table 2-2 makes the point. CO_2 emission per GDP (kg/US$) ranges from 0.11 recorded for France and Japan to 1.58 in China and 1.78 in Poland. Emission per capita (t/person) ranges from 0.19 for India and 0.52 for China to 5.79 in the USA.

What is more, the differentials seem to be widening over time. Figure 2-2, which depicts the energy intensity of GDP in recent years, shows that the world as a whole has become increasingly energy efficient. In developed countries, the amount of energy (in petroleum equivalent) required to produce 1000 dollars worth of GDP has declined from 0.45 tons in 1970 to 0.33 tons in 1989. In contrast, in developing countries, it has increased from 0.50 tons in 1970 to 0.56 tons in 1989, which is about 1.7 times larger than in the developed world. The situation in the formerly planned economies seems even worse, although evaluation of GDP in world prices is quite difficult.

If the present trend continues, it is estimated that CO_2 emission would amount to 12.4 billion tons of carbon equivalent in the year 2025 as compared with 5.2 billion tons in 1985 (Figure 2-3). Of this total amount in the year

Table 2-1 Per capita GNP and population growth

Country	GNP per capita US $, 1988	Population growth, average of '85-'90 %
Mozambique	100	2.65
Ethiopia	120	2.67
Nigeria	290	3.30
China	330	1.45
India	340	2.07
Pakistan	350	3.44
Indonesia	440	1.93
Egypt	660	2.39
Thailand	1,000	1.53
Mexico	1,760	2.20
Poland	1,860	0.65
Malaysia	1,940	2.64
Brazil	2,160	2.07
Hungary	2,460	(-)0.18
Korea	3,600	0.95
Saudi Arabia	6,200	3.96
Singapore	9,070	1.25
Australia	12,340	1.37
U.K.	12,810	0.22
Italy	13,330	(-)0.03
France	16,090	0.35
Denmark	18,450	0.08
Germany (West)	18,480	0.10
U.S.A.	19,840	0.81
Japan	21,020	0.43

Note: Data from United Nations, *World Population Monitoring 1989*; World Bank, *World Development Report 1989*.

2025, 44% would be attributable to less developed countries, considerably larger than the recent figure of about 26%.

From an egalitarian point of view, which is reflected in the emission per capita, the developed world is consuming too much energy and depriving the less developed world of their chance for growth. From a technological point of view, which is reflected in emission per GDP, overburdening of the global environment is attributable to less efficient technology in the less developed world, especially when we consider the larger growth potentials there. A part of such differentials is attributable to natural factors such as climatic conditions in particular regions. Obviously, in colder regions human habitation requires a much larger energy consumption for heating. But a larger portion is due to different technological practices. As a world community as a whole, we are being pressured to find a way of allowing economic development to take place in a more resource-efficient way.

In order to avoid global warming, we need to control fossil fuel burning. It would not be practical, however, to put the issue as a choice between growth or no growth, especially when serious income differentials persist in the world.

2.2 Policy options for the global community

Choices open to the global community seem to include the following:

(1) Nuclear option,
(2) Solar option,
(3) High-tech option,
(4) Lifestyle option, and
(5) Structural option.

In the following, we examine the feasibility of each of them from technical as well as socio-political points of view.

Nuclear option: Faced with a growing demand due to the rapid growth of some of the developing economies, on top of the continued growth of the developed economies, the demand for energy and other resources will continue to grow. Formerly planned economies are now going through a period of restructuring and are suffering from the adjustment they have to go through, but they are expected to start catching up. Although there seems to be room for improvement of efficiency in energy and resource use in these countries, the total demand will continue to grow in the long-run. If we take growing demand for granted and try to cope with the situation from the supply side, we are tempted to try the nuclear option.

Incidentally, this is believed to be the most economical. Japanese data on the cost of electric power generation is as follows [Institute of Energy Economics , 1990, pp.443-444]:

Figure 2-2 Energy intensity trends by region

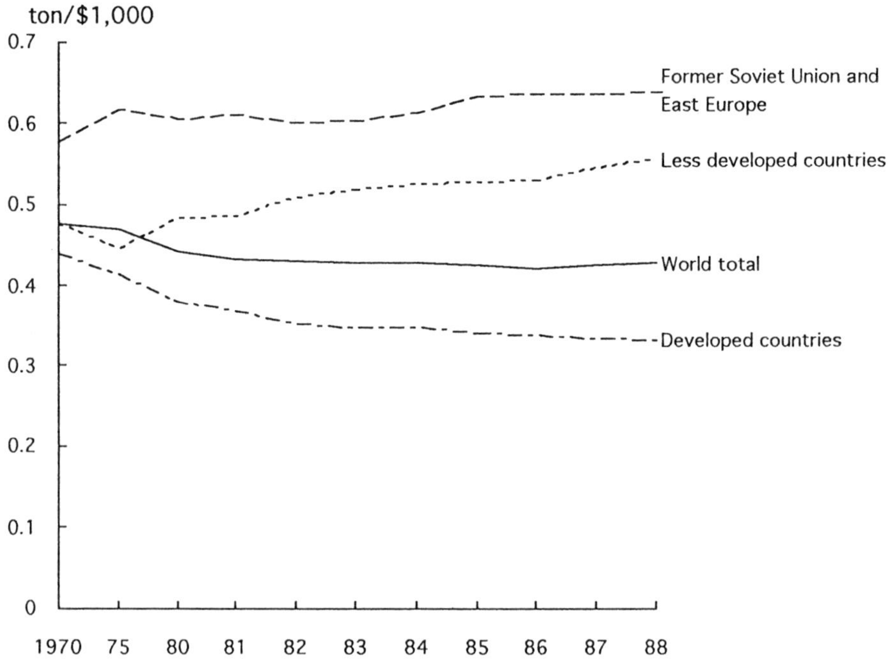

Note: The figures refer to primary energy consumption converted to petroleum per one thousand dollar GNP (1988 prices).

Source: Economic Planning Agency, *World Economic White Paper*, 1990.

	Construction cost (per KW)	Power generation cost (per KWH)
Hydro	¥640,000	¥13
Petroleum thermal	¥190,000	¥11
Coal thermal	¥230,000	¥10
LNG thermal	¥200,000	¥10
Nuclear	¥310,000	¥ 9

In the case of nuclear power, cost increase including fuel fabrication and reprocessing is assumed to rise 0 to 0.1% annually. The cost of decommissioning the nuclear power plant is included in the generation cost and estimated at about 0.2 yen/KWH based on a report by the Nuclear Branch of the Comprehensive Energy Council (Sogo Enerugi Chosakai) in 1985. Decommissioning a plant of the 110,000 KW class is estimated to cost 30 billion yen in 1984 prices.[1]

The fuel cycle of a nuclear power plant consists of three main stages [NEA, 1985, p.20]:

(1) the so-called front-end which extends from the mining of uranium ore until the delivery of fabricated fuel elements to the reactor site;
(2) fuel use in the reactor, where fission energy is employed to produce electricity; and
(3) the so-called back-end, which starts with the shipping of spent fuel to away-from-reactor storage and ends in the final disposal of reprocessing wastes or of the spent fuel itself.

A report by the Nuclear Energy Agency writes, "The cost of the decommissioning option may be a very important factor. At the end of the plant operations, the immediate dismantling of the facility (Stage 3, i.e., removal of all contaminated equipment and structures) would require a much larger expenditure than would be needed to accomplish Stage 1 (i.e., continuing surveillance and maintenance) only. If funds have been set aside for decommissioning during plant operation, then the immediate cost may be less important. The situation is different if the decommissioning has to compete for funds with the owner's other capital needs. Although the prediction of future decommissioning costs involves large uncertainties, plant owners should plan the financing of the chosen decommissioning alternative so that this item does not become a controlling factor." [NEA, 1986, p.19]

"Decommissioning of power reactor generates significant quantities of low-level waste from neutron-activated materials and from the surface contaminated materials. A small amount of intermediate-level waste comes from the reactor and its internals. Considering the total radioactivity of decommissioning waste , it should be noted, however, that more than 90 per cent of the radioactivity in a reactor at shutdown is contained in the spent fuel element which will be removed and generally shipped off-site." Based on estimates from

Table 2-2 World CO2 emission due to fossil fuel burning

Country	Fossil fuel (share, %)	Emission per GDP (kg/US $)	Emission per capita (t/person)
World total	5,888.8 (100.0)	—	1.15
United States	1,426.2 (24.2)	0.29	5.79
Soviet Union	1,101.9 (18.7)	—	3.86
China	569.7 (9.7)	1.58	0.52
Japan	276.1 (4.7)	0.11	2.25
Germany (West)	200.1 (3.4)	0.18	3.25
United Kingdom	168.5 (2.9)	0.23	2.95
India	157.5 (2.7)	0.57	0.19
Canada	132.4 (2.2)	0.30	5.10
Poland	124.4 (2.1)	1.78	3.29
Italy	117.5 (2.0)	0.15	2.04
France	100.3 (1.7)	0.11	1.79
Germany (East)	87.5 (1.5)	—	5.25
Netherlands	63.9 (1.1)	0.30	4.33
Korea	53.7 (0.9)	0.36	1.28
Spain	52.5 (0.9)	0.17	1.35

Note: Emission data are in terms of carbon. Emission data are estimates for 1988, and may not be accurate due to statistical concepts and the methods of data compilation.Original sources of data are United Nations, etc.

Source: *Environment White Paper, 1991, Overview*, p.14.

Canada, FRG, Sweden, and the USA, the report writes that "for all reactor types considered the volume of decommissioning waste is about the same order of magnitude as the volume of waste from their operation during a 25-year period." [NEA, 1986, p.34]

"After removal from the reactor, the spent fuel will normally be cooled in ponds at the reactor site and may be transferred for a period to an interim store. After a period of cooling the fuel assemblies may be encapsulated directly or may be disassembled using remote handling techniques so that the fuel pins can be packed together more closely prior to encapsulation. The encapsulated fuel in appropriate containers can be disposed of in a range of ways paralleling those for the vitrified high level waste from reprocessing. The most elaborate plan is that which has been put forward by Sweden, which involves disposal of the containers in deep rock formations. They will be embedded in a buffer material which will prevent any flowing ground water coming into contact with the container."[NEA, 1985, p.34] The time lag from the time of decommissioning of the plant to completion of waste disposal is generally assumed to be 40 years or longer.[NEA, 1985, p.86] Compared with the front-end cost, the back-end cost may amount to about 38% based on data provided by NEA [1985,p.62].

Apart from the fear of nuclear accidents and public resentment, there seems to be large economic cost to be incurred at the back-end.[2] Since the whole process takes quite a long time to complete, inflation may intervene as it did in the past, and if it does, financing of decommissioning may become a managerial headache. A question remains of whether the cost advantage of nuclear energy as presented above is realistic. At the same time, there are great uncertainties involved in the decommissioning of plants, giving rise to a wide ranging cost calculation. In areas where earthquakes and faults abound, disposal of the encapsulated fuel may also pose a serious problem.

We have examined the nuclear option as an alternative to increased fossil fuel burning, and the guiding principle was 'no regret' for the future generation. Judged against this principle, the nuclear option seems to create another source of regret in an attempt to escape from the one involving continued CO_2 emission and global warming .

Solar option: From the "hydrocarbon hill" to the "solar plateau" was the basic scenario in IIASA's prospectus on eco-restructuring.[Ayres and IIASA, 1990] In fact, many people put great hopes in solar energy as a typical case of the soft-energy path. People clearly prefer increased use of renewable energy such as solar. Sunlight will strike the earth at an almost constant rate far into the future, regardless of human activities. Converting sunlight into electricity would not upset the global environment. And such technology is becoming available as the efficiency of photovoltaic cells improves and its cost declines. Solar energy on the earth's surface has a rather low energy intensity of about 1KW per square meter, and its usefulness is limited by climatic conditions. However, it is clean, renewable, and immense. Its use as a heat source is already widespread in heating and cooling. In Japan, there are about 330,000 units of solar system and 4,000,000 units of solar heater for household use. Photovoltaic cells convert solar energy directly into electricity. Around 70% of the cells are currently used for calculators and watches, and the remaining 30% are for electricity generation which is used as the power source for street

Figure 2-3 Projection of CO2 emission by region

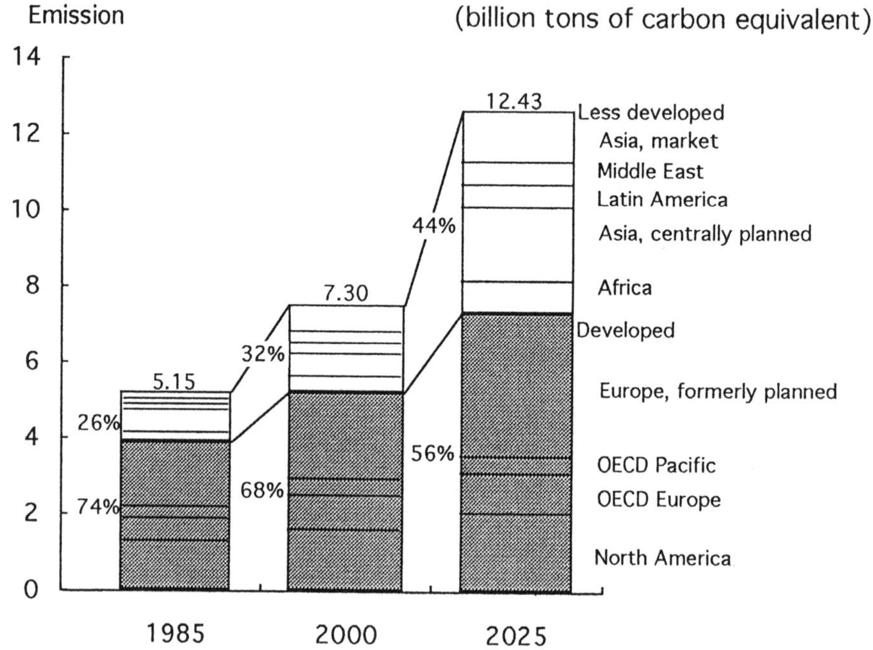

Note: Data refer to energy-originating emission only.
Source: IPCC.

lamps, lighthouses, etc. If they become available at a lower cost, the market is expected to spread to various household and social facilities.

The fact remains, however, that solar energy alone will not be able to serve as the alternative to fossil fuels because of its low energy intensity. "The solar plateau" in IIASA's document in reality refers to various nonfossil energy sources.

High-tech option: We need to explore the possibility of technological change , both at the micro-level where energy efficiency is attained through improved production processes and at the macro-level where it is attained through structural changes of the economy. The latter is often termed as a shift from smokestack industry to high-tech and services. Also important is to consider lifestyle options. This aspect pertains to household activities and transportation, often calling for the development of new technologies.

In response to the energy price hike in the early 1970s, industries reacted by improving the efficiency of their energy use. Figure 3-4 in the next chapter shows the case in Japan. In the iron and steel industry, the energy input per unit production has declined to 80.8 percent during 1973 and 1989 (fiscal year). This is attributable to improvement in plant operation, improvement in the production processes , recovery of waste heat and efficient use of energy. Introduction of a continuous casting process, power generation by blast furnace top gas pressure, and dry fire extinguishing of coke are concrete examples. In the case of ethylene production in the petrochemical industry , recovery of waste heat and rationalization of production processes such as high-efficiency compressors were the contributing factors. In cement production, introduction of the new suspension preheater kiln (NSP) and medium-to-low temperature waste heat power generation were the typical improvements. In the paper and pulp sector, introduction of a continuous production process, recovery of waste heat, increased use of recycled paper, and improvement in the production processes contributed to energy conservation.[3]

More recently, however, the leading sector of the economy has shifted from heavy industry to services and information. With energy conservation and structural shift in manufacturing on the one hand and higher levels of consumption and transportation activities on the other, a new approach is required to attain economy-wide energy efficiency.

Examples are co-generation (energy efficiency can reach 75% compared with 35% in the conventional method)[4], power generation combined with refuse incinerator, use of heat generated by waste incineration and power stations; recovery of energy from LNG; recovery of waste heat from the subway system, underground power cables, and river water, among others. Better insulation of houses and offices should also help.

Lifestyle option: Quality of life depends on the available stock of consumer durables, housing, and other social stock, rather than the annual purchase of durables or annual construction of houses. Stock, rather than flow, is also relevant from the point of view of energy consumption. The policy implication seems to be that, with the improved durability of products, people can attain the desired economic well-being, while conserving the energy required for producing them. In the same vain, recycling of wastes contributes to energy saving. This is true of steel, aluminum, plastics, paper, and other resources, the production of which involves a large energy consumption. Rather than focusing on the production phase alone, therefore, we should keep track of the services they render during their life span and recycle them at the end. In many instances, this presupposes the participation of consumers as an important linkage point in the system.

Despite energy efficiency being attained in industry, household activity and transportation seem to be tending toward increased energy requirements. People have more consumer durables requiring energy, have larger houses with air conditioning, and are shifting from public transportation such as railways to automobiles. Higher diffusion rates reflecting increased income levels more than offset the energy saving attained through improved technology. There seems to be no easy way out of the situation in a society which is market- and democracy-oriented. Lowering the room temperature in the winter and tolerating a higher room temperature in the summer, taking fewer showers or baths, turning off the pilot burner of gas water heaters, turning off unused lights, and replacing light bulbs by fluorescent lamps should all help. More fundamental measures, however, would be to provide a social infrastructure which reduces energy wastes. Examples are provision of efficient mass transit; prevention of road traffic congestion; optimal allocation of distribution channels. These improvements will enable us to maintain the levels of quality-of-life, or even improve it, with less resource input.

It is always desirable to change our wasteful lifestyle to one compatible with sustainable development . However, one should not forget that there is another side to the conservation effort. Ayres writes, "Energy is currently used very inefficiently to create final services, as compared to the first stage of energy conversion and distribution. The problem is that energy is lost and wasted at each step in the chain of successive conversions, from crude fuels to intermediates, to finished goods, to final services. Incandescent lights (converting electricity to white light) are only 7% efficient (fluorescent lights are better). Moreover, lighting fixtures are typically deployed very inefficiently (c. 10%) so that the final service (illumination where it is actually needed) is probably less than 1%."[Ayres and IIASA, 1990]

Lifestyle itself is changing. We may call it service orientation. Household expenditure on services is increasing faster than for hardware, reflecting higher income and, presumably, changing values. Although the total energy implication is not known, partly due to difficulties in measuring the indirect energy requirement to provide such services, there is a possibility that service orientation would reduce the total energy requirement.

Structural option: It is sometimes taken for granted that economic growth necessarily involves energy consumption. In economics jargon, the elasticity of demand for energy is believed to be close to unity. According to the Japanese experience in the past fifteen years, the relationship between the two variables is not that simple. Energy efficiency for the economy as a whole has improved considerably in the 1970s and 1980s in Japan. The energy price hike has triggered energy-saving technology, and at the same time, the industrial structure has been shifted to avoid energy consumption.

There is a method, based on input-output tables, to break down changes in total energy consumption into the following five parts:

(1) due to changes in unit consumption coefficients;

(2) due to changes in input coefficients which describe the structure of the intermediate inputs in an industry;

(3) due to changes in the levels of final demand;

(4) due to changes in final energy consumption; and

(5) due to the cross-product term.

In a sense, this provides a comprehensive picture including improved energy efficiency in individual sectors (first factor), bundle of industrial sectors (second factor), and pattern of final demand (third factor). Empirical examination of the economic adjustment in Japan which followed the first oil crisis revealed that for the decade as a whole, the usage of energy-saving technology was reflected in lower unit consumption coefficients, and the shifting industrial structure worked toward a lower energy requirement for the economy.[5] In reality, however, there are not many countries where the economic policy can influence the economic structure, apart from expecting individual response of the private sector to a price signal.

2.3 Policy issues

We have precluded the non-growth option in order to cope with the acute poverty which prevails, regrettably, in large portions of the world population. The nuclear option seems to be self-defeating if we want a no-regret strategy. Solar option becomes only feasible after extensive technological development. Thus, it seems we are left with a combination of the high-tech, lifestyle, and structural options.

We have seen that not all countries are following the best technological practice available. This seems to remain true even when we take account of climatic differences and the size of the countries. The implication is that we should implement a policy whereby technological knowledge can be diffused among countries.

Technological differentials have arisen for various reasons, including the following:

(1) Lack of financial resources to implement capital formation which embodies the best technological practice. Also, where market interest rates tend to be higher, reflecting scarcity of capital or greater country risk, it may not be economically feasible to adopt the best technology.
(2) Lack of knowledge pertinent to environmentally sound technology.
(3) Lack of foresight about events such as environmental deterioration and global warming. Sharp discounting of the future, in turn, is often a manifestation of acute poverty at present or major economic disequilibrium accompanying transition of the economy.
(4) Lack of proper economic incentive which is often attributable to the fact that many environmental issues are not properly valued due to externalities (market failures).

(5) Lack of proper political incentives due to the fact that the cost of the environmental policy has to be borne by a particular industry or a country, whereas its benefit is diffused among the population at large (political failures). In this case, rent is generated by not implementing the policy required to nullify the damage done, and the rent can be exploited by the private firm for higher profits or by the political machinery for economic and political purposes. This may take any form ranging from export drive by a firm to growth promotion by a country.

(6) In addition, reflecting the stage of economic development, medium-income economies tend to have larger shares of resource-intensive sectors such as the steel and petrochemical industry, resulting in high resource intensity on the aggregate. This may be called the structural factor.

Official development assistance (ODA) in the environmental field is increasing steadily. Also, in carrying out ODA for development purposes, care is being taken not to exert a hazardous effect on the environment of the recipient country, as well as for the world community as a whole. ODA is provided based on a request from the governments of the recipient countries. However, environment-related projects usually do not bear direct and immediate economic benefit and have not enjoyed high priority. If appropriate measures are not taken to prevent environmental harm, permanent damage may be done, or it may turn out to be extremely costly to repair the damage. It is necessary to promote dialogue between the provider and recipient of ODA. Necessary measures include provision for training of personnel who engage in environmental monitoring and consultation, provision of manuals, guidelines, standards, etc. to deal with environmentally sensitive projects, diffusion of technological knowledge through database, among others. In view of the fact that the global environmental issue is closely related to energy strategies, projects aimed at reducing the use of fossil fuel and at improving energy efficiency should also enjoy high priority in ODA.

Equally important is the role of direct foreign investment by the private sector. This is much larger in magnitude than the ODA. During the past decade, some of the less developed countries have 'graduated' from being the recipients of ODA. However, concern for the environment does not have top priority in private investment in less developed countries, just as is the case in the developed world.

"Foreign investment, including investment through multinational enterprises, plays a major role in global economic development and trade, both in OECD economies and in those of the Third World. During the past decade, 60% of the industrial investment in the Third World countries originated outside those countries, and much of this was supplied by multinational enterprises. Multinationals are also responsible for a major part of the growing volume of trade in minerals, manufactures and technology between OECD and the Third World." "Environmentally unsound development confronts many of these countries with difficult trade-offs between the immediate benefits of development and its long-term costs to human health and the environment. Coupled with the lack of effective means to manage resources and the development proposals, the result can be extensive and often irreparable damage to the ecological basis both for the development concerned and for future economic development."[OECD, 1982b, pp.64-65] Similar trade-offs prevail in the restructuring of the formerly planned economies. The problem is now compounded by the fact that, in contrast to the pollution problems in the 1960s and 1970s when the issue was localized and hence covered by a single policy body having a set of consistent laws and regulations, today's issue extends across national boundaries in its geographical span and far into the future in its time span. No single policy body can be responsible for an issue of global implication, but an institutional framework on an international level has yet to be devised.

2.4 Policy tools

Let us turn here to the examination of the policy tools employed in the past in environmental field. In many countries, such a policy has evolved out of necessity, but in retrospect they may be classified as follows.[6]

(1) Charges: Charges may be considered as a "price" to be paid for pollution. Polluters have to pay for their implicit claim on environmental "services" , which thereby enters at least in some part into private cost-benefit calculations. Various types of charges are reported:
Effluent charges: charges based on the quantity and/or quality of discharged pollutants.
User charges: payments for collective or public treatment of effluents.

Product charges: charges on the price of products which are polluting.

Administrative charges: payments for authority services for control and authorization.

Tax differentiation: provides more favorable prices for environmentally friendly products and vice versa.

(2) Subsidies: Various forms of financial assistance for the purpose of providing incentive for polluters to change their behavior or to comply with imposed environmental standards include the following:

Grants

Soft loans: the interest rates are set below the market rate.

Tax allowances: accelerated depreciation or tax exemptions for anti-pollution measures.

(3) Deposit-refund systems

(4) Market creation: Artificial markets are created where participants buy pollution rights for pollution or where they can sell their rights.

Emission trading: if a discharger releases less pollution than is allowed by standards, the firm can sell or trade the differences between its limits and actual discharges to another firm. The latter then has the right to release more than its initial limits.

Market intervention: price intervention (price guarantees, etc.) for the purpose of creating a market for, for example, potentially valuable residuals.

Liability insurance: creation of a market in which risks of damage penalties are transferred to insurance companies. Premiums will be lowered for more secure processes, lower damage, fewer wastes, etc., providing incentive for better performance on the side of actual or potential polluters.

(5) Financial enforcement incentives: This may be considered a legal instrument rather than economic.

Non-compliance fees: fees imposed when polluters do not comply with certain regulations.

Performance bonds: refunding takes place when compliance has been achieved.

Traditional environmental policy relied on the command-and-control form of regulation which specifies standards and requirements in detail. In contrast, these alternative methods are intended to provide incentive through the market mechanism and increase the room for private decision-making. The strength of economic instruments is believed to be as follows.[Boland, 1989; OECD, 1989]

a. Bring about incremental improvements in environmental quality over and above those attainable through direct regulation. b. Facilitate self-financing pollution control programs. c. Create additional stimuli for technical progress. d. Reduce total cost burden on polluters. e. Provide flexibility in enforcement without sacrificing environmental goals.

In reality, however, their application has been limited. Reporting on the situation in the USA, Boland writes: "Comparatively little use is made of economic instruments in the US. Pollution control policy continues to rely on a 'command and control' philosophy, admitting the possibility of economic incentives only so long as they do not interfere with conventional regulation. Applications that do exist tend to be narrowly circumscribed with respect to geography, environmental media, and potential participants."[Boland, 1989, p.9] A similar situation prevails in the UK. Barett observes: "The United Kingdom has a bevy of environmental regulations, but not one of them makes use of economic incentives."[Barett, 1989, p.1] Japanese environmental policy, which is believed to be successful, consists mainly of the enactment of environmental quality standards, adoption of stringent regulations, and implementation of pollution control programs at regional levels which coordinate private and public efforts with a focus on particular regions. "Environmental standards were to be achieved through various laws which stipulate direct regulatory measures."[Uno, 1987]

Direct regulatory instruments include the following[IEA, 1989a, pp.38-41]:

(1) Environment quality standards: Environmental quality standards are aimed at the protection of human health or of the ecosystem. Indicators of "quality" are defined precisely in operational terms and specify allowable average concentrations over a specific time for a given pollutant in a particular region.

(2) Fuel quality regulations: In the case of fuels, standards are structured around the types of fuels in use (e.g., coal, various types of refined oil products, motor fuels). They are limited by the technical possibilities and the costs of the cleaning process for the different fuels.

(3) Fuel use regulation: Control of fuel use has been employed as a strategy for air pollution reduction or to respond to general environment concerns.

(4) Emission standards: Emission standards set a maximum allowable rate of pollution output for each generic type of source (transport, power plant, industry) by type of pollutant.

(5) Prescriptive technology standards: They provide precise definitions of which type of control technology or method should be applied in a particular instance. Rarely used because of their inherent lack of flexibility.

(6) Licensing: It is a key component in the siting of new facilities, along with environmental impact assessments. A license can be revoked if the polluters fail to satisfy a certain level.

(7) Zoning: Geographical restriction of the location of industrial facilities and applies mainly to the siting of stationary sources.

(8) Safety regulation: It is designed to minimize hazards associated with energy or other activities, both in terms of occupational risks and in terms of risks incurred by the public.

Technological change is a preferred way to do away with environmental pollution and still continue the desired activity without incurring additional economic and ecological costs. Precisely on this point, we have noted above that economic instruments are said to be conducive to innovation. It is interesting to note counterargument on this point. "All regulation imposes a constraint to which industry tries to react in the most efficient fashion, i.e. at the least cost. In many cases, the most economic solution is an innovatory one."[OECD, 1985a, p.34] In Japan, for example, more stringent regulations are imposed for new firms than the existing ones in order to stimulate new technologies. In addition, environmentally sensitive regions opted to establish stricter standards than the national stipulation, forcing business firms to "invent or vanish."

In order to overcome the obstacles and narrow the technological differentials among countries enumerated earlier, we have at our disposal a set of economic instruments and regulations. The pros and cons of both approaches have been debated, but their relative effectiveness, especially in an international context, is not altogether clear. Given the nature of the environmental concerns, which is global as well as regional, the world community is probably getting ready to accept universal rule just as it has agreed in the case of limiting CFC emissions. Given the fact that many of the financial resources and new technologies originate in OECD countries, the responsibility of the developed world seems to be large.

This does not mean that new technologies should be transferred to the rest of the world free of charge. This is detrimental to innovative activities in the private sector. It would be advisable to create a market for such technological knowledge where it can be traded at a market price and its diffusion protected by intellectual property right. Rather than leaving the transaction to private hands, the provider of ODA and international aid agencies should promote the creation and functioning of the technology market and exercise economic instruments as deemed necessary from the policy point of view.

FOOTNOTES

[1] The calculation assumes that the operation starts in 1989. The costs of construction and power generation have been calculated for model plants for each energy category as follows.

Hydro	10,000	to	40,000 KW
Petroleum thermal	600,000	KW	class x 4 units
Coal thermal	600,000	KW	class x 4 units
LNG thermal	600,000	KW	class x 4 units
Nuclear	1,100,000	KW	class x 4 units

Plant life is assumed to be 40 years for hydro, 15 years for thermal, and 16 years for nuclear, based on figures which are commonly used among OECD countries. Rate of capacity utilization is assumed to be 45% for hydro and 70% for thermal and nuclear. Energy cost reflects the future, petroleum and LNG prices are assumed to increase to $26-30 per barrel in the year 2000, and the coal price is assumed to rise at an annual rate of 1 to 1.5%. Energy efficiency in thermal plants is assumed to be around

40%.

[2] An OECD report points out long lead times as a cost factor. "With the fall in fossil fuel prices, particularly since 1985, and expectations for these price levels to continue into the medium term, nuclear power no longer enjoys in some OECD regions undisputed cost advantages over its alternatives for the generation of baseload electricity. Safety concerns following the Three Mile Island and particularly the Chernobyl accidents have produced a climate siting and licensing have become more difficult; lead times and hence interest payments have increased, and the cost of nuclear plants themselves has risen substantially. ... In this uncertain environment, many utilities have, for the time being, been opting out of nuclear expansion on the grounds of risk aversion." [IEA 1990b]

[3] Energy consumption is found to be sensitive to price levels. Price elasticity of energy consumption was estimated at the disaggregated level, based on the following specification.

$$\begin{aligned} \ln QECi = {} & b0 + b1 \ln VQREALi + b2 \ln (PMFUEL/Pi) \\ & + b3 \ln QECi(-1) \end{aligned}$$

where	QECi:	Energy consumption
	VQREALi:	Gross output, constant prices
	PMFUEL:	Import deflator, mineral fuels
	Pi:	Sectoral deflator

Empirical data were obtained from the Japanese experience during the mid-1950s to 1980. [uno 1987] According to the result, price elasticity is found to be (-)0.1034 for manufacturing as a whole, (-)0.0714 for transportation, and (-)0.1014 for commercial and home use. It is seen that energy conservation was prompted by higher energy prices.

[4] In Japan, the co-generation capacity has increased from 61 MW in 1970 to 1372 MW in 1990 for industrial use and 0.3 MW in 1974 to 213 MW in 1990 for commercial use. Refer to *Comprehensive Energy Statistics,* Agency of Natural Resources and Energy [1991], pp.192-193. The capacity of refuse power generation has increased from 10 KWH in 1970 and 34 KWH in 1975 to 342 KWH in 1989.

[5] Despite the growing size of the Japanese economy and an increase in final energy consumption, which when taken together meant an additional

energy requirement worth about 8835 billion yen (1980 prices), the actual increase remained at 4548 billion yen, or about 1/2 of the would-be level.

[6] OECD [1989], pp.14-16. This source provides a survey of country practices.

Chapter 3

Energy Demand and Supply

3.1 Energy and CO_2 emission

The world consumption of primary energy has been on the increase ever since the Industrial Revolution . The energy consumption in 1860 is estimated to have been less than 100 million tons of oil equivalent, which rose by 1990 to 800 million tons. The increase has been particularly marked since WWII when the sources of primary energy shifted from coal to petroleum. Although hydro and nuclear power are now among the primary sources of energy, about 90% of the total supply is still attributable to fossil fuels .

The developed countries consume 50% of the world's primary energy supply, while constituting only 15% of the world population. The energy consumption of developing countries amounts to 19% and that of formerly planned economies, 32%. Per capita energy consumption in developed countries is 10 times as great as that of developing countries and 3 times that of formerly planned economies.

The demand for energy is expected to increase into the future. Assuming 2.7% annual growth of the world economy, the world primary energy consumption in 2010 will reach 14.5 billion tons of oil equivalent by 2010, or 1.8 times the 1989 level. If the economic growth is accompanied by efficiency gains in energy consumption amounting to 1.2% per annum, energy consumption will

Table 3-1 Projection of CO_2 emission by world

Region	Cases	CO_2 emission (million tons of C) 1989	2000	2010 (%)	Annual growth (%)	Ratio to 1989
Developed	1	2996	4072	5227(42.1)	2.7	1.7
	2	2996	3598	4150(46.3)	1.6	1.4
Formerly planned	1	2160	2804	3994(34.6)	3.0	1.8
	2	2160	2283	2609(18.0)	0.9	1.2
Developing	1	1166	1715	2400(23.3)	3.5	2.1
	2	1166	1570	2055(35.7)	2.7	1.8
World total	1	6322	8591	11621(100.0)	2.9	1.8
	2	6322	7451	8814(100.0)	1.6	1.4
Asia	1	1119	1868	3164(38.6)	5.1	2.8
	2	1119	1473	2048(37.3)	2.9	1.8

Notes: Asia includes China but excludes Japan. Case 1 refers to the situation where unit energy input remains at the present level, while in case 2 conservation efforts are made. Case 2 takes account of the shift among primary energy sources in calculating the CO_2 emission. The extent of energy conservation and the composition of primary energy are based on relevant data provided in reports by Working Party 3. The estimation was carried at the Economic Planning Agency based on BP statistics and IPCC reports.

Source: Economic Planning Agency [1991b].

still amount to 11.4 billion tons, or 1.4 times the present level. This directly translates into CO_2 emission. The shares of nonfossil energy such as hydro and nuclear will increase from 12% of the total to 15%; however, at the same time, the share of coal will rise from today's 28% to 30% according to a realistic scenario. Thus, the CO_2 emission in the year 2010 will reach 11.6 billion tons (in terms of carbon) if the technology remains unchanged, which is about 1.8 times the present level. In the energy conservation scenario, it will be 8.8 billion tons, or 1.4 times the present level.

The trends in developing countries are particularly noteworthy because of their high growth potentials. By 2010, it is estimated that the primary energy consumption and CO_2 emission will expand by 2.1 times in a constant technology case and 1.8 times in an energy conservation scenario. The impact on global energy demand will further increase beyond 2010 due to continued economic expansion in the developing countries.

Table 3-1 also reveals the trends in Asia as a whole (which includes China but excludes Japan), which is the growth center in the global community. During the past decade, primary energy consumption has increased 1.6 times in the area, whereas the world average was 1.2 times. By the year 2010, energy consumption is expected to double. Emission of CO_2 will further increase by 1.8 times during the same period due to a structural shift in energy consumption such as reduction of coal (from 58% to 50% of total supply) and increase in natural gas (from 5% to 11%), hydro and nuclear (from 7% to 11%) power.

The situation in formerly planned economies (including China) is rather confused at present. Assuming that the growth until the year 2000 is modest and that the room for improvement in energy consumption is relatively large (1.7% per annum; for China, 2.5% per annum), the primary energy consumption will rise by a modest margin of 1.2 times. If the economic reform is carried out smoothly, leading to higher growth, the demand for energy may be that much higher.

The Japanese economy has attained the highest level of energy efficiency in the world, which is attributable not only to individual energy-related technology but also to comprehensive system design including production facilities and production management. This implies that the transfer of energy-efficient technology will require a lot of money and time. Considerable energy savings will be realized if the Japanese practice is transferred to other countries. According to an estimate by the Economic Planning Agency, energy consumption and CO_2 emission can be reduced by 20% in developed countries and as much as 50% in developing countries and formerly planned economies.[1] The amount of capital required is 50 to 150 trillion yen in developing countries and formerly planned economies in order to replace the existing capacity in 1989. Total requirement until the year 2010 will be 100 to 300 trillion yen if one includes the financing of capacity expansion during the period. The net amount will be smaller if an economic benefit of cost reduction is taken into account. In any event, considering the rather limited financial resources available in these areas, it seems imperative to transfer technology and money in order to achieve energy conservation on a global scale.

Japanese consumption of primary energy in the world amounts to 5% of the world total, as is shown in Table 3-2. Her consumption of petroleum amounts to 7.3% of the world total. This results in CO_2 emission which is estimated to be 4.5% of the world total. Considering the fact that her GDP is around 15%

Table 3-2 CO_2 emission in OECD countries

	Primary energy (MTOE)	CO_2 emission (mil.ton C)	GDP (mil.$)	Population (thousand)
USA	1943.4	1480.0	5392200	248777
Japan	404.6	280.2	2977725	123116
Germany	271.6	281.9	1487958	78665
France	219.0	110.8	1190837	56160
Italy	153.5	117.3	1090755	57525
UK	211.3	164.0	975455	57236
Canada	219.6	135.5	570039	26248
Spain	86.4	63.2	491240	38888
Australia	85.5	76.3	296315	16807
Netherlands	65.1	53.5	279138	14849
Sweden	48.1	21.4	228102	8493
Switzerland	23.5	12.1	224878	6723
Belgium	47.3	33.6	192392	9938
Austria	23.8	17.0	157378	7624
Finland	29.0	18.4	137234	4964
Denmark	17.9	16.2	130950	5132
Turkey	49.6	43.8	108409	55255
Norway	22.8	10.6	105822	4227
Greece	22.0	22.7	65956	10033
Portugal	16.1	13.0	59678	10337
New Zealand	12.9	7.9	42768	3343
Ireland	9.8	8.1	42468	3515
Luxembourg	3.4	2.7	8723	378
Iceland	1.3	0.6	5856	253
EEC	1123.4	887.0	6015550	342656
OECD	3987.3	2998.6	16262279	848486
World	7815.7	6255.9	n.a.	5292200

Note: Data refer to 1989 except GDP figures which refer to 1990. World total of CO_2 emission refers to 1988. Countries are listed in the order of GDP size.

Source: Primary energy supply and CO_2 emission data are from OECD [1991b]. GDP data are from OECD, *National Accounts.*

of the world, we can say that Japan has a lenient socio-economic structure as far as energy is concerned.

For example, the US GDP is about 1.8 times larger than Japan while consuming 4.8 times more energy and emitting 5.3 times more CO_2. The German GDP is about 50% of Japan, but energy consumption is about 67%, and CO_2 emission is just about the same. Such comparisons only provide a rough picture. The relative size of GDP itself cannot be defined with any precision.[2] Energy consumption reflects various factors such as the industrial structure, geographical span, and natural climate, and may not directly reflect the efficiency in energy use. The composition of the primary energy sources has to do with CO_2 emission. Nevertheless, the fact remains that Japan represents a significant share of the world economy and that she has attained considerable energy efficiency.

This makes Japan's performance in CO_2 emission relatively good. Japan's CO_2 emission per GDP stood at 0.09 T-C/$1000 compared with the OECD average of 0.18, and CO_2 emission per person stood at 2.27 T-C/person compared with the OECD average of 3.53.[3]

We thought that we had put the pollution problem behind us. We have contained air pollution, water contamination, and waste disposal by removing lead from gasoline, prohibiting the use of heavy metals in production processes and final products, removing the sulphur content from crude oil and from fume gas, increased the provision of sewerage and water purification plants, among others.

But global warming is a relatively new concern to which a proper response has yet to be devised, and Japan is no exception. Considering the internationally agreed policy target of restraining the CO_2 emission in the year 2000 to the current level, energy efficiency and reduction of greenhouse gases is a big challenge. It is in this context that we take up the Japanese experience.

3.2 Energy demand and the economy

The Japanese energy scene is detailed in Table 3-3a, which shows the energy matrix identifying sources of primary energy, energy conversion, and final consumption at five-year intervals. Among the sources, one can identify coal, petroleum, natural gas, hydro, and nuclear. Nonconventional sources such as solar, waste incineration, etc. are also listed. Table 3-3b shows the annual series of total energy supply and demand in the same format but without identifying sources of energy. The data are expressed in physical terms rather than value.

Figure 3-1, which is expressed in terms of petroleum equivalent, provides an overview of the energy flow from primary supply to final consumption.

Figure 3-1 Energy flow in Japan, 1990

(unit: million kl of petroleum equivalent, %)

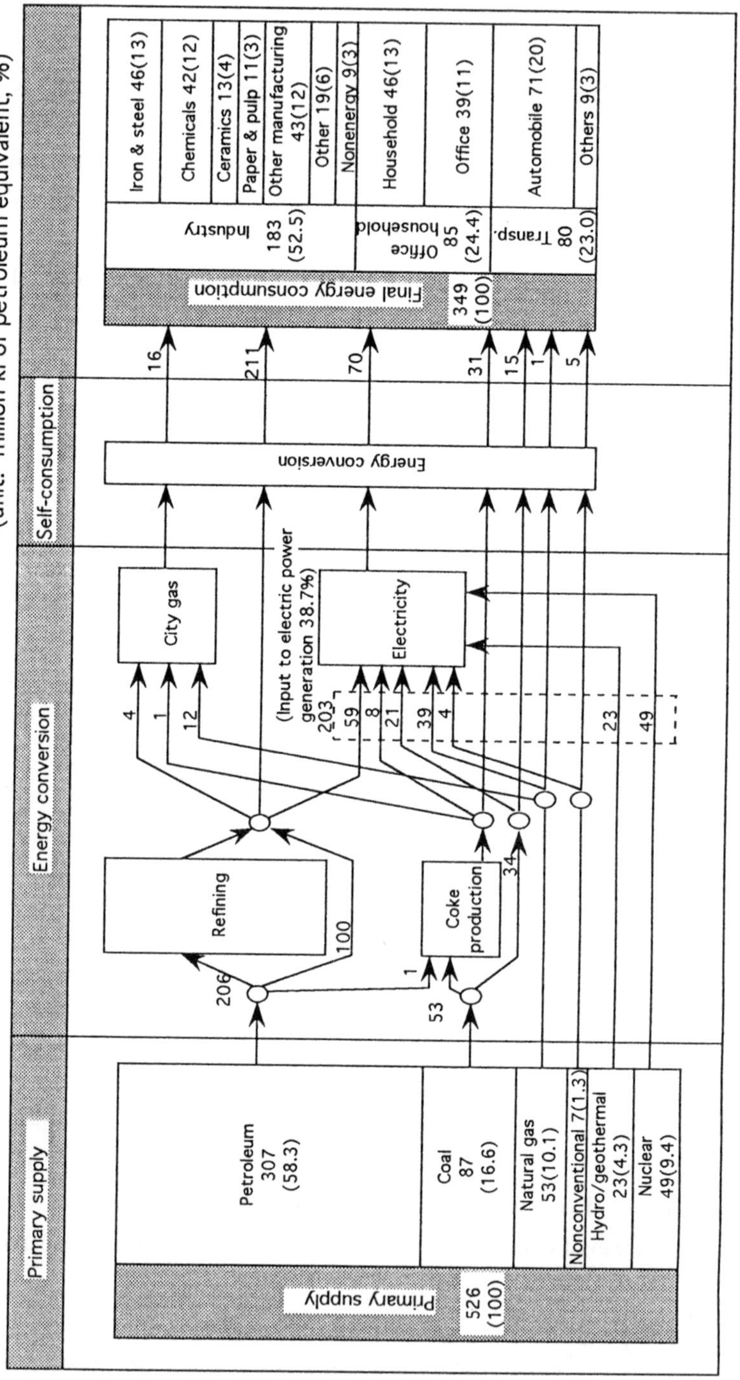

Source: MITI, *Energy '92*.

Let us start by looking at the demand side. After the first oil crisis, the demand for energy has been stable in Japan, reflecting energy conservation measures, structural change of the industry from heavy industry to high tech and services, stagnant economic activity due to appreciation of the yen, among others. Higher relative prices of energy also had a role to play.[4] The energy consumption had remained almost constant during 1973 and 1986. In contrast, from 1987 onwards, energy consumption began to increase, reflecting economic recovery. According to Table 3-3b which shows the time series data on energy, the final consumption showed an increase of 4.8% in 1987, 5.7% in 1988, 3.5% in 1989, and 3.8% in 1990.

In most industrialized countries, the share of industry use stands at around 34% (33.8% in USA, 34.6% in FRG., 31.3% in UK, and 34.7% in France), whereas the Japanese figure has traditionally been higher (according to Table 3-3a, 49.8% in 1990). Note that IEA statistics give lower figures than Japanese statistics. The rest is equally divided into domestic use (24.4%) and transportation (23.0%).

Let us move from the snapshot of the recent year to the time trend by looking at the energy consumption per unit GDP. Energy demand is induced by economic activity, but energy efficiency can and actually does change over time, reflecting technological progress and the shifting structure of the economy. Figure 3-2 depicts the ratio (energy consumption/GDP) for selected countries. We see clearly that the energy efficiency of the economy as a whole has been improving since the mid-1970s. The pace may have slowed down more recently, apparently reflecting a relative cheapening of energy. Still, the energy efficiency of the Japanese economy is one of the best among the industrialized eonomies, as noted earlier. It is interesting to see that the energy intensity of the economy has been on the increase in the 1960s and early 1970s in some countries, including Japan. This was the time when the industrialization drive led to rapid growth of heavy industry. Subsequently, the economy has shifted to electronics and services.[5]

Industrial growth is seen to inevitably accompany in proportion an increased consumption of energy and raw materials. Some of the ex-LDCs are now entering the phase of rapid industrialization, and if the past pattern is repeated, it will certainly lead to resource exhaustion. In some developed marked economies, however, the demand for energy has levelled off despite

Table 3-3a Energy supply and demand, the structure; 1970

Supply and demand sectors	Sector Code	Coal	Cokes
Primary energy:			
1. Domestic production		24946	-
2. Imports		38564	62
3. Total supply of primary energy	00	63509	62
4. Exports		- 7	-69
5. Changes in stock		-311	-18
6. Total supply for domestic use		63191	-25
Energy conversion:			
7. Electric power companies	70-	-10916	-1386
8. Electric power for own use	70-	-186	-1170
9. Heat supply companies	70-	-	-
10. City gas	70-	-	-1936
11. Cokes production	27-	-46907	40268
12. Petroleum refinery	27-	-	-
13. Petrochemical industry	26-	-	-
14. Others		-	-
15. Own consumption and loss		- 572	-3301
16. Statistical discrepancy		- 1451	-2059
Final energy consumption:			
17. Total		3159	30390
18. Industrial sector total		1629	28580
19. Agriculture, forestry, and fisheries	01&	-	-
20. Mining	10&	91	-
21. Construction	15	-	-
22. Manufacturing total	0M	1538	28580
23. Food processing	18	148	12
24. Textiles	20&-	47	-
25. Paper and pulp	24	217	-
26. Chemicals	26-	228	522
27. Nonmetallic mineral products	30	352	298
28. Iron and steel	31	297	27158
29. Nonferrous metals	32	61	415
30. Machinery	34&=34+35+36+37	63	138
31. Other manufacturing	38&&	126	36
32. Domestic sector total		1207	1061
33. Household	0F	756	987
34. Business	75	451	74
35. Transportation sector total	60T	323	749
36. Non-energy use		-	-

Source: Resource and Energy Agency, *Comprehensive Energy Statistics.*

(unit: 10**10 kcal)

	Crude oil	Petroleum products	Gasoline	Naphtha	Jet fuel	Kerosene	Light oil
1.	847	14	-	-	-	-	-
2.	192580	36452	2	5356	284	158	257
3.	193427	36466	2	5356	284	158	257
4.	-	-18137	-59	-3	-1560	-122	-316
5.	-973	-3518	-18	-247	-30	-1167	-299
6.	192454	14811	-75	5106	-1306	-1131	-358
7.	-6805	-33917	-	-	-	-	-66
8.	-	-7239	-	-	-	-	-
9.	-	-	-	-	-	-	-
10.	-1211	-1879	-	-1283	-	-	-
11.	-	-1227	-	-	-	-	-
12.	-182392	178490	17857	18140	2273	15648	11611
13.	-	385	-	192	-	-	-
14.	-	-	-	-	-	-	-
15.	-	-8108	-107	-133	-	-53	-45
16.	-1479	-36	-24	-2276	55	-1464	-235
17.	566	141280	17651	19748	1022	13000	10906
18.	566	82755	393	19748	-	4431	2792
19.	-	6463	-	-	-	1734	440
20.	-	554	-	-	-	84	142
21.	-	2834	-	-	-	681	1344
22.	566	72904	393	19748	-	1932	866
23.	-	3053	-	-	-	18	-
24.	-	4576	-	-	-	54	-
25.	-	4422	-	-	-	88	-
26.	566	29864	-	19707	-	342	-
27.	-	10259	-	-	-	194	-
28.	-	10944	-	-	-	306	-
29.	-	1415	-	-	-	91	-
30.	-	2860	-	-	-	914	-
31.	-	5512	393	41	-	-75	866
32.	-	20259	-	-	-	8569	80
33.	-	8403	-	-	-	5717	-
34.	-	11856	-	-	-	2852	80
35.	-	32228	17258	-	1022	-	8035
36.	-	6038	-	-	-	-	-

	Heavy oil	LPG	Natural gas and LNG	City gas	Hydro	Nuclear	Geo-thermal
1.	-	14	2702	-	17894	1054	66
2.	24769	3419	1269	-	-	-	-
3.	24769	3433	3970	-	17894	1054	66
4.	-15481	-51	-	-	-	-	-
5.	-1355	-284	-4	-	-	-	-
6.	7933	3097	3967	-	17894	1054	66
7.	-33851	-	-991	-	-16634	-1054	-19
8.	-7144	-	-212	-	-1260	-	-37
9.	-	-	-	-	-	-	-
10.	-	-211	-783	5191	-	-	-
11.	-	-	-	-	-	-	-
12.	103852	4193	-	-	-	-	-
13.	-	-	-	-	-	-	-
14.	-	-	-	-	-	-	-
15.	-4998	-523	-69	-619	-	-	-
16.	-1391	1167	-137	-1	-0	-	-
17.	64400	7723	1774	4571	-	-	9
18.	51959	2641	1749	486	-	-	4
19.	4289	-	-	-	-	-	4
20.	328	-	-	-	-	-	-
21.	809	-	-	-	-	-	-
22.	46532	2641	1749	486	-	-	-
23.	3035	-	-	72	-	-	-
24.	4522	-	-	32	-	-	-
25.	4335	-	-	0	-	-	-
26.	8583	1232	1341	42	-	-	-
27.	9726	27	-	61	-	-	-
28.	10579	-	-	60	-	-	-
29.	1324	-	-	29	-	-	-
30.	1946	-	-	126	-	-	-
31.	2484	1382	408	63	-	-	-
32.	8187	3423	25	4085	-	-	5
33.	-	2686	-	2943	-	-	-
34.	8187	737	25	1142	-	-	5
35.	4254	1659	-	-	-	-	-
36.	-	-	-	-	-	-	-

(unit: 10**10 kcal)

	Nonconventional energy				Elect-ricity	Heat supply	Total
		Solar	Waste in-cineration	Others			
1.	3260	-	10	3250	-	-	50782
2.	-	-	-	-	-	-	268926
3.	3260	-	10	3250	-	-	319708
4.	-	-	-	-	-	-	-18213
5.	-	-	-	-	-	-	-4823
6.	3260	-	10	3250	-	-	296672
7.	-3	-	- 3	-	26058	-	-45667
8.	-690	-	- 7	-683	4468	-	-6327
9.	-	-	-	-	-	-	-
10.	-	-	-	-	-	-	-619
11.	-	-	-	-	-	-	-7865
12.	-	-	-	-	-	-	-3902
13.	-	-	-	-	-	-	385
14.	-	-	-	-	-	-	-
15.	-	-	-	-	-3617	-	-16286
16.	-	-	-	-	-	-	-5164
17.	2567	-	-	2567	26910	-	211226
18.	1906	-	-	1906	18576	-	136251
19.	-	-	-	-	87	-	6554
20.	-	-	-	-	167	-	812
21.	-	-	-	-	214	-	3048
22.	1906	-	-	1906	18108	-	125837
23.	-	-	-	-	675	-	3960
24.	-	-	-	-	799	-	5454
25.	1906	-	-	1906	1503	-	8049
26.	-	-	-	-	4292	-	36854
27.	-	-	-	-	1102	-	12073
28.	-	-	-	-	4624	-	43082
29.	-	-	-	-	1900	-	3819
30.	-	-	-	-	1406	-	4593
31.	-	-	-	-	1808	-	7953
32.	661	-	-	661	7360	-	34664
33.	661	-	-	661	4579	-	18329
34.	-	-	-	-	2782	-	16335
35.	-	-	-	-	973	-	34273
36.	-	-	-	-	-	-	6038

Table 3-3a Energy supply and demand, the structure; 1975

Supply and demand sectors	Sector Code	Coal	Cokes
Primary energy:			
1. Domestic production		12758	-
2. Imports		47195	40
3. Total supply of primary energy	00	59953	40
4. Exports		-23	-387
5. Changes in stock		253	-35
6. Total supply for domestic use		60184	-381
Energy conversion:			
7. Electric power companies	70-	-4031	-3259
8. Electric power for own use	70-	-	52 -1276
9. Heat supply companies	70-	-	19 -
10. City gas	70-	-	-1756
11. Cokes production	27-	-52703	45261
12. Petroleum refinery	27-	-	-
13. Petrochemical industry	26-	-	-
14. Others		-	-
15. Own consumption and loss		-279	-3755
16. Statistical discrepancy		-1458	-1290
Final energy consumption:			
17. Total		1642	33545
18. Industrial sector total		1342	33167
19. Agriculture, forestry, and fisheries	01&	-	-
20. Mining	10&	-	-
21. Construction	15	-	-
22. Manufacturing total	0M	1342	33167
23. Food processing	18	-	-
24. Textiles	20&-	-	-
25. Paper and pulp	24	34	-
26. Chemicals	26-	146	227
27. Nonmetallic mineral products	30	241	211
28. Iron and steel	31	209	32378
29. Nonferrous metals	32	162	219
30. Machinery	34&=34+35+36+37	83	94
31. Other manufacturing	38&&	466	38
32. Domestic sector total		280	377
33. Household	0F	163	340
34. Business	75	117	37
35. Transportation sector total	60T	20	-
36. Non-energy use		-	-

(unit: 10**10 kcal)

	Crude oil	Petroleum products	Gasoline	Naphtha	Jet fuel	Kerosene	Light oil
1.	650	26	-	-	-	-	-
2.	244284	23682	-	4751	441	-	15
3.	244934	23708	-	4751	441	-	15
4.	-	-22090	-	-86	-1588	-0	-36
5.	-1380	925	-59	227	24	72	87
6.	243554	2543	-59	4892	-1123	72	66
7.	-21874	37230	-	-1951	-	-	-
8.	-	8689	-	-	-	-	-
9.	-	-45	-	-	-	-33	-
10.	-59	-3524	-	-2293	-	-	-
11.	-	-1207	-	-	-	-	-
12.	-221717	219027	24557	21722	2907	19426	15102
13.	-	556	-	278	-	-	-
14.	-	-	-	-	-	-	-
15.	-51	-11466	-118	-466	-	-73	-34
16.	233	8506	-21	545	7	333	-303
17.	86	168472	24360	22726	1791	19726	14830
18.	86	87727	368	22726	-	4879	3645
19.	-	7053	-	-	-	1029	506
20.	-	498	-	-	-	47	161
21.	-	2963	-	-	-	333	1803
22.	86	77213	368	22726	-	3471	1174
23.	-	3646	-	-	-	18	-
24.	-	5602	-	-	-	21	-
25.	-	4709	-	-	-	66	-
26.	86	29530	-	20721	-	237	-
27.	-	10360	-	-	-	132	-
28.	-	11007	-	-	-	444	-
29.	-	1051	-	-	-	48	-
30.	-	1625	-	-	-	477	-
31.	-	9683	368	2005	-	2027	1174
32.	-	29382	-	-	-	14847	139
33.	-	12226	-	-	-	7568	-
34.	-	17156	-	-	-	7279	139
35.	-	45079	23992	-	1791	-	11046
36.	-	6285	-	-	-	-	-

	Heavy oil	LPG	Natural gas and LNG	City gas	Hydro	Nuclear	Geo-thermal
1.	-	26	2681	-	19237	5653	96
2.	9924	7031	6550	-	-	-	-
3.	9924	7057	9231	-	19237	5653	96
4.	-20109	-11	-	-	-	-	-
5.	489	31	-5	-	-	-	-
6.	-9696	7078	9226	-	19237	5653	96
7.	-35279	-	-4556	-	-17980	-5648	-30
8.	-8533	-	-138	-	-1256	-5	-53
9.	-12	-	-	-43	-	-	-
10.	-	-675	-2696	7860	-	-	-
11.	-	-	-	-	-	-	-
12.	123777	5316	-	-	-	-	-
13.	-	-	-	-	-	-	-
14.	-	-	-	-	-	-	-
15.	-5593	-1017	-806	-747	-	-	-
16.	1765	1254	442	-0	-	-	-
17.	66429	11956	1471	7070	-	-	13
18.	51381	4359	1470	892	-	-	6
19.	5517	-	-	-	-	-	6
20.	290	-	-	-	-	-	-
21.	827	-	-	-	-	-	-
22.	44746	4359	1470	892	-	-	-
23.	3628	-	-	105	-	-	-
24.	5581	-	-	32	-	-	-
25.	4643	-	-	1	-	-	-
26.	7198	1374	1087	56	-	-	-
27.	9864	31	-	77	-	-	-
28.	9476	1073	-	140	-	-	-
29.	916	87	-	70	-	-	-
30.	1148	-	-	233	-	-	-
31.	2292	1794	383	178	-	-	-
32.	8762	5633	1	6178	-	-	8
33.	-	4658	-	4544	-	-	-
34.	8762	976	1	1634	-	-	8
35.	6286	1964	-	-	-	-	-
36.	-	-	-	-	-	-	-

(unit: 10**10 kcal)

	Nonconventional energy	Solar	Waste in-cineration	Others	Elect-ricity	Heat supply	Total
1.	3372	136	34	3202	-	-	44473
2.	-	-	-	-	-	-	321752
3.	3372	136	34	3202	-	-	366224
4.	-	-	-	-	-	-	-22500
5.	-	-	-	-	-	-	-241
6.	3372	136	34	3202	-	-	343483
7.	-10	-	-10	-	35539	-	-59079
8.	-778	-	-25	-753	5312	-	-6934
9.	-	-	-	-	-2	67	-42
10.	-	-	-	-	-	-	-176
11.	-	-	-	-	-	-	-8649
12.	-	-	-	-	-	-	-2690
13.	-	-	-	-	-	-	556
14.	-	-	-	-	-	-	-
15.	-	-	-	-	-4719	-	-21822
16.	13	-	-	13	0	-10	6435
17.	2597	136	-	2462	36130	57	251083
18.	2114	-	-	2114	22135	-	148940
19.	-	-	-	-	107	-	7166
20.	-	-	-	-	145	-	643
21.	-	-	-	-	85	-	3048
22.	2114	-	-	2114	21798	-	138083
23.	-	-	-	-	889	-	4640
24.	-	-	-	-	859	-	6494
25.	2114	-	-	2114	1805	-	8663
26.	-	-	-	-	4213	-	35346
27.	-	-	-	-	1303	-	12192
28.	-	-	-	-	5964	-	49698
29.	-	-	-	-	2387	-	3889
30.	-	-	-	-	1783	-	3818
31.	-	-	-	-	2594	-	13344
32.	483	136	-	347	12798	57	49563
33.	483	136	-	347	7808	11	25574
34.	-	-	-	-	4990	46	23989
35.	-	-	-	-	1197	-	46295
36.	-	-	-	-	-	-	6285

Table 3-3a Energy supply and demand, the structure; 1980

Supply and demand sectors	Sector Code	Coal	Cokes
Primary energy:			
1. Domestic production		13181	-
2. Imports		54146	-
3. Total supply of primary energy	00	67327	-
4. Exports		-53	-1489
5. Changes in stock		1128	-3
6. Total supply for domestic use		68402	-1492
Energy conversion:			
7. Electric power companies	70-	-5533	-3684
8. Electric power for own use	70-	-160	-1693
9. Heat supply companies	70-	- 21	-
10. City gas	70-	-	-1348
11. Cokes production	27-	-53532	45398
12. Petroleum refinery	27-	-	-
13. Petrochemical industry	26-	-	-
14. Others		-	-
15. Own consumption and loss		-171	-3675
16. Statistical discrepancy		-2702	158
Final energy consumption:			
17. Total		6282	33665
18. Industrial sector total		5973	32930
19. Agriculture, forestry, and fisheries	01&	-	-
20. Mining	10&	-	-
21. Construction	15	-	-
22. Manufacturing total	0M	5973	32930
23. Food processing	18	-	-
24. Textiles	20&-	-	-
25. Paper and pulp	24	112	-
26. Chemicals	26-	164	250
27. Nonmetallic mineral products	30	4337	214
28. Iron and steel	31	253	32057
29. Nonferrous metals	32	213	224
30. Machinery	34&=34+35+36+37	28	59
31. Other manufacturing	38&&	867	126
32. Domestic sector total		309	735
33. Household	0F	67	241
34. Business	75	242	494
35. Transportation sector total	60T	-	-
36. Non-energy use		-	-

(unit: 10**10 kcal)

	Crude oil	Petroleum products	Gasoline	Naphtha	Jet fuel	Kerosene	Light oil
1.	448	17	-	-	-	-	-
2.	230952	31019	-	5329	377	253	295
3.	231400	31036	-	5329	377	253	295
4.	-	-16043	-	-	-1911	-	-172
5.	-290	-151	-64	126	56	221	-40
6.	231110	14841	-64	5455	-1478	474	84
7.	-14910	-37071	-	-1101	-	-	-112
8.	-	-7986	-	-	-	-	-
9.	-	-50	-	-	-	-26	-
10.	-0	-3583	-	-1399	-	-	-
11.	-	-1082	-	-	-	-	-
12.	-212210	209805	29065	16756	3921	21069	19791
13.	-982	982	-	727	-	-	-
14.	-	-	-	-	-	-	-
15.	-48	-11867	-100	-624	-	-91	-39
16.	-2894	3152	115	-232	138	-396	-89
17.	66	167142	29016	19582	2581	21029	19635
18.	66	76975	103	19582	-	6050	4969
19.	-	8478	-	-	-	1468	1158
20.	-	459	-	-	-	60	232
21.	-	3831	-	-	-	648	2513
22.	66	64207	103	19582	-	3874	1066
23.	-	3239	-	-	-	-	-
24.	-	3932	-	-	-	30	-
25.	-	3881	-	-	-	23	-
26.	66	27021	-	18499	-	193	-
27.	-	6962	-	-	-	114	-
28.	-	5596	-	-	-	432	-
29.	-	1407	-	-	-	144	-
30.	-	1587	-	-	-	444	-
31.	-	10581	103	1082	-	2493	1066
32.	-	29557	-	-	-	14980	65
33.	-	13630	-	-	-	8565	-
34.	-	15927	-	-	-	6414	65
35.	-	53693	28913	-	2581	-	14602
36.	-	6917	-	-	-	-	-

	Heavy oil	LPG	Natural gas and LNG	City gas	Hydro	Nuclear	Geo-thermal
1.	-	17	2350	-	20481	18583	290
2.	10781	11982	21814	-	-	-	-
3.	10781	11999	24164	-	20481	18583	290
4.	-13548	-5	-	-	-	-	-
5.	-84	-298	0	-	-	-	-
6.	-2851	11696	24165	-	20481	18583	290
7.	-34975	-883	-17446	-	-19177	-18452	-198
8.	-7821	-	-142	-	-1305	-131	-51
9.	-24	-	-	-70	-	-	-
10.	-	-1673	-5090	10065	-	-	-
11.	-	-	-	-	-	-	-
12.	104412	4950	-	-	-	-	-
13.	-	-	-	-	-	-	-
14.	-	-	-	-	-	-	-
15.	-4545	-1043	-593	-737	-	-	-
16.	-416	556	102	-0	0	-0	-
17.	53780	13602	996	9257	-	-	41
18.	39530	5743	974	1448	-	-	17
19.	5852	-	-	-	-	-	17
20.	167	-	-	-	-	-	-
21.	670	-	-	-	-	-	-
22.	32840	5743	974	1448	-	-	-
23.	3239	-	-	188	-	-	-
24.	3902	-	-	48	-	-	-
25.	3858	-	-	-	-	-	-
26.	6271	2058	731	371	-	-	-
27.	6529	36	-	97	-	-	-
28.	4354	801	147	207	-	-	-
29.	1101	163	-	113	-	-	-
30.	1144	-	-	275	-	-	-
31.	2443	2685	96	149	-	-	-
32.	8405	6107	21	7809	-	-	23
33.	-	5065	-	5649	-	-	-
34.	8405	1043	21	2160	-	-	23
35.	5845	1752	-	-	-	-	-
36.	-	-	-	-	-	-	-

(unit: 10**10 kcal)

	Nonconventional energy	Solar	Waste in-cineration	Others	Elect-ricity	Heat supply	Total
1.	3918	370	107	3440	-	-	59267
2.	-	-	-	-	-	-	337931
3.	3918	370	107	3440	-	-	397198
4.	-	-	-	-	-	-	-17585
5.	-	-	-	-	-	-	685
6.	3918	370	107	3440	-	-	380298
7.	-31	-	-31	-	43999	-	-72502
8.	-1390	-	-76	-1313	5459	-	-7399
9.	-	-	-	-	-8	121	-28
10.	-	-	-	-	-	-	44
11.	-	-	-	-	-	-	-9216
12.	-	-	-	-	-	-	-2405
13.	-	-	-	-	-	-	-
14.	-	-	-	-	-	-	-
15.	-	-	-	-	-5425	-	-22516
16.	469	-	-	469	0	-19	-1735
17.	2967	370	-	2597	44024	102	264541
18.	2398	-	-	2398	25320	-	146102
19.	-	-	-	-	104	-	8599
20.	-	-	-	-	143	-	602
21.	-	-	-	-	91	-	3922
22.	2398	-	-	2398	24982	-	132978
23.	-	-	-	-	1134	-	4561
24.	-	-	-	-	911	-	4891
25.	2398	-	-	2398	2081	-	8472
26.	-	-	-	-	4041	-	32645
27.	-	-	-	-	1748	-	13357
28.	-	-	-	-	6506	-	44766
29.	-	-	-	-	2454	-	4411
30.	-	-	-	-	2704	-	4653
31.	-	-	-	-	3404	-	15222
32.	568	370	-	198	17395	102	56519
33.	564	365	-	198	10313	23	30486
34.	5	5	-	-	7082	79	26034
35.	-	-	-	-	1309	-	55003
36.	-	-	-	-	-	-	6917

Table 3-3a Energy supply and demand, the structure; 1985

Supply and demand sectors	Sector Code	Coal	Cokes
Primary energy:			
1. Domestic production		10975	-
2. Imports		67836	-
3. Total supply of primary energy	00	78810	-
4. Exports		- 1	-1647
5. Changes in stock		-196	27
6. Total supply for domestic use		78613	-1620
Energy conversion:			
7. Electric power companies	70-	-13622	-5108
8. Electric power for own use	70-	-1034	-2115
9. Heat supply companies	70-	-24	-
10. City gas	70-	-	-1133
11. Cokes production	27-	-54217	46990
12. Petroleum refinery	27-	-	-
13. Petrochemical industry	26-	-	-
14. Others		-	-
15. Own consumption and loss		-178	-4107
16. Statistical discrepancy		301	-1761
Final energy consumption:			
17. Total		9839	31147
18. Industrial sector total		9637	30144
19. Agriculture, forestry, and fisheries	01&	-	-
20. Mining	10&	-	3
21. Construction	15	-	-
22. Manufacturing total	0M	9637	30141
23. Food processing	18	-	-
24. Textiles	20&-	5	-
25. Paper and pulp	24	528	-
26. Chemicals	26-	1242	380
27. Nonmetallic mineral products	30	5359	382
28. Iron and steel	31	1476	28470
29. Nonferrous metals	32	178	226
30. Machinery	34&=34+35+36+37	0	501
31. Other manufacturing	38&&	849	183
32. Domestic sector total		202	1002
33. Household	0F	43	147
34. Business	75	159	855
35. Transportation sector total	60T	-	-
36. Non-energy use		-	-

(unit: 10**10 kcal)

	Crude oil	Petroleum products	Gasoline	Naphtha	Jet fuel	Kerosene	Light oil
1.	619	15	-	-	-	-	-
2.	181810	45597	445	11809	2047	899	219
3.	182429	45612	445	11809	2047	899	219
4.	-	-11427	-	-	-3132	-6	-173
5.	1468	-177	175	-161	-17	63	-131
6.	183897	34008	620	11648	-1102	956	-85
7.	-12137	-21820	-	-290	-	-	-140
8.	-	-6438	-	-335	-	-	-
9.	-	-42	-	-	-	-15	-
10.	-	-3155	-	-384	-	-	-
11.	-	-554	-	-	-	-	-
12.	-167036	165116	30307	8641	3710	22117	23708
13.	-1917	1917	-	1725	-	-	-
14.	-	-	-	-	-	-	-
15.	-28	-9366	-123	-250	-	-90	-99
16.	-2766	2443	22	-999			
17.	13	162109	30826	19755	2659	22508	23575
18.	13	66124	55	19755	-	6386	6296
19.	-	8355	-	-	-	1761	1470
20.	-	306	-	-	-	35	198
21.	-	4260	-	-	-	949	2687
22.	13	55203	59	19755	-	3641	1941
23.	-	2502	-	-	-	-	-
24.	-	3069	-	-	-	22	0
25.	-	2719	-	-	-	28	2
26.	13	25574	-	19024	-	232	2
27.	-	3273	-	-	-	95	12
28.	-	2593	-	-	-	284	36
29.	-	1104	-	-	-	151	4
30.	-	1987	-	-	-	411	52
31.	-	10383	59	731	-	2418	1833
32.	-	31199	-	-	-	16123	28
33.	-	16678	-	-	-	10959	-
34.	-	14521	-	-	-	5164	28
35.	-	57478	30768	-	2659	-	17251
36.	-	7308	-	-	-	-	-

	Heavy oil	LPG	Natural gas and LNG	City gas	Hydro	Nuclear	Geo-thermal
1.	-	15	2328	-	19081	35905	398
2.	12800	14154	35885	-	-	-	-
3.	12800	14169	38213	-	19081	35905	398
4.	-7836	-36	-	-	-	-	-
5.	158	-286	2	-	-	-	-
6.	5122	13848	38215	-	19081	35905	398
7.	-20658	-731	-28812	-	-17867	-35771	-283
8.	-5347	-50	-143	-	-1214	-134	-53
9.	-23	-5	-	-98	-	-	-
10.	-	-2386	-8086	12726	-	-	-
11.	-	-	-	-	-	-	-
12.	60953	5134	-	-	-	-	-
13.	-	-	-	-	-	-	-
14.	-	-	-	-	-	-	-
15.	-2804	-334	-601	-752	-	-	-
16.	490	-58	343	-1	-	-	-
17.	37734	15416	916	11875	-	-	61
18.	24688	6612	898	2309	-	-	26
19.	5123	-	-	-	-	-	26
20.	73	-	-	-	-	-	-
21.	624	-	-	-	-	-	-
22.	18868	6612	898	2309	-	-	-
23.	2502	-	-	334	-	-	-
24.	2930	100	-	70	-	-	-
25.	2566	48	-	57	-	-	-
26.	3516	2580	627	275	-	-	-
27.	2385	227	-	221	-	-	-
28.	1767	495	187	426	-	-	-
29.	710	193	-	102	-	-	-
30.	1133	390	-	791	-	-	-
31.	1359	2579	85	33	-	-	-
32.	8266	6782	18	9566	-	-	35
33.	-	5719	-	6743	-	-	-
34.	8266	1063	18	2823	-	-	35
35.	4779	2022	-	-	-	-	-
36.	-	-	-	-	-	-	-

(unit: 10**10 kcal)

	Nonconventional energy				Elect-ricity	Heat supply	Total
		Solar	Waste in-cineration	Others			
1.	4875	878	227	3770	-	-	74195
2.	-	-	-	-	-	-	331127
3.	4875	878	227	3770	-	-	405323
4.	-	-	-	-	-	-	-13076
5.	-	-	-	-	-	-	1123
6.	4875	878	227	3770	-	-	393370
7.	-81	-	-81	-	51401	-	-84100
8.	-2229	-	-146	-2084	5850	-	-7510
9.	-	-	-	-	-13	164	-15
10.	-	-	-	-	-	-	353
11.	-	-	-	-	-	-	-7780
12.	-	-	-	-	-	-	-1920
13.	-	-	-	-	-	-	-
14.	-	-	-	-	-	-	-
15.	-	-	-	-	-6408	-	-21440
16.	1060	-	-	1060	-	-26	-406
17.	3624	878	-	2747	50830	138	270552
18.	2607	-	-	2607	26767	-	138526
19.	-	-	-	-	225	-	8606
20.	-	-	-	-	205	-	514
21.	-	-	-	-	88	-	4348
22.	2607	-	-	2607	26249	-	125058
23.	-	-	-	-	1376	-	4212
24.	43	-	-	43	918	-	4105
25.	2512	-	-	2512	2175	-	7991
26.	41	-	-	41	4220	-	32372
27.	-	-	-	-	1713	-	10947
28.	-	-	-	-	6280	-	39431
29.	11	-	-	11	1516	-	3137
30.	-	-	-	-	3966	-	7244
31.	-	-	-	-	4086	-	15619
32.	1017	878	-	139	22661	138	65838
33.	981	842	-	139	12693	28	37313
34.	36	36	-	-	9968	109	28525
35.	-	-	-	-	1402	-	58880
36.	-	-	-	-	-	-	7308

Table 3-3a Energy supply and demand, the stucture; 1990

Supply and demand sectors	Sector Code	Coal	Cokes
Primary energy:			
1. Domestic production		6149	-
2. Imports		74605	-
3. Total supply of primary energy	00	80754	-
4. Exports		-	-1192
5. Changes in stock		952	-48
6. Total supply for domestic use		81706	-1241
Energy conversion:			
7. Electric power companies	70-	-16533	-4636
8. Electric power for own use	70-	-2892	-2667
9. Heat supply companies	70-	-18	-
10. City gas	70-	-	463
11. Cokes production	27-	-49465	42818
12. Petroleum refinery	27-	-	-
13. Petrochemical industry	26-	-	-
14. Others		-	-
15. Own consumption and loss		-80	-3645
16. Statistical discrepancy		957	-1894
Final energy consumption:			
17. Total		13675	28273
18. Industrial sector total		13619	27499
19. Agriculture, forestry, and fisheries	01&	-	-
20. Mining	10&	-	-
21. Construction	15	-	-
22. Manufacturing total	0M	13619	27499
23. Food processing	18	-	-
24. Textiles	20&-	19	-
25. Paper and pulp	24	1099	-
26. Chemicals	26-	1031	259
27. Nonmetallic mineral products	30	5819	469
28. Iron and steel	31	5161	25899
29. Nonferrous metals	32	101	212
30. Machinery	34&=34+35+36+37	128	330
31. Other manufacturing	38&&	260	329
32. Domestic sector total		56	774
33. Household	0F	12	73
34. Business	75	44	701
35. Transportation sector total	60T	-	-
36. Non-energy use		-	-

(unit: 10**10 kcal)

	Crude oil	Petroleum products	Gasoline	Naphtha	Jet fuel	Kerosene	Light oil
1.	604	8	-	-	-	-	-
2.	219737	63187	1786	16867	3847	3376	4588
3.	220341	63195	1786	16867	3847	3376	4588
4.	-	-14518	-41	-367	-4672	-441	-618
5.	-4591	-557	-231	15	-1	255	-258
6.	215751	48120	1514	16515	-827	3189	3721
7.	-20685	-24665	-	-121	-	-	-142
8.	-	-9635	-	-696	-	-	-
9.	-	-63	-	-	-	-	38 -
10.	-	-3408	-	-235	-	-	-
11.	-	798	-	-	-	-	-
12.	-190088	187763	36101	9467	4093	21115	30821
13.	-4339	4339	-	4155	-	-	-
14.	-	-	-	-	-	-	-
15.	-24	-9713	-125	-123	-	-80	-114
16.	-598	3643	128	-3252	-13	-460	211
17.	17	195584	37617	25710	3254	23725	34488
18.	17	79587	98	25710	-	8874	8427
19.	-	11612	-	-	-	2918	2702
20.	-	398	-	-	-	34	221
21.	-	5392	-	-	-	1236	3424
22.	17	62185	98	25710	-	4685	2080
23.	-	2543	-	-	-	-	-
24.	-	2274	-	-	-	18	1
25.	-	3041	-	-	-	31	2
26.	17	31726	-	25558	-	276	2
27.	-	3429	-	-	-	109	17
28.	-	3292	-	-	-	344	28
29.	-	1880	-	-	-	184	4
30.	-	2007	-	-	-	423	76
31.	-	11992	98	152	-	3302	1950
32.	-	34550	-	-	-	14851	76
33.	-	17360	-	-	-	10925	-
34.	-	17191	-	-	-	3926	76
35.	-	72675	37519	-	32541	-	25985
36.	-	8773	-	-	-	-	-

	Heavy oil	LPG	Natural gas and LNG	City gas	Hydro	Nuclear	Geo-thermal
1.	-	8	2078	-	20512	45511	465
2.	11886	17208	47206	-	-	-	-
3.	11886	17216	49284	-	20512	45511	465
4.	-8070	-2	-	-	-	-	-
5.	-283	12	-8	-	-	-	-
6.	3532	17226	49276	-	20512	45511	465
7.	-23332	-1070	-36452	-	-19211	-45316	-329
8.	-7630	-93	-37	-	-1301	-196	-58
9.	-17	-7	-	-154	-	-	-
10.	-	-2804	-11422	15879	-	-	-
11.	-	-	-	-	-	-	-
12.	69500	5459	-	-	-	-	-
13.	-	-	-	-	-	-	-
14.	-	-	-	-	-	-	-
15.	-2036	-59	-378	-468	-	-	-
16.	1299	-104	-281	-0	-	-	-
17.	41316	18548	705	15256	-	-	78
18.	26318	8006	616	3977	-	-	34
19.	5991	-	-	-	-	-	34
20.	144	-	-	-	-	-	-
21.	732	-	-	-	-	-	-
22.	19452	8006	616	3977	-	-	-
23.	2543	-	-	555	-	-	-
24.	2092	136	-	103	-	-	-
25.	2790	68	-	215	-	-	-
26.	2586	2721	514	596	-	-	-
27.	2771	240	-	256	-	-	-
28.	2097	776	42	789	-	-	-
29.	1404	238	-	219	-	-	-
30.	991	517	-	1244	-	-	-
31.	2178	3310	61	-	-	-	-
32.	11373	8250	88	11279	-	-	45
33.	-	6434	-	7764	-	-	-
34.	11373	1816	88	3515	-	-	45
35.	3625	2292	-	-	-	-	-
36.	-	-	-	-	-	-	-

(unit: 10**10 kcal)

	Nonconventional energy	Solar	Waste incineration	Others	Electricity	Heat supply	Total
1.	6098	1050	407	4642	-	-	81426
2.	-	-	-	-	-	-	404734
3.	6098	1050	407	4642	-	-	486161
4.	-	-	-	-	-	-	-15710
5.	-	-	-	-	-	-	-4252
6.	6098	1050	407	4642	-	-	466198
7.	-130	-	-130	-	64294	-	-103663
8.	-3281	-	-276	-3005	8572	-	-11494
9.	236	-	-	-	-29	236	-29
10.	-	-	-	-	-	-	586
11.	-	-	-	-	-	-	-7445
12.	-	-	-	-	-	-	-2324
13.	-	-	-	-	-	-	-
14.	-	-	-	-	-	-	-
15.	-	-	-	-	-7765	-	-22074
16.	1300	-	-	1300	-	-34	3093
17.	3987	1050	-	2937	65072	202	322848
18.	2831	-	-	2831	32704	-	160883
19.	-	-	-	-	273	-	11918
20.	-	-	-	-	218	-	616
21.	-	-	-	-	122	-	5514
22.	2831	-	-	2831	32091	-	142835
23.	-	-	-	-	1785	-	4883
24.	57	-	-	57	999	-	3452
25.	2692	-	-	2692	2829	-	9876
26.	30	-	-	30	4976	-	39150
27.	-	-	-	-	1976	-	11950
28.	-	-	-	-	6995	-	42178
29.	51	-	-	51	1512	-	3975
30.	-	-	-	-	5842	-	9551
31.	-	-	-	-	5178	-	17819
32.	1156	1050	-	106	30654	202	78804
33.	1117	1010	-	106	16353	31	42708
34.	40	40	-	-	14301	171	36095
35.	-	-	-	-	1714	-	74389
36.	-	-	-	-	-	-	8773

Table 3-3b Energy supply and demand

Energy supply, time series

	Demand sectors	Sector code	1955	1956
Primary energy:				
1.	Domestic production			
2.	Imports		13311	16920
3.	Total supply of primary energy	00	64129	71105
Sources of primary energy:				
a.	Coal		30286	34143
b.	Cokes		—	12
c.	Crude oil		8949	12022
d.	Petroleum products		2322	2139
e.	Natural gas and LNG		244	294
g.	Hydro		17465	17606
h.	Nuclear		—	—
i.	Geothermal		—	—
j.	Nonconventional energy		4865	4889
j1.	Solar		—	—
j2.	Waste incineration		—	—
j3.	Others		4865	4889

Table 3-3c Energy supply and demand

Energy demand, time series

17.	Final consumption total		41102	46440
18.	Industrial sector total		22859	26710
19.	Agric., forestry, and fisheries	01&	1283	1537
20.	Mining	10&	165	232
21.	Construction	15	97	140
22.	Manufacturing total	0M	21314	24801
23.	Food processing	18	1416	1522
24.	Textiles	20&-	1733	1949
25.	Paper and pulp	24	2087	2338
26.	Chemicals	26-	4788	5527
27.	Nonmetallic mineral products	30	3060	3694
28.	Iron and steel	31	5485	6404
29.	Nonferrous metals	32	855	1035
30.	Machinery	34&=34+35+36+37	897	1158
31.	Other manufacturing	38&&	992	1172
32.	Domestic sector total		10248	10742
33.	Household	0F	7631	7781
34.	Business	75	2617	2962
35.	Transportation sector total	60T	7372	8200
36.	Non-energy use		624	787

Note: Numbers and symbols correspond to those in Table 3-3a.

(unit: 10**10 kcal)

	1957	1958	1959	1960	1961	1962	1963
1.	57106	53167	54625	57064	60554	56597	56886
2.	21673	20735	31235	43746	55389	64396	79678
3.	78779	73903	85860	100810	115943	120992	136564
a.	37545	32611	36207	41522	44812	41799	43221
b.	30	0	—	—	142	18	5
c.	14295	16227	23820	31318	37535	45231	59458
d.	3295	1682	3625	6611	9499	11847	12326
e.	425	536	700	939	1366	1763	2054
g.	18138	18228	16966	15780	17999	15892	16585
h.	—	—	—	—	—	—	—
i.	—	—	—	—	—	—	—
j.	5052	4619	4543	4640	4589	4444	2917
j1.	—	—	—	—	—	—	—
j2.	—	—	—	—	—	—	—
j3.	—	—	—	4640	4589	4444	2917

17.	49687	48846	56168	65271	74125	79846	88417
18.	28930	27545	33514	39815	45835	48353	54847
19.	1572	1554	1890	1947	2058	2374	2699
20.	251	244	260	272	332	357	342
21.	184	239	352	486	670	952	1130
22.	26923	25507	31013	37111	42775	44669	50676
23.	1632	1614	1747	1897	2024	2281	2394
24.	2005	1842	1925	2160	2243	2286	2331
25.	2580	2446	3048	3580	3836	3904	4276
26.	6063	5719	6916	8277	9535	10348	12538
27.	4143	3742	4549	5636	6097	6497	7065
28.	6944	7039	8900	10839	13373	13392	15462
29.	1011	815	1155	1328	1446	1571	1799
30.	1246	1085	1300	1576	1851	2121	2391
31.	1298	1205	1473	1819	2370	2269	2421
32.	11271	11159	11551	12657	13791	15344	15165
33.	8211	8025	8113	8801	9309	10165	9415
34.	3060	3134	3438	3856	4481	5179	5750
35.	8655	9204	9975	11503	12836	14286	16210
36.	831	937	1128	1295	1664	1863	2195

	1964	1965	1966	1967	1968	1969	1970
1.	55884	56193	56315	54984	55072	52832	50782
2.	94724	112717	128838	157402	188119	224693	268926
3.	150608	168910	185153	212386	243190	277526	319708
a.	44449	45619	47550	53095	57284	61416	63509
b.	1	35	16	227	107	130	62
c.	70381	83103	98731	118455	138847	164827	193427
d.	14792	17575	15664	19600	24274	27517	36466
e.	2027	2027	2112	2234	2409	2847	3970
g.	16194	17938	18336	15947	17096	17391	17894
h.	—	8	134	145	240	249	1054
i.	—	—	4	26	50	55	66
j.	2764	2604	2604	2657	2884	3095	3260
j1.	—	—	—	—	—	—	—
j2.	—	5	5	5	5	5	10
j3.	2764	2600	2600	2652	2880	3090	3250

17.	100299	108538	123604	141333	158642	186387	211226
18.	62920	67848	77761	89668	100494	119524	136251
19.	2771	3215	3892	4403	4793	5413	6554
20.	474	566	606	477	760	799	812
21.	1202	1330	1744	2105	2311	2635	3048
22.	58473	62738	71520	82683	92631	110677	125837
23.	2776	2974	3117	3363	3599	3842	3960
24.	2520	2767	3588	4093	4467	4828	5454
25.	4986	5205	5800	6349	6523	7364	8049
26.	15002	16484	18533	21392	24644	30922	36854
27.	8044	7737	8628	9779	10699	11355	12073
28.	17387	19210	22337	26981	30482	38318	43082
29.	1814	1860	2252	2599	2885	3443	3819
30.	2414	2383	2726	3023	3472	3989	4593
31.	3529	4119	4538	5105	5860	6617	7953
32.	16988	18700	21167	23679	26595	31026	34664
33.	9832	10696	11442	12791	14104	16570	18329
34.	7156	8004	9725	10889	12491	14455	16335
35.	17665	19055	21104	23944	27082	30520	34273
36.	2727	2934	3571	4041	4471	5319	6038

(unit: 10**10 kcal)

	1971	1972	1973	1974	1975	1976	1977
1.	48761	46137	40846	45071	44473	46970	43323
2.	276029	300900	344563	339608	321752	340362	343947
3.	324790	347037	385409	384679	366224	387333	387270
a.	55842	55872	59556	63512	59953	58620	55861
b.	11	—	31	179	40	22	2
c.	209479	230370	269140	257177	244934	256954	258312
d.	31018	31644	29095	29125	23708	30356	31367
e.	4002	4029	5914	7684	9231	10457	13860
g.	19219	19475	15772	18972	19237	19514	16958
h.	1802	2133	2184	4432	5653	7668	7123
i.	99	65	66	74	96	97	125
j.	3357	3448	3651	3523	3372	3643	3662
j1.	—	—	—	—	136	148	169
j2.	11	11	20	29	34	56	72
j3.	3345	3437	3631	3494	3202	3440	3421

17.	223535	240305	265234	257723	251083	266319	265244
18.	141647	150169	165663	158585	148940	157532	153406
19.	6652	6925	7634	7240	7166	7732	8141
20.	742	634	626	679	643	651	605
21.	3018	3637	3780	3270	3048	3109	3205
22.	131235	138974	153624	147395	138083	146040	141455
23.	4066	4320	4666	4641	4640	4699	4629
24.	5830	6175	6535	6405	6494	6646	6105
25.	8355	9047	9890	9225	8663	8856	8620
26.	38217	39855	41330	38307	35346	38580	38755
27.	12569	13624	14676	13252	12192	12909	13019
28.	44491	46553	54494	53587	49698	50647	46490
29.	3850	4191	4692	4350	3889	4080	4168
30.	4571	4628	5007	4383	3818	4226	4193
31.	9287	10581	12333	13244	13344	15397	15476
32.	38688	48276	48030	48281	49563	53929	54439
33.	19788	21542	23591	24377	25574	28098	28529
34.	18899	21734	24439	23904	23989	25831	25911
35.	36493	39454	43425	44127	46295	48617	50509
36.	6708	7406	8116	6885	6285	6242	6889

	1978	1979	1980	1981	1982	1983	1984
1.	48484	53965	59267	59040	60532	63909	65893
2.	337970	357170	337931	323128	303764	319631	337174
3.	386453	411134	397198	382168	364296	383540	403067
a.	51379	56677	67327	70406	67538	68921	75771
b.	—	—	—	—	—	—	—
c.	251353	257644	231400	212416	191543	196582	196771
d.	32052	36340	31036	31036	33459	39187	41790
e.	18012	21484	24164	24259	25238	28924	36962
g.	16469	18878	20481	20085	18686	19403	16419
h.	13346	15838	18583	19760	23047	25715	30210
i.	188	279	290	294	312	369	374
j.	3655	3994	3918	3913	4172	4438	4772
j1.	185	213	370	492	605	718	809
j2.	74	76	107	116	137	168	210
j3.	3396	3705	3440	3304	3430	3552	3753

17.	273313	279578	264541	257041	248815	260308	267406
18.	155066	157913	146102	138765	130073	132378	139003
19.	9078	9021	8599	8343	7599	8443	8832
20.	587	599	602	552	520	520	521
21.	3553	3744	3922	3754	3673	4013	4465
22.	141848	144549	132978	126116	118280	119403	125185
23.	4608	4620	4561	5075	4487	4462	4386
24.	6209	5300	4891	5848	5554	5190	4254
25.	9023	9502	8472	7985	7859	8029	8226
26.	39717	39649	32645	29440	28351	28907	31670
27.	14210	14389	13357	12466	11362	11425	11433
28.	44354	46743	44766	41607	38090	38077	40255
29.	4279	4606	4411	3562	2991	2946	3137
30.	4409	4454	4653	5789	5675	6250	6767
31.	15039	15288	15222	14343	13911	14116	15057
32.	57163	58488	56519	57213	57477	63677	63467
33.	30409	31917	30486	31802	31649	35609	35643
34.	26753	26572	26034	25412	25828	28068	27824
35.	53638	55690	55003	54312	54628	57212	57483
36.	7447	7487	6917	6752	6638	7041	7453

(unit: 10**10 kcal)

	1985	1986	1987	1988	1989	1990
1.	74195	73307	76241	77108	78004	81426
2.	331127	328909	346135	368259	383725	404734
3.	405323	402217	422377	445366	461729	486161
a.	78810	73285	76125	80539	79670	80754
b.	—	—	—	—	—	—
c.	182429	173554	174089	184926	195144	220341
d.	45612	53987	66228	70479	72285	63195
e.	38213	39592	40861	42593	46158	49284
g.	19081	18532	17148	20709	21119	20512
h.	38905	37869	42246	40198	41146	45511
i.	398	376	378	349	368	465
j.	4875	5022	5303	5573	5839	6098
j1.	878	935	979	1009	1030	1050
j2.	227	274	274	325	342	407
j3.	3770	3813	4050	4239	4466	4642

	1985	1986	1987	1988	1989	1990
17.	270552	271583	284536	300545	311102	322848
18.	138526	136212	142981	151754	156023	160883
19.	8606	9302	9796	10439	10834	11918
20.	514	523	537	549	605	616
21.	4348	4616	4851	5009	5095	5514
22.	125058	121772	127797	135758	139490	142835
23.	4212	4177	4450	4813	4749	4883
24.	4105	4100	4199	4325	4034	3452
25.	7991	8226	8911	9978	9963	9876
26.	32372	32885	34919	36371	38018	39150
27.	10947	10386	10633	11723	12141	11950
28.	39431	36621	38530	40328	41448	42178
29.	3137	3128	2834	2915	3223	3975
30.	7244	7171	7571	8302	8751	9551
31.	15619	15079	15750	17006	17163	17819
32.	65838	66517	70007	73805	75444	78804
33.	37313	37571	39981	41186	41881	42708
34.	28525	28946	30026	32619	33563	36095
35.	58880	60952	63427	66665	71226	74389
36.	7308	7901	8121	8321	8410	8773

Figure 3-2 Energy consumption per GDP in selected countries

(unit: TOE / million US$, 1985 prices)

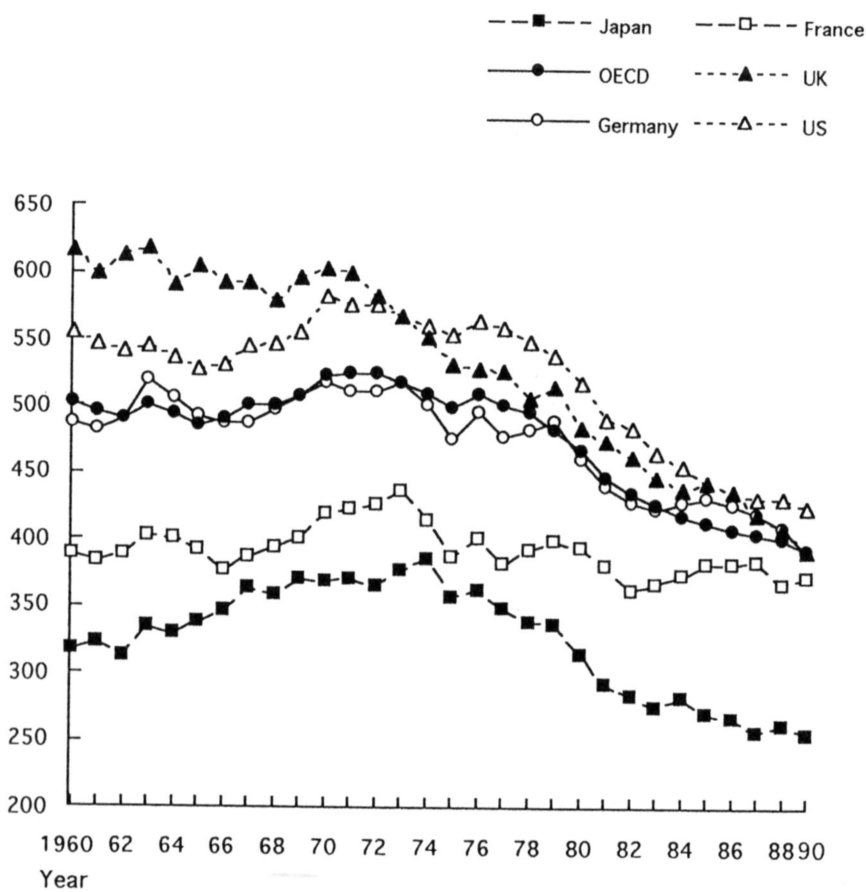

Source: OECD, *Energy Balances.*

continued economic growth. Such was the case also in Japan.

In this section we shall examine the technological change.

In the field of energy conservation, there were new developments in heat pumps, heat exchangers, waste heat recovery, energy storage, municipal solid waste systems, combined heat and power (CHP), and district heating (DH). New building design, building materials, and insulation materials have also been put into practical use. More energy-efficient engines and electric motors have appeared. Microelectronic sensor technologies have enabled us to control energy consumption precisely to meet actual requirements. We discuss this according to the end-use categories.

Energy conservation in industry: Energy consumption for industry amounts to about 50% of the total demand, making energy conservation in industry a matter of the utmost importance in an effort to conserve energy and reduce CO_2 emission. Price hikes of energy prompted the industrial sector to conserve energy by introducing energy-efficient production processes. As a result, energy consumption per unit production has shown remarkable improvement in the past. Figure 3-4 shows the trends in unit energy requirement (defined as physical weights of output/energy input in physical terms) in major energy-consuming sectors including iron and steel, paper and pulp, cement, and petrochemicals, which comprise about 70% of the energy requirement in manufacturing.

More recently, however, due to the relative cheapening of energy, the conservation effort has been relaxed somewhat, and even a reversal trend is seen such as a shift toward high-value added products with lower fuel efficiency.

Based on their previous study, IEA [1987, p.156, pp.236-237] lists the following technologies as important developments in the industrial use of energy:

* Energy-efficient production systems;
* Heat recovery systems (e.g. heat pumps);
* Recycling of energy-intensive products.

Regarding process heat, waste heat recovery is often economical, involving all types of fuels including oil, gas, and coal. It is especially true where heavy industries are concentrated. As for water heat, improved boiler designs and controls are currently economical for industrial and commercial use. Improved controls and insulation can be retrofitted to existing systems.

Figure 3-3 Energy conservation in industry

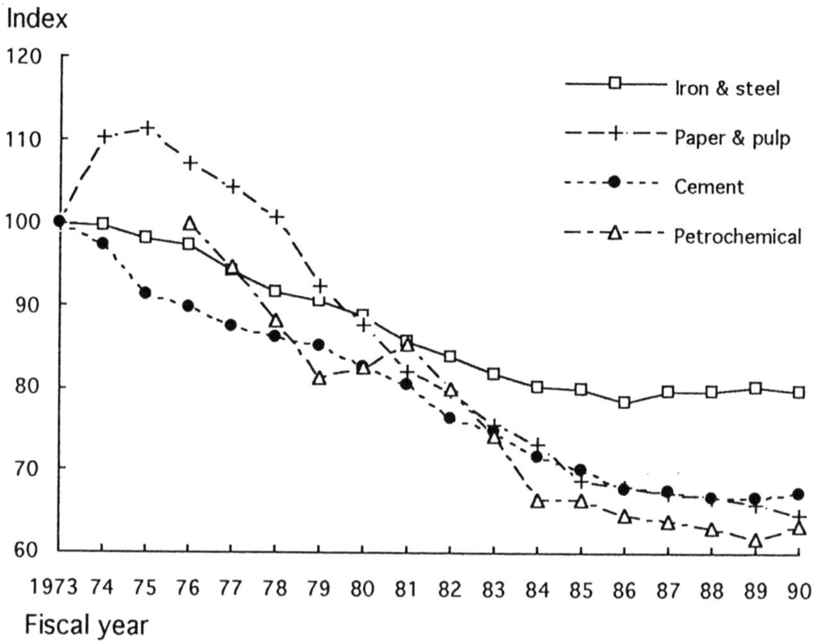

Note: For petrochemical industry, base year is fiscal 1977. The figures for iron and steel are adjusted to conform to the production condition in fiscal 1973. Other figures are apparent ratios (energy consumption/production).

Source: Original calculation conducted by the MITI. Reproduced here from *Energy Conservation Handbook.*

Electronic control technology has wide applications in all end-use sectors and is economical today. Its energy-saving potential is large for all types of fuel, although it may be more effective where electricity is involved. Some countries, however, lack strong electronics industries, resulting in insufficient information and advisory services.

In steel production, many factors influence the energy intensity such as coke production/importation, the method of refining pig iron, the use/nonuse of scrap, reheating, and various finishing treatments. Iron ore is first reduced to pig iron in a blast furnace using coke as the primary fuel, or is directly reduced to sponge iron. The pig iron from the blast furnace is further purified into steel in a basic oxygen furnace (BOF), an electric arc furnace (EAF), or using the open-hearth method which is the most antiquated, requiring much more energy than does the EAF or the BOF. The EAF can be charged with nearly all scrap, saving on the pig iron input. A BOF process requires approximately 19 gigajoules to produce a metric ton of crude steel. Using the EAF technology, it takes about 8.5 gigajoules [IEA 1991a, p.53]. The crude steel is moulded into ingots or is directly cast into shapes such as beams or bars.

Continuous casting of steel eliminates reheating step of ingots before breaking them down into billets or sheets. The diffusion of continuous casting is reported by IEA as follows:

	1980	1988
North America	21.1%	62.5%
Europe	38.5%	89.8%
Pacific region	56.2%	91.9%
IEA total	38.7%	77.5%

Energy intensity in steel production has thus improved considerably during the 1980s. In North America, it required 25.4 GJ to produce a metric ton of crude steel (excluding the production of coke) in 1980, which was reduced to 21.7 GJ in 1988. In the Pacific, an improvement from 19.5 GJ to 17.6 GJ was observed during the same priod [IEA 1991a, pp.54-55].

In the paper and pulp industry, wood is converted into pulp by a mechanical process or a chemical process, or a combination of both. In IEA countries, chemical processes account for 72%, mechanical processes 23%, and combined methods 5%. Pulp is then bleached, formed into a sheet and dried. The most energy-intensive processes are pulping and drying. Improved energy efficiency has been achieved through recycling of the steam used in making chemical and thermo-mechanical pulp. In the drying process, which typically consumes 1/2 of the energy requirement, hoods have been used in the production line since the early 1970s to retain the heat.

The use of waste paper reduces the energy requirement as well. Energy intensity per metric ton of paper varies, reflecting the fact that waste pulp can substitute for virgin wood pulp. The input of waste paper in paper production has reached nearly half (48.2%) of the total input in the Pacific. Some figures are cited here from IEA [1991a, p.56] regarding this point.

	1980	1986
North America	14.4%	18.1%
Europe	35.3%	35.2%
Pacific region	4.9%	48.2%
IEA total	18.5%	26.7%

Importation of pulp also contributes to energy efficiency. Thus, the IEA Pacific region was able to produce a metric ton of paper using 8.8 GJ in 1986, whereas in North America it took 19.1 GJ in the same year.

In the cement industry, energy accounts for about half of the total production cost. Clinker is produced in the kiln and is ground into cement. There are two basic ways of producing clinker, namely the wet process and the dry process. In 1980, the wet process required approximately 5.86 GJ per metric ton of clinker, whereas the dry process using a long rotary kiln required about 4.04 GJ per metric ton [IEA 1991a, p.57]. With the introduction of pre-heaters to the kiln feed, which uses exhaust heat, the energy requirement has been reduced to about 3.35 GJ per ton of clinker. The addition of pre-calciner further reduced the energy requirement to 3.14 GJ. These methods were introduced in the 1950s but did not find commercial application until the early 1970s, it was reported.

The Energy Council recommends the introduction of energy-efficient plants and equipment, promotion of effective use of waste heat, and recycling as necessary steps toward continued improvement in energy use. MITI's policies in this regard include:

a. Provision of financial measures for investment in plants and equipment

Figure 3-4 Fuel efficiency of passenger cars

a. Average

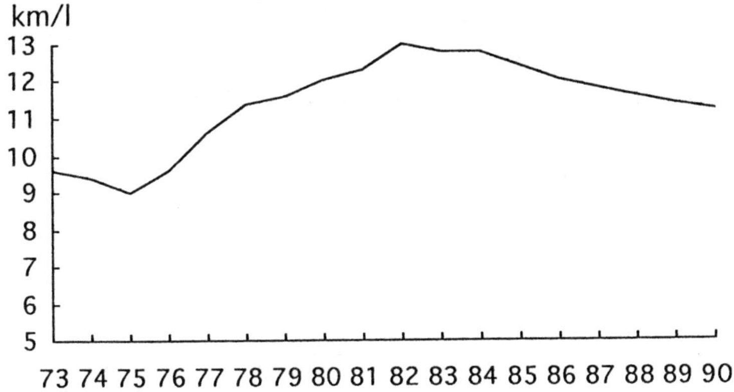

b. By engine size

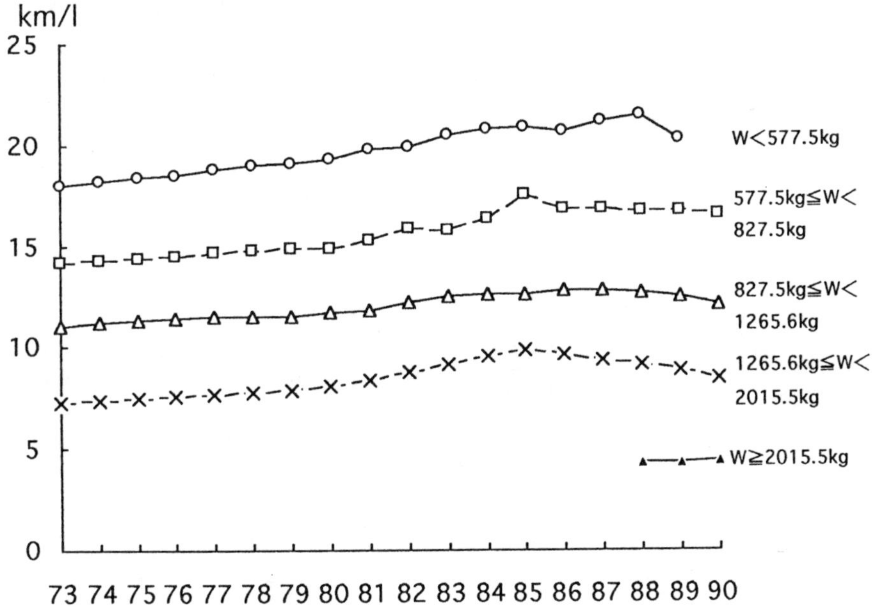

Source: MITI, *Energy '92*.

aimed at improvement of existing production technology and introduction of new ones.

b. Utilization of waste heat from factories for industrial purposes outside the factory sites. Office and household use of such hitherto unused energy sources in the local area should also be promoted.

c. Promotion of recycling of scrap iron, used paper, etc. which contributes to a reduction of the primary energy consumption.

d. Extending consulting services to medium and small-sized firms aimed at providing information on energy conservation. It is also useful to provide training for energy managers.[MITI, 1991, p.46]

Energy transformation sector: The technological developments in this field include more efficient power plants (e.g. combined cycle, fuel cells) and cogeneration systems.

Combined heat and power supply (CHP) coupled with district heating is described as economical in some situations. Primary energy is used for electricity generation. Availability of sites and connections with the grid are essential for commercial application. For this reason, existing power station sites in favorable locations should be reserved for future CHP use.

According to the Japanese experience, the average energy efficiency of power plants has nearly doubled in the post-WWII period as shown below [Resouces and Energy Agency, 1992]:

1953	20.7%
1955	24.0%
1960	31.9%
1965	36.5%
1970	36.9%
1975	37.4%
1976 and after	38.1%

This is partly attributable to the diffusion of state-of-the art technology, especially in the earlier years when some of the power plants were obsolete, and the rest is due to genuine technological progress.

The Electricity Utility Law prohibits the sale of surplus electricity by an enterprise. In order to promote cogeneration, the government is now adopting

Figure 3-5 Energy consumption in offices

a. Electricity consumption per floor space

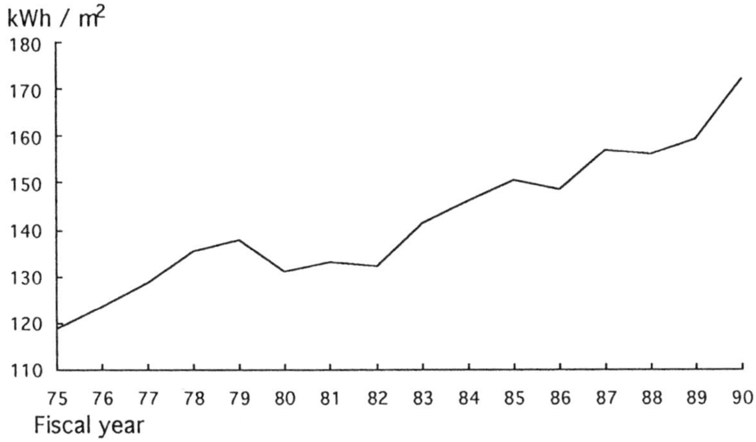

Source: MITI, *Energy '92*.

b. Electricity consumption by equipment type

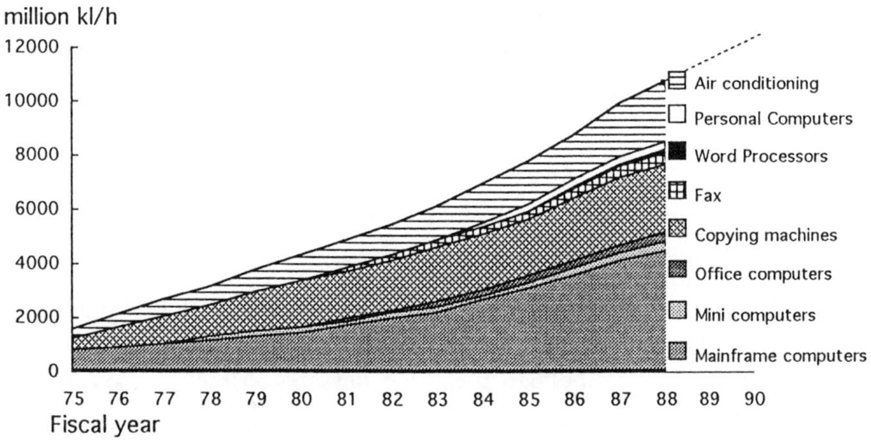

Source: MITI, *Energy '91*.

a new policy of allowing transactions between the utility companies and the nonutility producers of electricity.

Transportation sector: Improvement in the fuel efficiency of motor vehicles was observed, much the same as in the case of consumer durables, after the first oil crisis. As can be seen in Figure 3-4, automobiles produced in 1973 were able to travel 9.6 kilometers per liter, which improved to 13.0 kilometers in 1982 (figures are averages for cars produced in the years in question, and are not for identical types of cars). Reflecting the price reduction of fuels, however, the fuel efficiency of the Japanese car fleet has deteriorated somewhat since then. One contributing factor was the consumer demand which tended to shift toward larger, more luxurious cars. But if we look at the fuel efficiency of newly produced passenger vehicles classified by car weight, which is also shown in Figure 3-8, continued improvement was observed within each category. For instance, in the case of cars weighing between 1265 and 2014 kg, the distance travelled using one liter of fuel improved from 7.9 to 9.8 km between 1979 and 1985. This has somewhat deteriorated to 8.5 km in 1990.

One of the reasons for the high fuel efficiency of the Japanese car fleet is the compulsory automobile inspection which comes three years after purchase, and every 2 years thereafter, and every year after 10 years.

A factor which worked counter to fuel efficiency was the abolition of a commodity tax levied on luxuries including large-sized passenger cars. This was considered to be unduly discriminatory against imported cars and was criticized by the trading partners as constituting a nontariff trade barrier. The tax was abolished and integrated to a uniform 3% consumption tax (4.5% with temporary surcharge), which in effect substantially reduced the price of large-sized (hence less fuel-efficient) automobiles.

IEA summarizes the development of internal combustion engines as having a large energy-saving potential, particularly oil. Allowing for the introduction of new designs and rotation of car stock, they anticipate 10 to 15 years of lead time before the full effect is achieved. They identify low gasoline prices as the main impediments in the commercial application of new technologies. More efficient vehicles, such as those with ceramic engines, are also a possibility.

Government standards for higher fuel efficiency can act, often coupled with stricter environmental standards, as the proper stimulus. On this point, OECD [1988b, p.105] writes, "Many technologies which exist today to improve fuel

Figure 3-6 Energy consumption of household durables

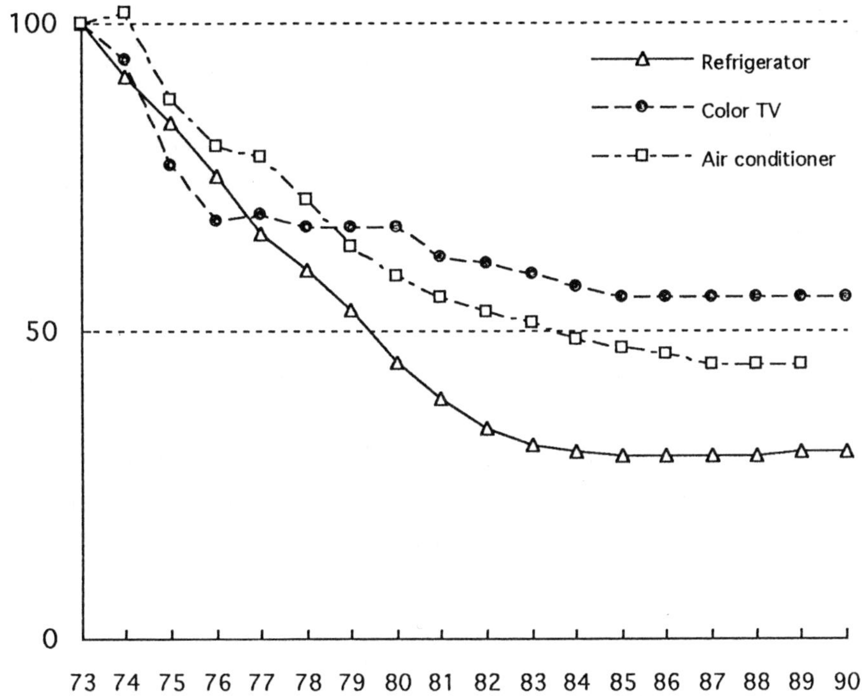

Note: For refrigerator, the data refer to monthly electricity consumption of a 170-liter class freezer-refrigerator with two doors. The data are for the refrigeration year (October in the previous year to September in the current year). The data for color television refer to 19/20 inches-type. The data for air conditioner refer to a 1600 kcal/h class, separate-type.

Source: MITI, *Energy '92*.

economy—for example, advanced air/fuel management systems such as fuel injection, electronic control of spark timing, advanced choke systems, improved transmission, etc.—can also result in significant exhaust emission benefits. In fact, some of the advances were developed as a direct result of tight emission control requirements." "Current experience indicates that vehicles may achieve very low exhaust emission levels for hydrocarbons, carbon monoxide and nitrogen oxides, alongside good fuel savings. A recent comparison between fuel consumption in IEA countries reveals that there need not be any fuel economy penalties due to more advanced emission control systems used in, for instance, Japan and the United States."

Electric battery vehicles will be entirely oil-saving. The technology is economically feasible now for some delivery vehicles. Wider application for cars and light vans will require 15 to 25 years. The public procurement of electric battery vehicles provides stimulus. Obstacles include the efficiency of the battery which limits driving distance between charges, high vehicle costs due to small market, and competition from increasingly efficient gasoline and diesel vehicles.

Household and commercial sector: In offices, energy efficiency had been improving in the case of air conditioning and elevators. Recently, however, the opposite trend has become apparent owing to such factors as service orientation of the economy, progress of information orientation, and 24-hour operation of business and social activities. The desire for a more comfortable office environment calls for the widespread use of air conditioning, and the progress of office automation (OA) requires equipment such as computers, copying machines, fax machines, word processors, and the like. Energy demand for the business field is described in Figure 3-5.

In Japan, energy efficiency has improved remarkably since the two oil crises in the 1970s, not only in the industrial field but also for consumer durables. Figure 3-6 shows energy consumption for typical consumer durables including refrigerators, television sets, vacuum cleaners and air conditioners. In the case of refrigerators, energy consumption for the identical type of equipment dropped to 1/3 of the 1973 level by 1988. In 1973, a 170-liter class refrigerator consumed 79.6 KWH/month; the 1988 figure stood at 26.0 KWH/month. In the case of color television sets and air conditioners, energy consumption dropped to about 2/3 during the same period. For the former, a 19.2-inch-type TV consumed 140W in 1973, which dropped to 83W by 1988; for the latter, a typical air conditioner consumed 847W of electricity, whereas the 1988 figure was 482W.

With rising real income, people are acquiring more consumer durables and opting for larger equipment. Energy consumption in the household has continued to increase as shown below. It is seen that energy consumption per household has increased nearly 10% in the past five years.[6] IEA [1987] has

Figure 3-7 Insulation of housing

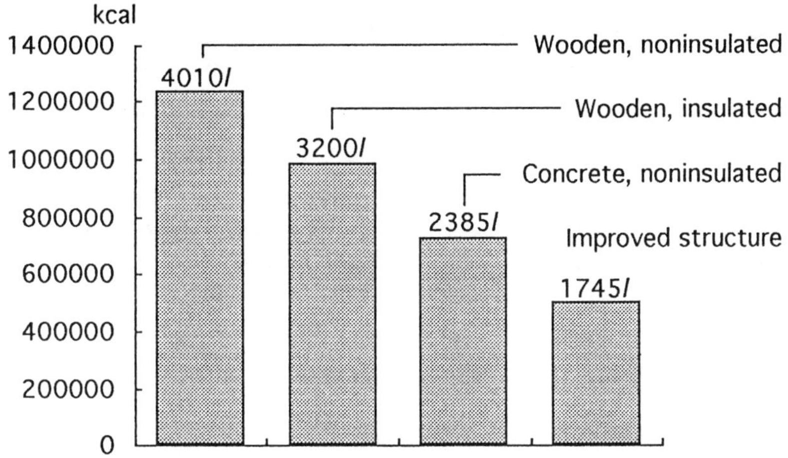

Note: Data refer to heating load during 4 months (December to March).

Source: Data provided by Daiwa House Co.

focused on technological developments such as building design, heating and cooling systems (e.g. district heating, heat pumps), and total energy management system.

The Law Concerning the Rational Use of Energy enacted in 1979 is a basic law providing the basis for government policy in Japan. MITI is responsible for the general energy conservation policy and implementing specific policies for the industrial, residential, and commercial sectors. MITI works with the Ministry of Construction and the Ministry of Transportation in relevant fields.

Heat pump technology is available today, with oil and electricity saving potentials. It is most competitive where the energy requirements are for heating and cooling and where air is the space-heating medium. Its attractiveness is limited to certain countries and regions, however.

One example of total energy management systems involves refuse collection and electric power generation. As to its energy-saving potential, IEA is rather cautious in saying that it is "local and minor, except in large conurbations."

The Japanese situation seems to be well fitted for this exceptional condition.

MITI has established efficiency standards for automobiles and air conditioners, and producers are encouraged to achieve the targets.

Concerning building design, IEA suggests higher standards for design and construction and insulation. Building design must vary according to the climatic variations and local building materials and design traditions. Considering the slow rotation of building stock, the total effect will not be felt until after the year 2000.

There seem to be fairly large potentials for energy saving in the residential and commercial sectors. The most energy-efficient technology is not necessarily diffused for practical use. The best available technology will achieve energy savings as listed below compared with stock average [IEA › 1ʕ ,p.100].

Residential electricity (U.S., Sweden)	78%
Building shell thermal efficiency (U.S., Sweden)	70 - 74%
Oil/gas heating system efficiency (U.S.)	26%
Central air conditioning system (U.S.)	50%
Refrigerator/freezers (U.S.)	50%
" (Germany, Japan)	20%+
Water heaters (U.S.)	57%
Commercial heating and cooling (U.S., Sweden)	75%
Energy use in large office buildings (U.S.)	63%
Commercial lighting (U.S.)	50%

IEA observes that "investment in energy savings today may be more cost-effective than investment in new energy production." A Canadian study conducted in 1984 compares the cost of energy conservation and the cost of energy supply and provides the following data on this point [IEA 1987, p.84]. Figures are in terms of 1983 Canadian dollars needed to conserve or produce energy of barrel of oil equivalent (BOE).

Table 3-4 Projection of energy demand

(unit: million kl, %)

	1989 actual	2000	2010
Industry	178(52.8)	193(49.5)	206(47.6)
Manufacturing part	151(44.8)	164(41.8)	173(39.9)
Office and household	82(24.3)	110(28.1)	134(30.9)
Transportation	77(22.9)	87(22.4)	93(21.5)
Final consumption	336(100.0)	391(100.0)	434(100.0)
Energy conversion	163	203	223
Total	499	594	657

Note: Conversion rate into crude oil is 9250 kl/liter.

Source: Energy Council, "Long-term Projection of Energy Demand and Supply."

Energy conservation:

New housing, super energy-efficient	$15 - 30
Existing housing, reduce average consumption	
by 30%	$13
by 40%	$28
High efficiency gas furnace	$10 - 13
Steam pipe insulation	$8 - 10
High efficiency commercial lighting	$15 - 30

Existing supply of energy:

Electricity	$35 - 60
Natural gas (export prices)	$15 - 35
Crude oil (at Montreal)	$22

New supply of energy:

Oil sands plant	$35 - 60
Offshore oil	$30
Offshore natural gas	$35
Nuclear electricity	$60

In Japan, new buildings are subject to insulation standards. New buildings with a floor space of more than 2000 square meters are required to submit an energy conservation plan at the time of application for a construction permit [IEA 1992a,p.183].

Table 3-5 Projection of energy supply

		1989 actual	2000	2010
Nonconventional	(mil. kl)	6.5(1.3)	17.4(3.0)	34.6(5.3)
Hydro	(bil.kwh)	88.0(4.6)	91.0(3.7)	105.0(3.7)
Geothermal	(mil. kl)	0.4(0.1)	1.8(0.3)	6.0(0.9)
Nuclear	(bil.kwh)	183.0(8.9)	330.0(13.3)	474.0(16.9)
Natural gas	(mil. kl)	49.9(10.0)	65.0(10.9)	80.0(12.2)
Coal	(mil. t)	113.6(17.2)	142.0(17.5)	142.0(15.7)
Petroleum	(bil. kl)	289.0(57.9)	305.0(51.3)	298.0(45.3)
Total	(mil. kl)	499.0(100.0)	594.0(100.0)	657.0(100.0)

Note: Conversion rate into crude oil is the same as in Table 3-4.

Source: Energy Council, "Long-term Projection of Energy Demand and Supply."

As to the future, the final use of energy for the year 2000 and 2010 is projected in Table 3-4 by the Advisory Committee for Energy affiliated with the Natural Resources and Energy Agency.

Energy demand in industry is expected to exhibit slow growth. Although it is difficult to foresee a drastic conservation effort like the one following the first oil crisis, energy consumption per unit output will exhibit steady improvement through the introduction of energy-conserving production processes. In most industrialized countries, the share of industry use stands at around 34% (33.8% in USA, 34.6% in Germany, 31.3% in UK, and 34.7% in France), whereas the Japanese figure has traditionally been higher.[7] The growth of energy consumption will be higher for office and household use (2.7% per year between 1989 and 2000 and 2.0% between 2000 and 2010). Transportation use will grow at 1.1% till 2000 and 0.7% thereafter. Industry use will grow at a slower pace (0.7% per year through 2010). Thus, the energy consumption pattern will approach that in other countries.

Industry refers here to the primary and secondary sectors except for the production (such as coal mining, petroleum and natural gas extraction, petroleum refining, coke production, etc.) and transformation of energy. Administration and self-consumption are excluded. Non-energy demand such as consumption in the petrochemical industry is included in the manufacturing sector. The household and commerce sector refers to the household sector excluding self-transportation, administration of industry, tertiary sector excluding electricity, gas, and transportation (i.e., included are wholesale and retail trade; finance, insurance, and real estate; communication; public administration; public services; and private services). Transportation demand includes transportation sector and self-transportation by industry and household. Conversion to crude oil is calculated as 9,250 kcal/l.

3.3 Sources of primary energy

Lacking an indigenous supply, Japan is at the consuming end of the global energy resource flow. The Japanese energy scene used to be dominated by coal, which was rapidly replaced by petroleum (Figure 3-8). It has placed great hopes on nuclear power, but the question remains whether this option provides the ultimate answer.

The primary energy supply in Japan stood at 499 million kl of petroleum equivalent. Petroleum is the major source of energy in every country, but its share is highest in Japan and Italy, amounting to about 60%. The USA consumes a large amount of natural gas which is available domestically. In Germany and the UK, the share of coal is relatively large, whereas in Canada, hydro-power plays a more important role.

Of the total energy supply, petroleum constituted 58.3% in 1990 (fiscal year)(see Table 3-3a). The long-term trends of petroleum in the total energy supply is shown:

actual:	1953		15.3%
	1955		17.6%
	1960		37.6%
	1965		59.6%
	1970		71.9%
	1973	(peak)	77.4%
	1975		73.4%
	1980		66.1%
	1985		56.3%
	1990		58.3%
projected:	2000		51.6%
	2010		46.0%

Earlier, the share of petroleum stood at around 15% (1953 figure), which

Figure 3-8 Sources of primary energy

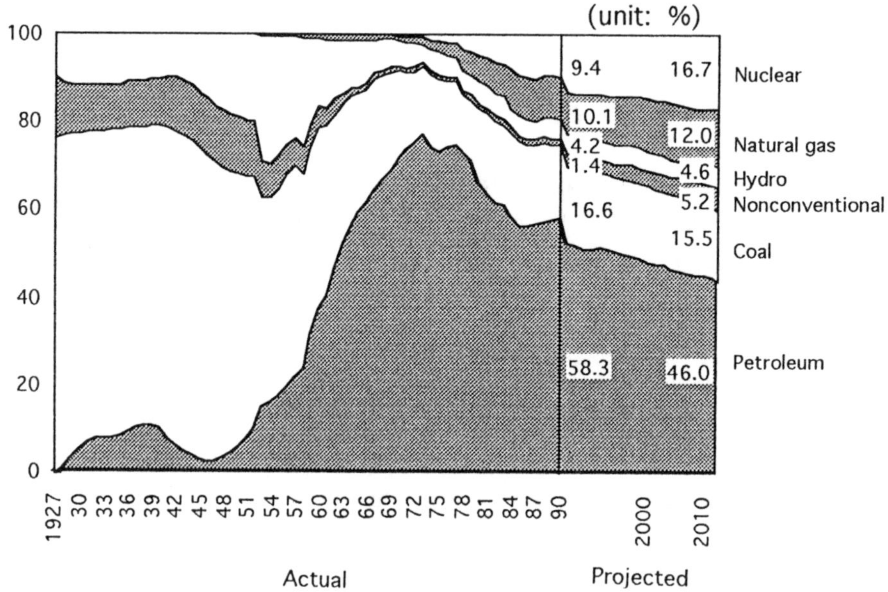

Source: Resource and Energy Agency.

started to climb rapidly, replacing coal (which then stood at 47.7%), hydro (29.0%), and fuel wood etc. (7.8%). The peak share was recorded in 1973, the year of the first oil crisis, at 77.4%. The share of oil in recent years stands at around 56 to 58%, showing a slight increase in the face of rapid economic expansion and lower oil prices.

For the future, the projection by the Advisory Committee for Energy expects larger shares for nuclear energy, natural gas, and new sources of energy such as solar (see table 3-5). Still, petroleum will have to provide 45% of the total energy supply in 2010; more so, if we try to be realistic about nuclear options, which entails such problems as (a) public opinion opposed it for various reasons, (b) treatment of radioactive waste, and (c) provision for decommissioning of reactors after use.

3.4 Policy implications

Lessons from the Japanese practice seem to be the following:

(a) Price does matter. The price hike at the time of the two oil crises and the relative reduction in price which followed have had their effect on energy efficiency.

(b) Technology does change.

(c) Investment is the key in the diffusion of new technology.

(d) There are divergent views on the policy tools to induce energy efficiency. Command-and-rule is out of the question. However, there is still room for government-industry interaction through establishing standards. A strategic shift to alternative energy sources is another example. In contrast, there are views which favor a market solution. In between, there is a school which favors energy tax and carbon tax, the effect of which is to alter the relative price of energy as a whole or of fossil fuels which generate CO_2.

Japan's policy measures in the energy field include the following according to a report in the Environment White Paper [1991]:[8]

(1) Energy conservation:
Systematization:

(a) Utilization of hitherto underutilized energy sources in the urban environment (such as river water, sewerage water, waste incineration facilities) for heat supply in local areas.

(b) Heat and electricity supply system based on fuel cells.

(c) Introduction of a heat pump in improving energy utilization within buildings.

(d) Recycling of resources such as used paper, etc.

Measures in individual fields:

(a) Energy-saving production facilities and processes in factories.

(b) Improved heat insulation in private homes.

(c) Introduction of energy-efficient devices such as inverter fluorescent lamp.

(d) Improvement in fuel efficiency of automobiles.

(e) Improvement in power generation efficiency by the introduction of combined-cycle power generation.

(2) Petroleum and natural gas: Increased provision of facilities aimed at security of supply.

(a) Improvement in competitiveness of the petroleum industry to be achieved through deregulation and improved autonomous business activities.

(b) Provision for security of supply including increased storage capacity, development of supply sources both at home and abroad, and emergency measures in order to cope with acute shortages and market disruption.

(3) Coal: High heat efficiency technologies such as coal gasification and combined cycle power generation technology by pressurized fluidized bed incinerator will be the major target of research and development leading to early utilization in practice.

(4) Nuclear: Improved safety regulations, development of related technology and safety measures; establishment of nuclear fuel cycle by back-end measures such as proper technology aimed at radioactive waste; and smoothening of development of new sites for nuclear power generation. More open public relations will be required in order to gain support among the population.

(5) Research and development:

(a) Sunshine Project including coal liquefaction, solar energy, geothermal energy, among others.

(b) Moonlight Project aimed at developing energy conservation technology.

(c) CO_2 fixation and utilization technology.

(d) Research into clean coal technology.

(6) International relations:

(a) Increased effort at developing common understanding that a stable supply-and-demand of petroleum in the global market is beneficial

Table 3-6 CO2 emission factors

(unit: Mt carbon/Mtoe)

	IEA total	Japan
Primary energy emission factors:		
Coal	1.09	1.10
Other solid fuels	0.89	0.89
Oil	0.84	0.84
Gas	0.64	0.64
Delivered fuel emission factors:		
Coal	1.14	1.34
Other solid fuels	0.89	n.a.
Oil	0.89	0.89
Gas	0.73	0.58a
Electricity	1.96b	1.50c

Note: a. Due to transformation of coal into gas.
b. From text on p.24 of IEA [1991a].
c. From table on p.208 of the same source.

Source: IEA [1991a], pp.23-24.

to both producing and consuming nations. To this end, intergovernmental exchange of views must be promoted.

(b) Cooperation in downstream production and distribution channel aimed at tightening of international relations.

(c) Cooperation in development of new oil fields.

Japan's policy option in essence would be to contain energy demand as much as possible while lowering dependency on oil and shifting toward nonfossil fuels. The statement above is based on the view that an increase in energy demand is inevitable in Japan, accompanying a better quality of life. In the long-run, the report recognizes that the oil price will tend to increase and that the global environment will continue to be a serious policy issue. With this background, the report foresees increased interaction between Japan and the global community in the energy field.

The supply targets of petroleum substitutes are determined at the government level based on the Law concerning the Development and Diffusion of Petroleum Substitute Energy. The latest version was published in October 1990 in conformity with the long-term demand-and-supply projection by the Energy Council in June of that year. The target year is 2010. The goals are in

Table 3-7 Carbon emissions in Japan

(unit: MT carbon, %)

	Coal	Oil	Gas	Elect-ricity	Total (% of total)
TPER	81.0	163.0	24.1	—	268.1(100.0)
Electricity generation	30.5	39.3	16.2	—	86.0(32.1)
TFC	50.5	123.7	7.9	86.0	268.1(100.0)
Industry	50.2	27.5	2.2	50.1	130.1(48.5)
Iron and steel	39.4	1.9	0.4	9.3	51.0(19.2)
Chemical	1.5	3.2	0.3	6.5	11.5(4.3)
Non-ferrous	0.2	0.4	0.0	1.7	2.3(0.9)
Non-metallic	6.4	2.9	0.0	2.2	11.6(4.3)
Transport equipment	0.0	0.0	0.0	2.6	2.6(1.0)
Machinery	0.0	0.0	0.0	3.5	3.5(1.3)
Textiles	0.0	2.4	0.0	0.9	3.3(1.2)
Food products	0.0	2.2	0.0	1.3	3.5(1.3)
Paper, pulp,& print.	1.2	0.7	0.0	3.6	5.5(2.1)
Others	1.6	13.7	1.5	18.4	35.2(13.1)
Transport	0.0	57.7	0.0	2.3	60.0(22.4)
Air	0.0	2.5	0.0	0.0	2.5(0.9)
Road	0.0	50.1	0.0	0.0	50.1(18.7)
Rail	0.0	1.0	0.0	2.3	3.3(1.2)
Others	0.0	4.2	0.0	0.0	4.2(1.6)
Others	0.3	38.5	5.7	33.5	78.0(29.1)
Agriculture	0.0	4.9	0.0	0.2	5.1(1.9)
Commercial/public	0.0	13.3	1.2	12.5	26.9(10.1)
Residential	0.3	10.6	4.4	20.9	36.2(13.5)
Others	0.0	9.7	0.0	0.0	9.7(3.6)

Note: The data refer to 1988.

Source: IEA [1991b], p.208.

conformity with the Action Plan for the Prevention of Global Warming which was implemented in October 1990. The summary is as follows [MITI, 1991, pp.12-13]:

(1) Promotion of energy conservation: Contain growth in energy demand which is projected as reflecting qualitative improvement of the people's life.

(2) Reduction of the shares of petroleum

(3) This is possible by increased supply of nonfossil fuels such as nuclear in the total energy supply follows.

Table 3-8 Accumulation of irradiated fuel from commercial plants

(unit: metric tons)

Country	1985	1990	2000
United States	12,601	21,800	40,400
Canada	9,121	17,700	33,900
Soviet Union	3,700	9,000	30,000
France	2,900	7,300	20,000
Japan	3,600	7,500	18,000
Germany	1,800	3,800	8,950
Sweden	1,330	2,360	5,100
Others	5,939	14,540	36,715
Total	40,991	84,000	193,065

Note: France and the United Kingdom (included in "other" category) totals do not include 16,500 tons and 25,000 tons, respectively, produced by dual-use military and civilian reactors. Canadian total is proportionately higher due to its use of natural uranium instead of enriched uranium in its CANDU reactor technology.

Source: Worldwatch Institute, *State of the World 1992.*

1973	5.6%
1989	14.8%
2000	20.2%
2010	26.8%

In the year 2010, the total energy supply is projected to be equivalent to 657 million kl of petroleum, of which 298 million kl (45.3%) is from petroleum. Nonpetroleum energy amounts will be equivalent to 359 million kl of oil.

The energy sources which need to be developed and distributed include the following:

Nuclear	30.8%	(111,000 kl)
Coal	28.6%	(103,000 kl)
Natural gas	22.2%	(80,000 kl)
Hydro	6.9%	(25,000 kl)
Geothermal	1.7%	(6,000 kl)
Other nonpetroleum energy	9.7%	(35,000 kl)
	100.0%	(360,000 kl)

One final question involves the consistency between the policy measures described here and the goal of reducing carbon emissions. The current situation of carbon emission in Japan is depicted in Table 3-7.

The United Nations Framework Convention on Climate Change, signed at the Earth Summit in Rio de Janeiro in June 1992, binds participants to cut greenhouse gas emissions to 1990 levels. Japan committed itself in 1990 to contain per capita emission of CO_2 at around the 1990 level after the year 2000 (which is around 2.6 tons).

This target cannot be met without increased dependence on nuclear energy. In order to meet the target, the Japanese government has been aiming at achieving a nuclear power generation capacity of 50.5 MKW by the year 2000. In 2000, Japan was to be generating 330 billion kWH in nuclear plants. It was reported recently that the capacity would remain 45 MKW, generating 300 billion kWH, or about 10% less than the target. The long-term plan already incorporates the maximum amount of energy conservation. Thus, any deficiency in supply would have to come from an increased dependence on thermal plants consuming fossil fuel. This implies that the emission of CO_2 in the year 2000 will exceed the target by 1 to 2%. [*Japan Economic Journal*, April 6, 1993]

The excess emission is relatively small, so that it may still be possible to absorb it by improving the energy efficiency of coal- and oil-burning power plants by the target year. The situation may even intensify the discussion for the introduction of an environmental tax on fossil fuels.

FOOTNOTES

[1] The values for unit energy consumption by industry used in the calculation are presented here. It should be noted that the energy efficiency varies widely among countries. The data for manufacturing refer to 1986 and those for electricity, 1988.

	Manufacturing (TOE/ton of output)			Electricity (TOE/100 mil. WH)		
	Steel	Chemi-cals	Non-metallic minerals	Coal	Oil	Gas
Former Soviet Union and East Europe:						
Former S.U.	0.4425	10.4993	0.3693	30.9	35.9	37.0
Poland	0.2860	3.0403	0.2536	37.9	45.3	45.0
Yugoslavia	0.3392	2.1157	0.1859	26.5	44.9	26.9
Other E.E.	0.4413	3.0180	0.3217	46.0	50.5	55.5
Developing countries:						
China	0.6706	24.6219	0.2318	27.2	23.9	26.1
India	0.4670	22.2733	0.1131	29.2	35.8	100.3
Brazil	0.2047	2.7724	0.0568	31.6	27.4	—
Korea	0.3136	2.2128	0.1972	23.7	22.1	23.3
Japan	0.1607	1.9853	0.0801	21.5	20.2	19.2

Note: Steel refers to pig iron+ferroalloy+crude steel production; chemical refers to ethylene and plastics production; nonmetallic minerals refers to cement production.

According to an estimate by the Economic Planning Agency, CO_2 emission of the countries listed here would be reduced from 1096.03 million carbon tons to 537.97 million carbon tons, or a reduction of 558.06 million carbon tons if the Japanese level of energy efficiency is attained. Economic Planning Agency [1991b], pp.64-65.

[2] The relative size of Japan's GNP is actually larger than stated here after the 1993 appreciation of the yen. Attempts are sometimes made to adjust by using the currency's purchasing power. I believe that such a practice is useful, for example, in comparing the levels of real consumption. When it comes to capital goods and electronic appliances, however, the Japanese price levels are lower than those among OECD countries, and if one adjusts the entire GDP by average prices for those goods, the Japanese economic size will be that much larger. Rather than resorting to artificial adjustment, which is good for one purpose but not for the other, we have used figures converted by the then-current exchange rate.

[3] Data are based on Table 3-2 where GDP figures are for 1990 and other data are for 1989.

[4] For an analysis of energy prices and energy demand, see Uno [1987]. Energy demand elasticity on output is estimated to be 0.608, and price elasticity price (ratio of price of energy/price of the sector in question) is estimated to be (-)0.104. Observation period is from 1956 to 1980. Parameters for individual sectors are also provided.

[5] See Uno [1989a] for an analysis of the economic structural change and energy demand. The analysis is based on input-output tables for 1970, 1975, and 1980, all in 1980 constant prices so that the effect of changing prices can be removed. It is shown that during the first half of the decade, demand increase attributable to economic expansion (factor 3 in the analysis) and increase in final consumption of energy (factor 4) predominated, whereas in the second half of the decade, energy conservation within every industrial sector (factor 1) played a major role in arresting the energy demand to a considerable extent despite expansion of the economy.

[6] For details, see "Lifestyle and environment" which is taken up in Chapter 4 of this volume.

[7] Note that IEA statistics give lower figures than Japanese statistics.

[8] For official statement, see "Global environmental problem calls for action" in *Environment Whitepaper*, 1991, General discussion, pp.113-156, which is summarized below.

International:

1. protection of ozone layer
2. prevention of global warming
3. strengthening of ODA in environmental field

National:

1. Government policy

   Action Plan for the Prevention of Global Warming

   Section 1. Background and the significance

   Section 2. Basic items to be considered in carrying out measures to prevent global warming

   Section 3. Goals

Section 4. Time schedule

Section 5. Policy measures

- CO_2 emission prevention
- Greenhouse gases emission prevention
- Absorption of CO_2 (reforestation)
- Scientific research, observation and control
- Research and development and diffusion of new technology
- Public relations and awareness
- Promotion of international cooperation

Section 6. Implementation of the plan

2. Local public bodies
3. Business circles
4. Local residents
5. Environmental education

Chapter 4

Lifestyle and Environment

4.1 Lifestyle as a variable

Although customs and social institutions seem invariable in a particular society, they do change in the long run. The changes are due to rising income, urbanization, service orientation, and new technology and products, among other influencing factors. The changes are sometimes induced by legal stipulation, various public policy, and provision of social infrastructure, but for the most part lifestyle is regarded as something that cannot be manipulated. Lifestyle reflects people's values, attitudes, and personal preferences and it should be honored as such. This has resulted in the neglect of empirical examination of changing lifestyle.

But different lifestyles translate into different industrial structure and different environmental impact. As we see in Figure 4-1, countries exhibit different patterns of energy use. For example, the share of industrial use is relatively high in Japan, whereas the opposite is true in the USA and the UK. In contrast, energy consumption by household is relatively low in Japan. Part of the explanation lies in the relatively modest winter in Japan, which reduces the need for heating fuel.

It should be noted, however, that household energy consumption is increasing and so is energy consumption in the transportaton sector where larger portion is also attributable to household activity.

We now have a consensus that global warming is a real threat, in addition

Figure 4-1 Energy consumption by use in selected countries

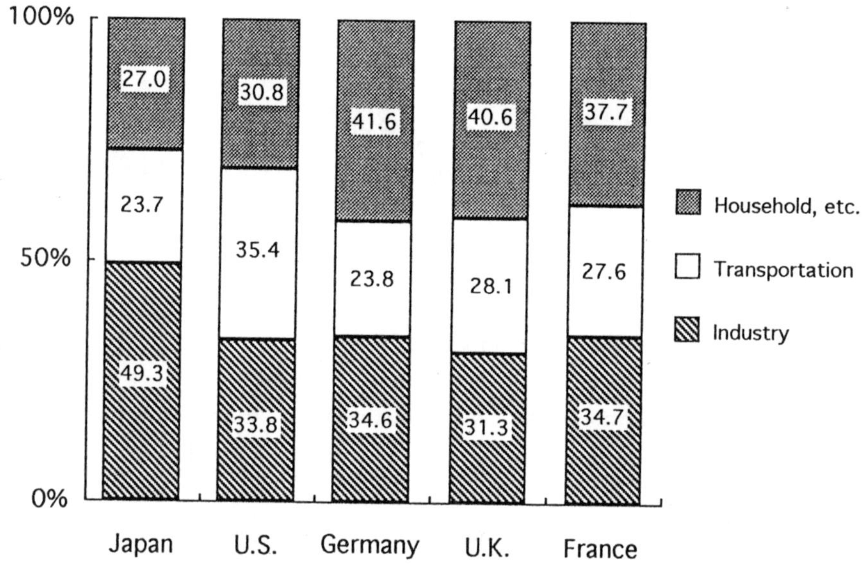

Source: Environmental Agency, *Environment White Paper, 1990,* Overview.

to the more or less traditional type of concern over air and water quality. Some limitations on the burning of fossil fuel is the key factor in controlling the quality of the air, preventing resource stock degradation, and arresting global warming. On the other hand, the growth of industrial production and improvement in the standards of living inevitably increase the consumption of energy. Logical conclusion which follows from this seems to be that economic growth would have to be stopped, or its pace slowed down, for the sake of safeguarding environmental quality.

The above argument leaves out two important factors, namely technology and lifestyle, or implicitly assumes that these two factors are constant over time. If, however, they are variables, especially if they prove to be amenable to policy means of control, we may be able to draw a different conclusion concerning the relation between energy consumption and economic growth. The discussion of lifestyle in this chapter is intended to shed some light on this point.

4.2 Household appliances and energy consumption

Let us first of all confirm the trends in household energy consumption. Table 4-1 relates to this point. The aggregate picture for the household sector as a whole can be derived from population size and the energy consumption attributable to household (data from Table 3-3b). It is calculated that per capita energy consumption has increased 3.7 times during 1960 and 1990. During the 15-year period from 1960 to 1975, it increased 2.4 times but the pace has slowed down to an increase of 1.5 times in the subsequent 15 year period from 1975 to 1990. Obviously, this is attributable to slower income growth following the oil crisis.

The table provides energy consumption per household. It can be seen that the number of household has grown much faster than the population size per se due to smaller number of persons in a family. Average family size is now less than three persons whereas it used to be about five. Thus, the number of household has more than doubled during the 30 year period from 1960 to 1990. This in itself is a manifestation of changing lifestyle as the economy develops. In addition to the fact that people are having less children now, extended family is more difficult to maintain in the urban area due to higher housing cost and increased mobility of the population. Increased attendance in higher education and increased labor participation of the female population are also contributing factors because universities and employment opportunities are found in urban areas, obliging younger population to be independent.

Let us look at table 4-2, which shows the diffusion rate of consumer durables. The changing lifestyle is observed in a shift from traditional Japanese life typified by 'tatami' mat, 'kimono' wardrobe, and 'hibachi' charcoal heating to Western style. Of the items covered by the survey, the ones listed here are

Table 4-1 Household energy consumption

	Population (1000)	Persons per house-hold (persons)	Number of house-hold (1000)	Energy consumption, household sector total (10**10kcal)	Energy consump-tion per household (10**6kcal)
1955	89276	4.97	17383	7631	43.9
1956	90172			7781	
1957	90928			8211	
1958	91767			8025	
1959	92641			8113	
1960	93419	4.54	19678	8801	44.7
1961	94287			9309	
1962	95181			10165	
1963	96156			9415	
1964	97182			9832	
1965	98275	4.05	23085	10696	46.3
1966	99036			11442	
1967	100196			12791	
1968	101331			14104	
1969	102536			16570	
1970	103720	3.41	30297	18329	60.5
1971	105145			19788	
1972	107595			21542	
1973	109104			23591	
1974	110573			24377	
1975	111940	3.28	33596	25574	75.8
1976	113089			28098	
1977	114154			28529	
1978	115174			30409	
1979	116133			31917	
1980	117060	3.22	35824	30486	85.1
1981	117902			31802	
1982	118728			31649	
1983	119536			35609	
1984	120305			35643	
1985	121049	3.14	37980	37313	98.2
1986	121672			37571	
1987	122264			39981	
1988	122783			41186	
1989	123255			41881	
1990	123611	2.99	40670	42708	105.0

Source: Population and household data are from the *Population Census*.

related to energy consumption. Even in the past twenty years or so, we have witnessed a rapid diffusion of consumer durables. Compared with traditional items such as carpets, sofa, and beds, the diffusion of energy-related items has been particularly marked. Also among the latter category, we notice a shift toward more energy-consuming items. Examples are the shifts from charcoal 'hibachi' and electric 'kotatsu' heaters to kerosene stoves and now to warm-air furnaces. The electric fan has been replaced by air conditioners (some of them are good for heating, too). Gas water heaters are quickly being replaced by electric water heaters. Moreover, many families now possess automobiles. Only 9% of Japanese households possessed automobiles as recently as 1965, but the ratio has reached 80% in 1990. According to a survey reported in Table 11-6 below, 1,090 cars are owned per one thousand family in 1989, which indicates that some families possess more than one car.

Some appliances, such as refrigerators and color television sets, seem to have reached a saturation point as far as the diffusion rate is concerned (although people can begin to have more than one unit in a family). Air conditioners leave much room for diffusion as can be witnessed by the rapid rate of increase even in the recent years. The figure has jumped from 55% in 1985 to 68% in 1990. The diffusion rate is much sharper in higher income brackets, indicating the room for further expansion. The same can be said of new items such as microwave ovens and water heaters (which typically utilize surplus, and hence cheap, electricity during the night).

It is clearly indicated that there is an economic reason (higher income and availability of consumer credit) as well as a technological one (emergence of new products) behind the increased use of consumer durables.

Table 4-3 itemizes the household electricity consumption. The diffusion rates listed here are slightly different from the ones provided in the previous table, but it is interesting to see the contribution of each item to total electricity consumption. Refrigerators has the largest, but declining, share (31% in 1980, 23% in 1990). Television sets also has smaller share now (15% in 1980, 11% in 1990). Lighting use stands at about 18% in 1990 which is slightly smaller than 10 years ago when it stood at 20%. Instead, the contribution of air conditioners (8% in 1980, 16% in 1990) and 'other appliances' (6% in 1980, 15% in 1990) is increasing at a fast pace.

In Japan, energy efficiency of consumer appliances has improved remark-

Figure 4-2 Temperature setting of air conditioning

a. Winter (heating)

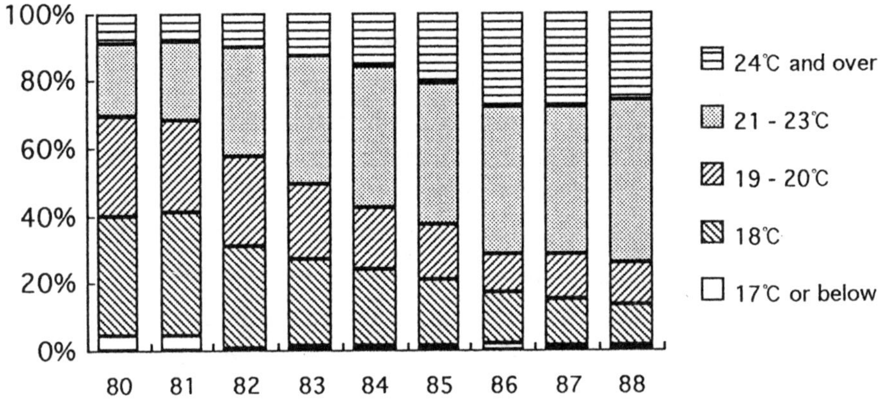

b. Summer (cooling)

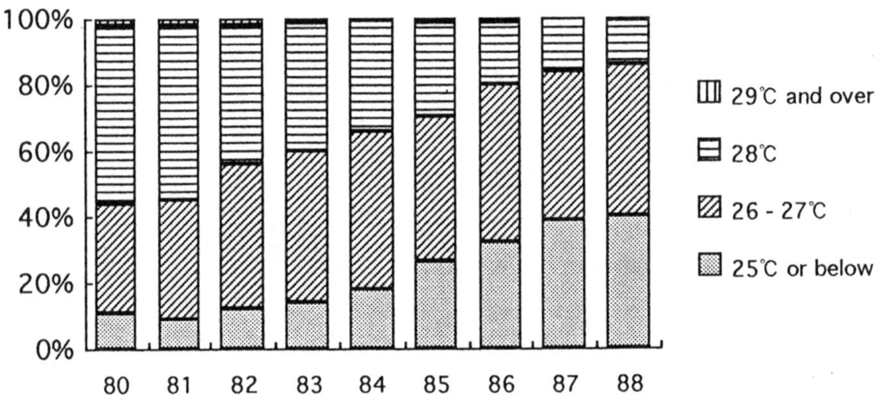

Note: Agency of Natural Resources and Energy, Survey on Efficient Use of Energy, 1988. Cited here from *Environmental White Paper, 1990*, Overview.

Table 4-2 Diffusion rate of consumer durables

(unit: percent)

	Refrigerators	Color television sets	Air conditioners	Water heaters	Microwave ovens	Passenger cars
a. Time series						
1965	51.4	'66 0.3	2.0	...	...	9.1
1970	89.1	26.3	5.9	...	...	22.1
1975	97.9	93.7	19.5	...	20.8	44.0
1980	99.2	98.5	41.3	...	37.4	58.5
1985	98.4	98.9	54.6	24.8	45.3	67.4
1990	98.9	99.3	68.1	37.0	75.6	79.5
b. Annual income group (million yen), 1990						
-3.0	96.7	98.5	52.1	19.9	64.3	57.7
3.0-4.0	98.0	99.0	65.2	26.9	77.0	77.8
4.0-5.5	97.8	98.9	72.2	29.7	80.5	82.2
5.5-7.5	98.5	99.3	76.6	36.2	86.6	85.6
7.5-	98.6	99.6	86.4	37.6	90.9	88.6

Source: Compiled from Economic Planning Agency, *Current Consumption Survey.*

ably since the two oil crises in the 1970s. We can see in Figure 3-6 above that the energy consumption for typical consumer durables including refrigerators, television sets, and air conditioners. The improving trend, however has tapered off in the latter half of the 1980s. Actually, the performance in 1990 was not any better than in 1985.

Figure 4-2 indicates that people now want warmer room temperature during the winter and cooler temperature during the summer. Relative cheapening of energy must be behind such trends in addition to higher income levels. People also want to live in larger houses. Based on the number of housing starts and total floor space (Table 9-8), one can calculate average floor space per unit. Starting from less than 60 square meters in 1950s and early 1960s, it reached 68 square meters in 1970 and 94 square meters in 1980. Recent figures are somewhat smaller than this, but this is a cyclical phenomenon reflecting an increase in units for renting which are smaller than owner-occupied ones. On the other hand, better insulation of housing units is a positive factor in energy conservation (Figure 3-7).

We may cite here one interesting estimate by the Energy Conservation Center.[*Environmental White Paper*, Overview, 1990, p.125] Energy consump-

Table 4-3 The structure of household electricity consumption

	Diffusion rate			Share in electric consumption		
Year	1980	1985	1990	1980	1985	1990
Electric appliances:						
Rice cooker	54.9	56.6	60.1	4.1	3.6	2.8
Refrigerators	107.1	109.0	109.8	31.1	26.4	23.4
"Kotatsu" warmer	108.7	112.9	114.5	7.2	5.8	5.4
Electric fans	151.4	152.3	154.1	0.8	0.7	0.7
Electric blankets	92.4	98.1	99.3	1.7	1.5	1.4
Room air conditioner	52.1	93.0	124.5	7.8	12.6	15.9
Washing machines	98.2	98.5	98.7	1.6	1.4	1.3
Vacuum cleaners	96.1	97.5	98.1	4.0	3.5	3.2
Microwave ovens	33.3	45.7	58.5	1.3	1.5	1.7
TV sets				15.0	11.3	10.6
of which first color	97.9	98.5	98.5	13.2	9.8	8.9
second color	36.2	55.1	69.4	1.3	1.5	1.7
Other appliances				5.8	13.1	15.2
Subtotal				80.3	81.4	82.2
Lighting:				19.7	18.6	17.8
Electricity total				100.0	100.0	100.0

Note: The data refer to fiscal year. TV sets for 1980 includes monochrome. Air conditioners for 1985 and 1990 include cooler/heater type.

Source: Resources and Energy Agency, *Summary Report on Electric Demand and Supply*. Cited here from Resources and Energy Agency [1990].

tion per household per year has increased from 7,317 thousand Kcal in 1973 to 9,615 thousand Kcal in 1986. This is an increase amounting to about 31%. Without the improvement in fuel efficiency, it is estimated that energy consumption would have reached 11,325 thousand Kcal. We may say that the difference between the two figures (1,710 thousand Kcal) represents energy conservation due to improved efficiency. On the other hand, if the diffusion rate of consumer durables remained at the 1973 level, and if 1986 energy efficiency is assumed, energy consumption per household in 1986 must have been 7,016 thousand Kcal. The difference between this figure and the 1986 actual (about 2,600 thousand Kcal) is attributable to higher diffusion rates, shift to large-sized products, shift to more sophisticated types, and emergence of new products.

Table 4-4 Production of passenger cars by engine size

(unit: 1000)

	Total	Light: 600cc or smaller	Small: 600cc-2000cc	Ordinary: larger than 2000cc
1960	165	36	129	—
1965	696	94	599	3
1970	3179	749	2378	52
1975	4568	160	4199	209
1980	7038	196	6439	403
1985	7647	161	6991	495
1990	9948	836	7361	1751

Source: Ministru of International Trade and Industry, *Current Production Statistics Survey.*

4.3 The case of automobiles

Table 4-4 reveals that consumers tend to prefer larger cars. Once popular light vehicles with engine displacement of 660 cc or smaller dropped its share in the 1980s. In 1990, the production has recovered considerably. Production of 'small' cars with engine size of less than 2000 cc continued to grow. The most marked change is the very rapid increase of 'ordinary' cars which have engine size of more than 2000 cc. In 1970, the production figure stood only at 52,000; in 1990, it was 1,571,000.

Fuel efficiency of passenger cars has improved. Automobiles produced in 1973 were able to travel 9.6 kilometers per liter, which improved to 13.0 kilometers in 1982 (Figure 3-8). Reflecting the cheapening of fuels, however, fuel efficiency has deteriorated somewhat since then. One contributing factor was the shift in consumer demand to larger, more luxurious cars.

Table 5-2 in the next chapter reveals that motorization proceeded at a rapid pace. In the case of freight transportation, the index (1970 as the base year) has jumped from 15.6 in 1960 and 36.0 in 1965 to 134.7 in 1980 and 154.1 in 1985. For passenger transportation, we have witnessed an explosion from 9.6 in 1960 and 26.4 in 1965 to 181.7 in 1980 and 212.2 in 1985. Non-business motor vehicle transportation recorded even faster growth, jumping from 3.8 in 1960 to 237.4 in 1985. It should be noted that the weight attached to motor vehicle transportation is distinctly larger than that for railways.

Table 4-5 Growth, wellbeing and energy consumption

	1955	1960	1965
Gross national expenditure	25.0	37.9	58.8
Economic wellbeing (NNW)	27.6	35.5	62.4
Final energy consumption	19.5	30.9	51.4

Source: GNE and NNW figures are from Table 11-1. Final energy consumption data are from Table 3-3.

Moreover, railway transportation exhibited an absolute decline in the case of freight transportation. Passenger railway transportation is still growing, but the pace is much slower than for motor vehicles. The rapid expansion of automobile transportation is consistent with what we have observed for the diffusion rate of passenger cars. It is known that in most countries, about 80% of energy for transportation use is attributable to automobiles. The Japanese case is compared with the OECD total in Table 5-1 in the next chapter.

4.4 Policy implications

"The issue is not just numbers of people, but how those numbers relate to available resources." "Human resource development is a crucial requirement not only to build up technical knowledge and capabilities, but also to create new values to help individuals and nations cope with rapidly changing social, environmental, and development realities." [The World Commission on Environment and Development, 1987, p.11] "The challenge is to ensure that these new values are more adequately reflected in the principles and operations of political and economic structures." [WCED 1987, p.28] "Perceived needs are socially and culturally determined, and sustainable development requires the promotion of values that encourage consumption standards that are within the bounds of the ecological possible and to which all can reasonably aspire." [WCED 1987, p.42.]

We may conclude that lifestyle is changing in the direction in which more energy is required. Despite the improved energy efficiency of individual equipment, people are now wanting larger, more luxurious types. Not only that, they tend to possess one or more cars, refrigerators, television sets, air conditioners, etc. per family. They are also willing to purchase new products such

(unit: index, 1970=100)

	1970	1975	1980	1985	1990
	100.0	124.2	155.6	187.6	250.0
	100.0	149.5	181.8	213.2	268.2
	100.0	118.9	125.2	128.1	152.8

as microwave ovens, warm-air furnaces, water heaters, audio equipment, and video tape recorders. People now want a warmer room temperature during the winter and cooler temperature during the summer. In homes as well as in offices, new equipment such as PCs, fax machines, and copying machines are widely used. This requires larger floor space. Emission of heat from the equipment also necessitates a larger cooling capacity. People are shifting from public transportation such as railways to automobiles. Despite energy efficiency being attained in industry, household activity and transportation seem to be tending toward increased energy requirement.

There seems to be no easy way out of the situation in a society which is market- and democracy-oriented.

Lifestyle is shifting toward services. Household expenditure on services is increasing faster than for hardware. Although total energy implication is not known, there is a possibility that service orientation would reduce total energy requirement.

Our Common Future calls for changing the quality of growth. "Sustainable development involves more than growth. It requires a change in the content of growth, to make it less material- and energy-intensive and more equitable in its impact." [WCED 1987, p.52] Table 4-5 compares the trends in energy consumption, GNE, and economic well-being as measured in the NNW framework. The methodology is detailed in Chapter 11, but in essence it boils down to modify the concept of GNE by imputing environmental damages (deduction), value of leisure, and the services of household durable goods. The table reveals that, taking 1970 as the base year (=100), energy consumption stood at 153 in 1990, while GNE reached 250. What is of interest here is the trend in NNW which stood at 268 in 1990. The implication is that improvement in economic well-being is being achieved without a parallel input of energy.

There is increasing recognition that production or raised value added does not automatically lead to improved well-being. Indeed, it is too much to ask national income statistics to represent welfare implications because it is geared toward the current market transaction. As such, there are important omissions in the concept of GNE of items which actually improve (or deteriorate) the quality of life. Examples are environmental pollution, increased leisure, the value of household production, and services of social capital stock and consumer durables.

Empirical examination indicates that the growth of such measures lagged behind economic growth, due to the 'cost' of growth such as pollution during the high growth era in the 1950s and 1960s. When proper measures were taken to cope with environmental disruption toward the end of 1960s and early 1970s, this negative effect of growth was removed to a considerable degree. With increased stock of consumer durables (plus increased leisure and other factors), economic well-being kept growing at a faster pace than GNE. Indeed, the quality of life depends on the available stock of cosumer durables, housing, and other social stock, rather than the annual purchase of durables or annual construction of housing. Stock, rather than flow, is also relevant from the point of view of energy consumption. The policy implication seems to be that, with improved durability of products, people can attain a desired economic well-being, while conserving the energy required for producing them. In the same vain, recycling of wastes contributes to energy saving. This is true of steel, aluminum, plastics, paper, and other resources, the production of which involves a large energy demand. Rather than focusing on the production phase alone, therefore, we should keep track of the services they render during their life span and recycling them at the end.

Chapter 5

Transportation

5.1 Transportation and environment

Transportation activities encompass both industry and household sectors. In the conventional definition of production boundary, some of the transportation services fall outside it. In dealing with the environment, however, one has to tackle the full scope of transportation, and this provides another reason why we have to go beyond the well-established economic account and start compiling an environmental account.

In a typical OECD country, 1/3 of energy consumption is attributable to transportation activities, implying that a large part of CO_2 emission is traceable to traffic sources. This in turn is likely to contribute to global climatic change through increased concentration of CO_2 in the atmosphere.

The total final energy consumption by the tranportation sector is shown in Table 5-1 for Japan and OECD as a whole. It can be seen from the table that, among various means of transportation, road is far more important in terms of energy consumption. The share of road tranport in total stood at 85% in Japan and 82% for OECD total in recent years, and for this reason our discussion in this chapter will focus on problems related to automobiles.

Fossil fuel combustion in the transportation activities also results in emission of air pollutants such as lead (Pb), carbon monoxide (CO), nitrogen oxides (NO_X), hydrocarbons (HC) and volatile organic compounds ($VOCs$), and particulate matter.[OECD, 1988b, pp.48-53]

Table 5-1 Energy consumption by mode of transportation

(unit: MTOE)

Modes		1970	1980	1985	1990
Japan:					
	Air transport	1.0	2.5	2.5	3.0
	Road transport	24.9	43.6	48.5	60.6
	Rail transport	3.7	2.6	2.3	2.6
	Total	32.9	54.2	57.5	69.6
OECD:					
	Air transport	69.0	87.0	96.3	118.2
	Road transport	445.0	607.4	636.6	732.7
	Rail transport	29.4	27.0	22.6	22.9
	Total	562.7	741.0	774.9	895.0

Source: OECD [1993], p.233.

Lead: Lead is known to cause damage to the kidney, liver, reproduction system, blood formation, basic cellular processes, and brain functions at relatively high levels in humans. Lead intake occcurs by absorption of some ingested lead from the gastrointestinal tract and by absorption of some inhaled lead from the lower respiratory tract. For this reason, children are found to be more susceptible to exposure to lead in dust and soils. It was found that children with high levels of accumulated lead in the body exhibit more behavioral problems, decreased ability to concentrate, and lower IQ. The average blood lead levels are correlated more strongly to gasoline lead than to lead in the air alone. More recently, it was revealed that even at a very low level, lead brings about adverse effects on health. Thus, reducing the lead content of gasoline has received much attention relatively early on in the history of pollution prevention in Japan.

Carbon monoxide: Carbon monoxide is one of the most directly acting toxic substances in high concentration in automobile exhaust gas. It causes adverse health effects with its interference with the absorption of oxygen. It combines with haemoglobin more than 200 times faster than oxygen and restricts the supply of oxygen by the red blood cells to body tissue. Carbon monoxide in the atmosphere is known to lower worker productivity and cause general discomfort. It can affect the central nervous system, thus affecting physical coordination, vision, and judgement. Carbon monoxide is also known for its synergistic action in conjunction with other air pollutants. The World Health Organisation (WHO) recommends a long-term goal of 10 mg per cubic meters of ambient CO as an eight-hour average.

Nitrogen oxides: Nitrogen oxides reduce gaseous exchange in the blood and increases respiratory symptoms, causing irritation and leading to oedema or emphysema. Most serious adverse effects on human health, especially for people with existing health problems, occur in combination with other air pollutants. The WHO established a maximum of one-hour exposure of 0.10-0.17 ppm as a level to safeguard human health, and exposure at this level should not occur more than once in every month.

Nitrogen oxides also have an adverse effect on vegetation. The effect is compounded when nitrogen oxide and sulphur dioxide are present simultaneously.

Hydrocarbons: Hydrocarbons and other organic compounds are contributing factors in photochemical oxidant formation. It has been revealed also that particular organic compounds (aldehydes, polycyclic aromatic compounds, benzene, or organic acids) have hazardous health effects while others (such as ethene) exert hazardous effects on vegetation.

Particulate matter: Particulate matter or aerosols consist of small solid or liquid particles of various chemical composition suspended in the atmosphere. They are emitted with the exhaust gas of motor vehicles. The wear and tear of tires and brakes and road dust caused by automobile traffic are also contributing factors. Fine particulate matter penetrates into the respiratory system, irritating lung tissue, which results in long-term disorder. Asbestos fibers are emitted from the wear and tear of brake-lining. Asbestos at high concentration is known to induce asbestosis, lung cancer, and other health hazards, although it is believed that the ambient asbestos level attributable to road traffic is generally low. Particulate matter may be toxic in itelf or may carry toxic trace substances on its surface. As in the case of diesel exhaust gas, particulate matter is visible as smoke in some cases. It also reduces visibility by scattering and absorbing light.

Photochemical oxidants: Photochemical oxidants are produced by a mixture of various pollutants after they have undergone transformation in the atmosphere. Photochemical oxidants are defined as "compounds with a strong

Table 5-2 Index of transportation activities

(unit: 1970=100)

	Total	Freight: Rail-ways	Motor vehicles	Passenger: Rail-ways	Motor vehicles Total	Non-business
Weights:	9836	31	1330	336	7985	7374
1955	11.2	67.2	7.1	44.9	4.4	1.0
1956	12.5	74.1	7.8	48.2	5.6	1.4
1957	14.1	79.0	9.7	50.0	6.7	1.9
1958	14.8	72.9	11.2	52.4	7.5	2.6
1959	16.3	79.2	13.2	54.8	8.1	2.7
1960	18.7	87.4	15.6	59.7	9.6	3.8
1961	21.3	92.5	18.7	64.0	12.0	5.5
1962	24.3	93.1	23.9	68.7	14.1	6.6
1963	28.2	94.1	29.4	73.0	18.1	10.6
1964	33.6	96.1	35.8	78.1	21.8	13.5
1965	36.0	92.0	36.0	82.2	26.4	18.1
1966	43.5	88.7	46.0	83.1	33.2	25.0
1967	54.1	94.2	58.1	85.9	43.6	35.8
1968	67.7	95.4	72.9	90.0	58.3	52.0
1969	84.4	96.2	88.7	92.4	78.6	75.1
1970	100.0	100.0	100.0	100.0	100.0	100.0
1971	110.0	99.4	106.5	98.9	116.8	120.2
1972	119.1	95.7	115.7	104.9	125.7	130.9
1973	121.5	93.7	112.3	109.2	131.0	137.3
1974	114.8	87.1	99.4	109.9	127.6	135.2
1975	118.7	75.4	96.9	109.9	138.9	149.5
1976	125.2	76.2	99.5	108.6	148.1	160.9
1977	126.4	66.4	106.2	104.5	147.8	160.7
1978	137.5	64.8	115.5	104.3	161.4	176.7
1979	150.4	69.1	128.0	104.5	176.8	194.4
1980	155.0	63.6	134.7	106.2	181.7	200.3
1981	155.2	55.9	136.0	106.7	181.7	200.3
1982	163.5	51.2	141.0	106.4	193.0	214.3
1983	168.3	47.5	144.9	108.4	198.8	221.5
1984	172.1	38.5	151.1	109.6	202.6	225.9
1985	179.2	36.7	154.1	111.2	212.2	237.4
1986	185.2	34.8	161.2	112.9	218.9	245.6
1987	197.9	33.2	166.9	115.7	235.8	265.4
1988	210.7	39.0	184.1	121.9	255.5	306.1
1989	228.8	42.1	196.7	124.2	279.5	338.6
1990	241.7	45.5	205.2	130.5	287.4	349.6

Note: The data refers to domestic transport only, although the weights are calculated including international transport.

Source: Ministry of Transport, *General Indexes of Transport Activities.*

oxidizing potential formed by chain reactions between unsaturated hydrocarbons and other reactive organic compounds, nitrogen oxides, and oxygen in the presence of the sunlight." [OECD, 1988b,p.52] Ozone (O_3) is the most prevalent photochemical oxidant. About 1/2 of HC and NO_X emissions are attributable to transportation activities. The WHO recommends a total oxidant concentration of 0.06 ppm as a best single value estimate of exposure limit. Oxidants cause adverse health effects such as increased susceptibility to infections, pulmonary disease, pulmonary and systemic biochemical changes, impairment of pulmonary functions, and irritation of eye, nose, and throat. Oxidants can cause damage to vegetation, including thin-leaved vegetables, agricultural crops, and forest trees.

Acid deposition: Sulphur and nitrogen oxides can undergo chemical transformation in the atmosphere, and acids and acid salts thus formed return to the earth. Approximately 1/3 of the acidity in rainfall is attributed to NOx emissions which are then transformed into nitric acid. Photochemical oxidants are suspected of playing a key role in the conversion of sulphur and nitrogen oxides into acids. Acid deposition is damaging to ecosystems (forest and aquatic) and corrosive materials.

Traffic noise: Noise and vibration caused by road, rail, and air traffic exert physical effects, physiological effects (directly measurable ones), and psycho-sociological effects (such as annoyance and other reactions).

Transformation of land use: Expansion of automobile traffic calls for increased provision of road infrastructure. More land has to be taken for infrastructures, often from farm land and forests. Building and improving roads in urban areas also tend to aggravate traffic noise unless strict zoning in land use is put into effect.

5.2 Trends in transportation activities

The long-term trends in transportation activities are shown in Table 5-2.[1]The data refer to the value of transportation and not the physical units. The index for the total jumped from 18.7 in 1960 to 100 in 1970, or a fivefold increase. The 1970s have seen much slower growth, the index reaching only 155.0 by 1980. Thus, on the aggregate level, tranportation is a reflection of economic growth. Focusing on the modal shift from railways to motor vehicles,

Table 5-3 Number of vehicles

a. Passenger transportation

	Buses for business		Passenger cars for business		Buses for private use	
Year	Vehicles (1000)	Distance travelled (mil.km)	Vehicles (1000)	Distance travelled (mil.km)	Vehicles (1000)	Distance travelled (mil.km)
1955	33.0	(1227)	45.7	(2728)	1.6	(20)
1956	37.0	(1375)	50.1	(3513)	1.8	(23)
1957	41.4	(1542)	54.7	(3954)	2.1	(27)
1958	45.4	(1715)	58.3	(4115)	2.2	(34)
1959	48.7	(1892)	63.2	(4444)	2.9	(40)
1960	53.1	(1946)	76.6	(4591)	4.2	(48)
1961	58.6	(2144)	98.6	(6281)	6.3	(62)
1962	62.2	(2343)	106.6	(7755)	9.1	(105)
1963	70.2	(2618)	123.4	(8880)	13.8	(157)
1964	74.4	(3008)	143.9	(10160)	20.9	(318)
1965	77.4	(3148)	151.0	(11721)	27.8	(442)
1966	79.6	(3345)	161.9	(13036)	37.6	(553)
1967	80.9	(3487)	174.9	(14897)	52.1	(790)
1968	83.2	(3581)	191.3	(16412)	70.1	(1126)
1969	84.8	(3602)	206.2	(18236)	90.8	(1411)
1970	85.9	(3674)	217.7	(19873)	105.1	(1720)
1971	84.1	(3629)	226.2	(20214)	112.7	(1749)
1972	83.9	(3670)	226.7	(19780)	121.8	(1997)
1973	84.4	(3625)	232.6	(19047)	129.3	(1849)
1974	84.6	(3576)	238.4	(17153)	131.1	(1742)
1975	86.7	(3623)	243.3	(17515)	133.1	(1828)
1976	86.7	(3695)	244.9	(17898)	135.5	(1774)
1977	87.0	(3730)	246.2	(17950)	137.0	(1993)
1978	87.2	(3786)	248.1	(18812)	139.4	(2110)
1979	87.6	(3864)	249.4	(19457)	140.7	(2097)
1980	88.4	(3890)	250.5	(19328)	140.9	(2156)
1981	88.7	(3954)	251.1	(19456)	140.8	(2137)
1982	89.2	(3982)	251.4	(19059)	140.4	(2096)
1983	89.3	(4026)	251.9	(19277)	140.2	(2138)
1984	89.8	(4066)	252.3	(19251)	140.2	(2247)
1985	90.0	(4112)	253.0	(19249)	141.0	(2237)
1986	91.0	(4136)	253.0	(19331)	141.0	(2319)
1987	92.0	(4252)	255.0	(19758)	143.0	(2375)
1988	93.0	(4392)	256.0	(19543)	146.0	(2345)
1989	94.0	(4502)	257.0	(19230)	148.0	(2460)
1990	95.0	(4609)	256.0	(19348)	151.0	(2502)

Note: Buses include those for route service and for charter. Figures in this table are based on Ministry of Transportation, *Statistical Survey on Motor Vehicle Transport*.

	Passenger cars for private use		Light vehicles	Two-wheeled vehicles
Year	Vehicles	Distance travelled		
	(1000)	(mil.km)	(1000)	(1000)
55	91.8	(1309)	530.9	50.6
56	110.9	(1334)	643.2	53.4
57	142.8	(1820)	768.6	56.3
58	149.2	(2526)	940.3	57.3
59	239.4	(2990)	1222.7	55.2
60	363.7	(4134)	1459.7	50.0
61	503.7	(5586)	1790.0	45.8
62	656.3	(7634)	2129.4	44.4
63	955.7	(12055)	2473.9	45.0
64	1317.8	(15753)	2747.0	44.4
65	1726.8	(22281)	3057.9	48.0
66	2312.9	(30453)	3395.6	55.3
67	3099.5	(43203)	3938.1	64.4
68	4100.7	(59795)	4566.5	76.9
69	5306.1	(78336)	5298.4	109.7
70	6559.1	(100709)	5968.4	171.5
71	7947.0	(117185)	6436.3	220.0
72	9738.6	(126492)	6736.5	238.1
73	11365.0	(144964)	6654.1	261.7
74	12840.3	(143857)	6552.6	276.7
75	14578.7	(158520)	5867.3	257.2
76	15961.2	(168184)	5953.7	277.0
77	17323.1	(187416)	6185.0	292.3
78	18937.5	(201550)	6331.9	327.5
79	20309.6	(215550)	6749.4	383.6
80	21292.9	(222131)	7297.1	444.9
81	22263.5	(227701)	8089.4	522.2
82	23137.7	(236908)	9036.0	617.3
83	24030.6	(240467)	10004.8	700.1
84	26785.8	(245845)	11035.8	775.6
85	25595.0	(256308)	12062.0	851.0
86	26435.0	(265963)	13133.0	912.0
87	27570.0	(288303)	14223.0	974.0
88	28720.0	(301953)	15259.0	1016.0
89	30625.0	(322144)	15975.0	1046.0
90	32177.0	(346249)	16769.0	1000.0

Note: After fiscal 1988, distance travelled for passenger cars includes light motor vehicles. Figures for light motor vehicles and two-wheeled cars are from *Motor Vehicles Ownership*.

b. Freight transportation

	Trucks for business		Light trucks for business		Trucks for private use	
Year	Vehicles (1000)	Distance travelled (mil.km)	Vehicles (1000)	Distance travelled (mil.km)	Vehicles (1000)	Distance travelled (mil.km)
1955	48.2	(1187)	43.8	(575)	111.8	(1710)
1956	53.5	(1376)	53.2	(729)	119.1	(1812)
1957	58.5	(1587)	62.0	(888)	127.6	(2055)
1958	60.6	(1734)	70.1	(1113)	131.5	(2236)
1959	67.7	(2050)	80.4	(1300)	141.7	(2379)
1960	76.3	(2732)	89.8	(1645)	155.2	(3817)
1961	89.3	(3231)	97.7	(1830)	182.2	(4026)
1962	99.0	(3844)	102.1	(2026)	205.2	(4707)
1963	115.6	(4937)	107.6	(2353)	233.9	(6005)
1964	132.4	(5811)	109.2	(2468)	256.6	(6666)
1965	143.5	(6080)	106.6	(2385)	281.8	(6751)
1966	164.4	(7557)	104.5	(2454)	331.6	(8517)
1967	191.4	(9189)	100.4	(2512)	396.5	(10569)
1968	217.0	(10934)	96.7	(2442)	454.2	(13072)
1969	238.8	(12432)	93.2	(2439)	508.8	(15114)
1970	258.6	(13175)	92.2	(2417)	555.2	(16285)
1971	277.7	(13725)	91.6	(2283)	594.9	(16470)
1972	303.0	(14747)	90.2	(2396)	664.5	(17917)
1973	333.3	(15722)	87.0	(2234)	754.8	(16802)
1974	345.7	(15203)	86.3	(2082)	797.3	(16079)
1975	353.0	(15780)	86.0	(2141)	822.4	(16422)
1976	368.9	(17269)	86.2	(2202)	844.7	(16774)
1977	385.0	(19033)	86.5	(2286)	877.6	(17698)
1978	408.2	(22081)	86.0	(2363)	951.1	(20721)
1979	434.1	(24234)	85.9	(2388)	1019.9	(22532)
1980	450.7	(24535)	86.6	(2347)	1051.6	(23155)
1981	464.1	(25045)	88.0	(2371)	1069.1	(23061)
1982	477.1	(26108)	89.3	(2447)	1076.1	(22969)
1983	498.0	(27549)	91.0	(2512)	1088.9	(23680)
1984	524.2	(30048)	92.4	(2570)	1105.5	(24076)
1985	537.0	(32055)	93.0	(2627)	1120.0	(24635)
1986	563.0	(34574)	91.0	(2668)	1140.0	(25504)
1987	590.0	(37287)	95.0	(4674)	1175.0	(27161)
1988	632.0	(40755)	95.0	(5014)	1250.0	(29191)
1989	675.0	(43667)	95.0	(5193)	1334.0	(30602)
1990	713.0	(45839)	94.0	(5349)	1431.0	(32288)

Note: Distance travelled figures include light vehicles starting in 1987. Total trucks and total motor vehicles series are from *Motor Vehicle Ownership*. Total motor vehicles include special purpose vehicles and heavy special vehicles.

	Light trucks for private use		Total trucks	Total motor vehicles
	Vehicles	Distance travelled		
	(1000)	(mil.km)	(1000)	(1000)
55	486.5	(3306)	693.0	1501.7
56	580.0	(3692)	809.3	1775.1
57	672.8	(4815)	925.3	2069.1
58	765.4	(5734)	1032.7	2404.1
59	876.9	(9396)	1172.3	2898.4
60	994.2	(9250)	1321.6	3403.7
61	1167.7	(12386)	1543.9	4135.8
62	1368.3	(14550)	1782.0	4922.0
63	1665.4	(19278)	2130.6	5937.2
64	1984.3	(26144)	2491.5	6984.8
65	2328.7	(29347)	2870.2	8123.0
66	2796.2	(38830)	3407.0	9639.3
67	3355.7	(47565)	4056.6	11690.7
68	3887.3	(56058)	4671.4	14021.9
69	4265.8	(62032)	5126.2	16528.5
70	4530.4	(68163)	5460.3	18919.0
71	4802.7	(68224)	5791.8	21222.7
72	5177.3	(72595)	6262.7	23869.1
73	5510.9	(71921)	6720.5	25962.8
74	5789.8	(66792)	7057.9	27870.4
75	6079.4	(70516)	7381.0	29143.4
76	6415.4	(81901)	7757.8	31048.1
77	6630.4	(92219)	8023.4	32965.0
78	6894.4	(89837)	8387.6	35179.5
79	7053.7	(91831)	8646.7	37333.2
80	7036.6	(91509)	8682.9	38992.0
81	6974.5	(90933)	8654.6	40834.0
82	6861.5	(89532)	8563.5	42687.4
83	6723.0	(89280)	8462.0	44558.8
84	6596.1	(87639)	8382.2	46362.8
85	6600.0	(87216)	8306.0	48241.0
86	6476.0	(87117)	8271.0	50223.0
87	6416.0	(165024)	8352.0	52646.0
88	6443.0	(172392)	8549.0	55136.0
89	6492.0	(172419)	8695.0	57994.0
90	6510.0	(172396)	8835.0	60499.0

Table 5-4 Traffic volume and fuel consumption

a. Passenger transportation

	Passenger-km (million km)		Distance travelled (million km)		Fuel consumption (1000 kl)	
Year	Buses	Passenger cars	Buses	Passenger cars	Gasoline	Light oil
1955	23320	4180	1246	4038	764	217
1956	27600	5700	1399	4852	874	280
1957	31300	7200	1575	5774	968	346
1958	34300	8400	1749	6641	1022	394
1959	39200	8800	1933	7434	1320	481
1960	43999	11532	1994	8725	1225	502
1961	49237	15958	2206	11867	1603	586
1962	54399	19622	2448	15389	2017	677
1963	62873	26491	2775	20935	2458	809
1964	76039	31922	3327	25913	2881	1002
1965	80134	40622	3590	34002	3581	1067
1966	83933	54595	3898	43490	4382	1120
1967	90476	74111	4278	58100	5837	1196
1968	95314	103612	4707	76208	7724	1259
1969	100192	141869	5013	96572	9529	1285
1970	102893	181335	5394	120582	11772	1323
1971	100843	211635	5378	137400	14238	1330
1972	108211	220346	5667	146272	14504	1368
1973	111713	225732	5474	164010	16812	1357
1974	115776	228400	5318	161010	16726	1360
1975	110063	250804	5451	176035	18237	1379
1976	98714	264499	5469	186083	19428	1410
1977	104639	263961	5723	205367	21739	1438
1978	107009	296044	5896	220363	23554	1540
1979	108317	319869	5960	235007	25614	1615
1980	110396	321272	6046	241459	26300	1666
1981	108827	328251	6091	247156	26359	1705
1982	104836	347219	6077	255968	26985	1711
1983	103418	360747	6164	259744	27294	1721
1984	103064	365631	6313	265096	27275	1735
1985	105000	384000	6352	275557	27827	1740
1986	102000	398000	6455	285294	28688	1746
1987	103000	456000	6626	308062	26978	3800
1988	107000	501000	6737	321496	27999	4083
1989	109000	556000	6376	341374	29658	4584
1990	110000	576000	7112	365597	31726	5222

Note: Figures are in terms of fiscal year. Okinawa is not included up to 1974.

Source: Ministry of Transport, *Statistical Survey on Motor Vehicle Transport*.

b. Freight transportation

	Ton-km (million km)		Distance travelled (million km)		Fuel consumption (1000 kl)	
Year	Business use	Private use	Business use	Private use	Gasoline	Light oil
1955	3700	5800	1762	5016	1629	285
1956	4500	6400	2105	5503	1994	372
1957	5500	7700	2475	6871	2264	484
1958	6300	9100	2847	7970	2504	503
1959	7900	10500	3351	8775	2772	686
1960	9638	11163	4377	13068	2634	919
1961	11700	14900	5060	16412	2928	1447
1962	14091	18338	5870	19257	3137	1813
1963	18541	23490	7290	25283	3856	2233
1964	20222	26993	8279	32810	4773	2720
1965	22385	26006	8465	36098	4913	2814
1966	30602	34310	10012	47347	5960	2723
1967	37189	43904	11701	58134	6892	4940
1968	46972	54480	13376	69130	7922	6121
1969	58135	61730	14871	77146	8541	7101
1970	67330	68586	15592	84448	9179	7816
1971	72050	70618	16007	84694	9079	8547
1972	76515	77095	17142	90512	9129	9048
1973	73367	67612	17956	88754	8990	9390
1974	72044	58726	17285	82871	8341	9289
1975	69247	60455	17922	86938	8684	9869
1976	72789	59830	19471	98675	9781	10672
1977	80020	63075	21319	109917	10525	11787
1978	86873	69192	24444	110558	9936	13417
1979	98190	74698	26622	114362	9651	14710
1980	103541	75360	26883	114664	9212	15226
1981	108276	73034	27416	113994	8881	15460
1982	116832	70887	28555	112502	8294	15882
1983	124680	68847	30060	112960	7776	16688
1984	132028	68786	32617	111716	7064	17785
1985	137000	69000	34682	111851	6473	18669
1986	148000	68000	37242	112622	5995	19685
1987	155000	71000	41961	192186	5658	21188
1988	171000	75000	45769	201583	5117	23179
1989	184000	79000	48860	203022	4602	24531
1990	194000	80000	51188	204684	4314	25867

Note: Figures are in terms of fiscal year. Okinawa is not included up to 1974. After fiscal year 1988, figures include light motor vehicles.

Source: Ministry of Transport, *Statistical Survey on Motor Vehicle Transport.*

the table does not include data on sea and air transport. It is shown that, in the case of freight transport, railways reached a peak in 1970 and have shown a rather steep decline since, reaching about 30% in 1990. Freight transportation by motor vehicles took its place. In the case of passenger transportation by rail, the trend up to 1970 seems more or less similar to that of freight, only that it had exhibited much faster growth. After 1970, it never experienced an absolute decline except for short-term fluctuations reflecting economic slowdowns following the two oil crises. Motor vehicles started to play a major role in passenger transportation by the 1960s, and especially noteworthy is the rapid increase of nonbusiness passenger transport by automobile. In the base year (1970), the share of this category had reached 73.7% of total.

The nature and magnitude of the environmental impact of transportation depend on the stock of motor vehicles in use and the volume of traffic. Let us look at these factors.

Table 5-3 provides an overview of the historical development in the number of motor vehicles in Japan.[2] "Business" in the table refers to commercial transport operators and "private" to company-owned fleets and household-owned cars. The total number of vehicles stood at 60.5 million in 1990, which consisted of 32.2 million passenger cars for private use, 16.8 million light motor vehicles [3], and 8.8 million trucks, among others. Of these, most notable is the passenger car for private use. In 1960, only 363 thousand of them existed; by 1970, this figure already exceeded 6 million and by 1980, 21 million. The growth of stock of passenger cars was not arrested in the 1980s, as can be seen in the 50% increase in passenger cars and more than twofold increase in the stock of light motor vehicles.

Table 5-4 is intended to establish linkages between traffic volume and fuel consumption.[4] Traffic volume is represented by passenger- and ton-km and by the distance travelled. There are several points to be made. First, in the case of passenger transportation, the role of buses has become rather minor. Towards the end of the 1960s, the passenger-km of bus transportation was taken over by passenger cars. Second, for freight transportation, business use is dominant as opposed to private use in terms of ton-km, whereas in terms of distance travelled, the opposite is true. Private use seems to involve small-sized trucks for deliveries. Third, in terms of fuel consumption, we observe an increased consumption of light oil reflecting a shift to diesel-powered vehicles.

The trend is more marked in the case of freight transportation where gasoline consumption has decreased in absolute terms. In fact, as early as 1977, the consumption of gasoline and light oil was nearly equal in proportion, and in 1990, the former has less than halved while the latter has more than doubled.

The trends in stocks of vehicles and traffic volume indicate a considerable slowdown in the last decade compared with the previous years.

5.3 Technology and policy

In order to interpret the environmental impact of the stock of motor vehicles, we need information regarding the technological change which affects fuel efficiency and emission of various pollutants and the policy measures intended to influence these factors. They include:

(1) the progressive introduction of fuel-efficient vehicles, consumer preference for larger cars, and an increase in commercial vehicles in the heavier classes;

(2) the diffusion of less polluting vehicles in terms of lead (Pb), CO, NO_X, HC, and noise emissions;

(3) the increased use of diesel vehicles, which is particularly marked in Japan as well as in Europe. Diesel-powered commercial vehicles account for about 70 to 80 per cent of the total in Japan and Europe but only about 30 per cent in the USA [OECD, 1988b, p.29]; and

(4) the average life-span of vehicles which tends to be longer for technical reasons such as increased durability and economic reasons such as slower growth of the economy.

Improvement in the network of roads is also a factor affecting fuel efficiency, traffic volume, and wear and tear of the motor vehicles.

Lead emissions have been closely connected to the amount of leaded fuel consumed. Introduction of catalyst converters starting around 1975 made it imperative to make lead-free gasoline available. The three-way catalyst system, which is designed to reduce HC, CO, and NO_X simultaneously, is known to be sensitive to the use of leaded gasoline, which permanently damages the capability of reducing the emission of pollutants. In Japan, over 95 percent of all gasoline sold was unleaded already in 1983. The lead content of gasoline in

Table 5-5 Gasoline lead content in selected countries

(unit: gl)

Country	Actual lead content	Target maximum lead content (in effect)
EEC	0.15-0.4	0.013(compulsory, 1989)
Australia	0.4 -0.8	0.013(1985)
Austria	0.15	0.013
Canada : unleaded	0.013	
leaded	0.77	0.29
Finland	0.4	0.15
Japan	0.004	
Spain	0.48-0.65	0.4 (1986)
Sweden	0.15	0.0 (1987)
Switzerland	0.15	0.013(1986)
United States : unleaded	0.013	
leaded	0.13	0.026(1986)

Source: OECD [1988b], p.92.

selected countries is presented in Table 5-5.[OECD, 1988b, p.55]

As a result of the reduction of the lead content in gasoline, lead emission from the transport sector has been reduced. For example, an index of lead emission (1975=100) has been reduced to 33 in the USA in 1983, 55 in Canada, 54 in The Netherlands, and 92 in the UK.[OECD, 1988b, p.56]

Carbon monoxide emission has decreased sharply in Japan during the 1970s. In 1982, it was reported that at all, except for one, roadside monitoring stations, the Japanese ambient air quality standard was met (8-hour standard: 20 ppm). The annual average of CO concentrations in Tokyo and Osaka decreased by 45 percent between 1973 and 1982, from 1.7 ppm and 1.9 ppm to 0.9 ppm and 1.1 ppm, respectively. See also Table 6-12 on the trend of air quality in Japan. Similar improvement has been achieved in the USA, whereas in Europe no clear trends can be discerned, according to an OECD report. Starting in 1988, new environmental policy measures went into effect regarding car exhaust emissions. The measures were started in 1985 on a voluntary basis and were made mandatory in October 1988. [5]

Nitrogen oxide remained a stubborn problem for a long time, and Japan was no exception. The sequence of regulations concerning NO_X emission of motor vehicles is listed in Table 5-6. As can be seen from the table, the

Table 5-6 NOX reduction rate of mobile source

(unit: index, no regulation = 1.00)

	Passenger cars gasoline & LPG		Trucks and buses diesel (direct injection type)	
1955	No regulation	1.00	No regulation	1.00
1960				
1965				
1970				
1971				
1972				
1973	'73 regulation	0.71		
1974			'74 regulation	0.80
1975	'75 regulation	0.39		
1976	'76 regulation	0.20		
1977			'77 regulation	0.68
1978	'78 regulation	0.08		
1979			'79 regulation	0.56
1980				
1981				
1982				
1983			'83 regulation	0.49
1984				
1985				
1986				
1987				
1988			'88 regulation	0.42
1989			'89 regulation	0.42
1990				

Note: Exhaust level prior to environmental regulation is taken as 1.00. The 1976 regulation for passenger car was stipulated as 0.20 for vehicles below 1 ton in EIW (equivalent inertia weight) and 0.27 for vehicles exceeding 1 ton in EIW. Regulation for diesel-powered trucks and buses listed here applies to direct injection type with gross vehicle weight of more than 2.5 tons. Other stipulations apply to direct injection type of lesser weight and for indirect injection type, although the general trend of regulation is more or less the same. The 1988 regulation for diesel applies to vehicles with gross weight of 3.5 tons or less, and the 1989 regulation, to vehicles exceeding 3.5 tons.

Source: Environment Agency, *Environment White Paper.*

regulation was tightened much more quickly for passenger automobiles using gasoline. By 1978, taking the pre-regulation level as 1.00, the stipulated level was 0.08 for passenger cars, whereas for trucks and buses using diesel engines,

Table 5-7 New car price increases to comply with emission

(unit: in 1981 US $)

Model year	HC/CO/NOX (g/km)	Initial cost increase($)
1968-69	5.9/50.8/–	30
1970-71	3.9/33/3/–	50
1972	3.0/28.8/–	70
1973-74	3.0/28.0/3.1	100
1975-76	1.5/15.0/3.1	150
1977-79	1.5/15.0/2.0	175
1980	0.41/7.0/2.0	225
1981	0.41/3.4/1.0	350

Source: OECD [1988b], p.106.

the regulation remained at more than 0.50. NO_X emissions from the transport sector as a whole has been reduced somewhat. Taking 1975 as the base year (=100), it reached 71 in 1980 in Japan. In the USA, it was 103 in 1980 and 99 in 1983; in France and Germany, it increased to 123 and 142 in 1983.[OECD, 1988b, p.56]

As a result, secondary pollutants, such as photochemical pollution, have also decreased in Japan. The average number of days exceeding 0.12 ppm O_3 at roadside monitoring stations decreased from 2.8 in 1975 to 0.3 in 1982. The number of people who reported adverse health effects attributable to photochemical pollution has declined from a peak number of 46,000 in 1975 to about 2,000 or less in early 1980s.

Reduction in HC emissions was also a factor behind this improvement. It is reported that urban O_3 trends are more closely related to HC than to NO_X concentrations. Between 1977 and 1882, HC emissions were curtailed by 25% in Japan.[OECD, 1988b, p.58]

NO_X emission contributes to atmospheric acidity, which leads to an increased acid deposition in the soil. No clear time trends can be delineated in this regard because of insufficient data.[OECD, 1988b, p.58]

Tighter exhaust emission control technologies cannot be adopted without cost. Table 5-7 shows the cost estimates based on the US data. No comparable figures are available for Japan. The cost estimates listed here are based on the retail price equivalent of each component in the emission control system used in gasoline-fuelled cars. A cost model was established based on these data which shows new car price increases as a function of various emission standards. The model is used in calculating the initial cost paid by the consumers to comply with the US emission standards. All emission standards are based on the 1975 US test procedure. The cost estimate for 1981 stood at $350. There are other sources of estimates, which put the cost at $300 to $1,000. Extra service cost was estimated at $10 per year. The cost is arrived at by netting out extra service cost for the new system and the reduced cost due to improved engine oil consumption and spark plug life.[OECD, 1988b, p.106]

5.4 Further remarks

One important aspect of road transportation is the provision of roads, which often requires changes in land use with inevitable consequences on the environment. This will be taken up in Chapter 9 where the issue of land use is discussed.

The energy requirement in transportation activities accounts for nearly 1/4 of the total energy demand in Japan (1/3 elsewhere) and directly contributes to the greenhouse gases such as CO_2, a topic which is taken up in the total energy demand-supply structure in Chapter 3. The diffusion of privately owned cars played an important role in household expenditures and asset formation together with other consumer durables. This aspect is taken up in Chapter 4 in conjunction with lifestyle and again in Chapter 11 as a contributing factor to the quality of life.

Motorization also has negative effects such as automobile accidents, which must be considered when assessing the relation between environment and quality of life.

The energy efficiency of various means of transportation is an important issue in redesigning the socioeconomic system to make it amenable to the environment. As a measure of energy efficiency, we can calculate the unit energy consumption which is defined as energy consumption/traffic volume for each mode of transportation.

Energy consumption can be measured in terms of kcal and traffic volume in passenger-km and ton-km for passenger and freight transportation, respectively. Seen in this context, passenger cars are the least efficient means of transportation, consuming 5 times more energy than railways and 3 times more than buses. For freight transportation, the energy efficiency is markedly good for coastal sea transportation and railways, followed by trucks for business use. Freight transportation by privately owned trucks is the least efficient means, consuming 17 times more per ton-km than coastal sea transportation and railways, and 3 times more than commercial trucks.

However, the shift toward private cars constitutes consumer preference for privacy, comfort, and increased flexibility and cannot be dealt with simply from the point of view of energy efficiency. The increased use of company-owned fleets in freight transportation indicates a changing need for delivery service, which tends toward more timely delivery and probably in small quantities. There has been a major shift from postal package to private delivery service. In industry, there has been a shift toward just-in-time delivery in order to streamline production and logistics. In wholesale and retail trade, sales and delivery are directly linked by computer. Introduction of POS (point of sales) reduces the need for excessive inventory at the sales outlet and increases the efficiency of the trade sector but will demand a more flexible delivery system. The situation is probably different in the case of urban transit: heavy dependence on private cars will lead to traffic congestion, worsening of air and noise pollution, and destruction of the natural environment by increased provision for highways.

These considerations point to the fact that improving fuel efficiency and reducing pollutant emission of individual automobiles are necessary steps but will not provide the ultimate answer to the relation between transportation and the environment.

FOOTNOTES

[1] Comparable long-term figures for transportation activities are available in input-output tables for the Japanese economy. It should be noted that passenger and freight transportation by privately owned cars has been included in the tables up to the 1985 edition but was omitted thereafter. Thus, 1975-1980-1985 link tables and annual extension tables for the re-

cent period contain discontinuities at this disaggregated level. Details of the transportation sector include those of national railways, passenger other than urban transit; national railway, freight; national railway, urban transit; private lines, passenger; private lines, freight; buses taxis and rental limousines; road freight transportation; delivery services; ocean freight; coastal passenger; coastal freight; harbor transport; international air; domestic air, passenger; domestic air, freight; services using airplanes; warehouse; packing; road-related services; sea transport services, public; sea transport service, business; air transport-related services, national and public; air transport-related services, business; and other transport-related services.

[2] Data from the *Motor Vehicles Ownership* published by the Ministry of Transport. The category motor vehicles owned consists of registered motor vehicles entered in the motor vehicle registration files in accordance with the Road Transport Vehicle Law and light two-wheeled vehicles and light motor vehicles. Registered motor vehicles are classified into buses, trucks, passenger cars, special use vehicles and heavy special vehicles. Special use vehicles refer to vehicles used for special use such as tank lorries, concrete mixers, etc. Heavy special vehicles are those specially built such as road rollers, forklift trucks, etc. Light two-wheeled vehicles and light three- and four-wheeled vehicles are subject to compulsory motor vehicle inspection, whereas light motorcycles (not included in the table) are subject to reporting only.

[3] Defined in Japan as ones with engines of less than 800 cc and with physical dimensions falling under certain specifications; the definition was altered from time to time. In addition to fuel economy, light motor vehicles were exempt from laws requiring registration of parking place. The taxes for car ownership, as distinct from purchase, are also considerably lower. Commodity tax rates have been equalized recently.

[4] Statistics on motor vehicle transport volume and fuel consumption are from Statistical Survey on Motor Vehicle Transport. The present survey was organized in April 1960 as one of the designated statistics. Sample size is about 39,000 vehicles for May and October and about 11,000 vehicles for other months. Buses are counted completely. Light motor vehicles were newly covered by the survey from fiscal 1987.

[5] OECD [1988b] reports that in the USA, the ambient air quality standards (8-hour standard : 9 ppm) was met by 35 % of the monitoring stations in 1975 and 70% in 1983. cf. ibid., p.57. For details of legislation governing car exhaust emission in the EC, see pp.89-91, p.129.

Part III

Coping with Environmental Disruption

Chapter 6

Pollution Prevention Investment

6.1 Quality of life and environmental damage

Economic growth is being achieved without a parallel input of energy, and well-being is being enhanced at a faster pace than production. Environmental damage is being contained as far as traditional types of pollution are concerned.

We should not be optimistic, however, because we have witnessed an emergence of new types of environmental concerns. The major issues in the 1960s and 1970s had been deterioration of air and water quality and disposal of wastes. In the 1980s, we were concerned about such diverse topics as global warming, acid rain, ozone hole, toxic chemicals, deforestation, nuclear wastes and disaster, and fossil fuel burning, in addition to more or less traditional issues. The problems promise to be with us throughout the 1990s and into the 21st century.

Environmental concerns encompass quite a large scope of issues. Rather than trying to deal with the whole array of problems, the purpose of this chapter is limited to assessing trends in prevention investment. We will look deeply into pollution prevention investment and its effectiveness. The scope is limited to air pollution, water pollution, and disposition of industrial and household wastes. For these aspects of the problem, we have by now a fairly reliable statistical database on which we can draw some lessons as to the

Table 6-1 Pollution abatement investment in selected countries

(unit: index, 1985=100)

	Japan	USA	Germany	France	Netherlands
1972	178	71	68	—	—
1973	240	86	69	—	—
1974	339	87	68	—	—
1975	341	96	66	—	85
1976	263	97	61	—	71
1977	132	94	56	—	106
1978	104	92	51	—	82
1979	86	95	46	—	78
1980	86	93	54	—	87
1981	110	91	57	138	81
1982	122	85	67	119	104
1983	124	82	68	108	61
1984	95	95	63	109	70
1985	=100	=100	=100	=100	=100
1986	74	97	129	111	128
1987	68	102	134	110	—
1988	81	—	138	—	—
1989	101				
1990	105				

Notes: Index based on data expressed in national currencies and in 1985 prices. Deflator used refers to price index for gross fixed capital formation. Germany refers to West only. Japanese figures for 1988, 1989, and 1990 have been calculated by the present author from the MITI's data on pollution prevention investment by industry (Table 6-4 below), adjusted by deflator on private investment in plant and equipment from SNA.

Source: OECD, *OECD Environmental Data 1991.*

time lags involved in recognizing the problem, technological development and the diffusion process, the magnitude of economic resources to be diverted for this purpose, and the changes in environmental quality. An effort is made to establish environmental accounting, bridging pollution prevention investment (by industry and pollution category) to improvement in environmental quality.

6.2 Trends in pollution prevention investment

The statistical data on pollution prevention investment are rather numerous among the new fields of social concern, where data collection tends to lag in relation to the emergence of the issue. In fact, environmental pollution was the major concern already in the mid-1960s, prompting various public as well as private organizations to start collecting some figures. Table 6-1 reveals the fact that the pollution prevention investment was activated much earlier in Japan compared to other countries. However, the data on new types of environmental concern are only beginning to be collected. Another problem is the lack of a coherent framework within which such data can be fitted.[1] In this chapter, we trace the trends in pollution prevention investment, link it to the treatment capacity of various pollutants, and then to environmental quality.

Table 6-2 represents the production of pollution prevention equipment in Japan classified by the purchasing sectors.[2] This data set is particularly useful for our purpose in that it provides figures for major equipment categories that cope with particular environmental issues. The data can also be used in conjunction with capital formation data, enabling us to know how many real resources are devoted to this purpose.[3] Environmental accounting should include a tabulation of a similar nature, although it may be necessary to add a new pollution category to reflect today's concern. In particular, it seems imperative to add energy conservation effort to the list.[4]

Pollution prevention equipment in the table are classified into the following:

A. Air pollution prevention equipment

- (1) Dust collection
- (2) Heavy oil desulphurization
- (3) Flue gas desulphurization
- (4) Flue gas nitrogen oxide removal
- (5) Exhaust gas treatment
- (6) High stack
- (7) Related equipment

B. Water pollution prevention equipment

- (1) Industrial waste water processing
- (2) Sewerage treatment
- (3) Toilet water processing
- (4) Sludge processing

  (5) Sea water pollution prevention

  (6) Related equipment

C. Waste treatment equipment

  (1) Household waste processing

  (2) Industrial waste processing

  (3) Related equipment

D. Noise and vibration prevention equipment

  (1) Noise abatement

  (2) Vibration prevention

  (3) Related equipment

Purchasing sectors are divided into private (13 manufacturing sectors and 3 nonmanufacturing), public (local public bodies and others, the latter including public corporations and the central government), and exports. The industrial classification is based on the standard scheme and is amenable to the one adopted for the multisectoral econometric model and input-output framework (although the data here are on a business firm basis, whereas the classification in the input-output tables is on an activity basis).

The table reveals that the largest amount of money is being spent for household waste treatment, in 1990, (29.6% of total value of production of pollution prevention equipment), followed by sewerage treatment (18.9%). On the whole, water pollution prevention is the most expensive in terms of operation (50.0%). The waste treatment and air pollution prevention shares are 19.6% and 29.6%, respectively. Air pollution prevention is almost exclusively the responsibility of the private sector. An exception is dust collection equipment for which local public bodies are also customers in order to purify exhaust air from the incineration of wastes. In the case of water pollution prevention and waste treatment, the private industry is responsible for the treatment of their share, whereas the treatment of sewerage and wastes from household and commercial activities are generally the responsibility of local public bodies. Of the total production of pollution prevention facilities, 70.0% is purchased by the public sector and 26.3% by the private sector in 1990. Exports comprise a small fraction.

Based on a time-series compilation of such data, one can obtain statistical series on the pollution prevention effort for individual sectors by pollution category. Table 6-3 is a summary for total domestic demand (i.e., excluding exports), divided into private and public sectors.

Table 6-4 compares the data we have used in this paper with those from alternative sources. For private industry, probably the most frequently cited data are those compiled by the Ministry of International Trade and Industry. The MITI figures are based on a survey covering establishments with a capital of 100 million yen (for mining, 50 million yen) or more which are under its jurisdiction, i.e., mining, manufacturing, and electricity and gas.[5] There are some limits in the usefulness of the data, which include the following. First, the survey publishes only for recent years investment data coping with air pollution, water pollution, and industrial wastes, without providing further details. Second, the source does not include the public sector, making it difficult to judge the relative importance of private and public sectors. Third, the survey does not cover small businesses.[6] One clear advantage of this source is that it provides the ratio of pollution prevention investment to total fixed capital formation in individual sectors. This makes it easier to come up with an estimate of total investment for this purpose, coupled with the sectoral capital formation data.[7] We list SNA data on investment in plant and equipment in the private sector for comparison.

As for the public sector, 'administrative investment' data are published annually by the Ministry of Home Affairs. This source identifies public investment in 25 categories, of which environment sanitation and public sewerage more or less correspond to our household waste treatment and sewerage processing, respectively. The survey covers national government as well as cities and towns, adjusted for duplication among them. The business units covered by administrative investment are, in principle, those related to the gross public fixed capital formation in the system of national accounts, with minor exclusions. The amount of investment includes expenditures on the maintenance and repair of facilities, improvement projects including cost of land and compensation, office expenses, and planning and surveys.[Ministry of Home Affairs, various issues] We list SNA data on fixed capital formation in the public sector for comparison. The Economic Planning Agency estimates capital stock in the public sector with similar classification to the administrative

Table 6-2 Production of pollution prevention equipment by purchasing sector

a. 1970

Purchasing sectors:	Food	Textiles	Pulp, paper
Sector code	18	20&	24
A. Air pollution prevention equipment			
1. Dust collection	554	180	1344
2. Heavy oil desulphurization	0	0	0
3. Stack fume desulphurization	338	32	12
4. Stack fume nitrogen oxide removal	-	-	-
5. Exhaust gas treatment	80	40	68
6. High stack	0	204	0
7. Related equipment	1	72	17
Subtotal	973	528	1411
B. Water pollution prevention equipment			
1. Industrial water processing	3037	1146	5091
2. Sewerage processing	52	22	19
3. Toilet water processing	4	0	0
4. Sludge processing	18	27	263
5. Sea water pollution prevention	-	-	-
6. Related equipment	252	10	296
Subtotal	3363	1205	5669
C. Waste treatment equipment			
1. Household waste processing	34	0	0
2. Industrial waste processing			
3. Small-sized incinerators	-	-	-
4. Related equipment	0	0	0
Subtotal	34	0	0
D. Noise and vibration prevention equipment			
1. Noise abatement	0	14	12
2. Vibration prevention	1	1	0
3. Related equipment	0	0	0
Subtotal	1	15	12
Total	4371	1748	7122
Percentage	(2.2)	(0.9)	(3.7)

Note: Samples are 187 firms, of which 155 (82.9 %) responded. The data here refer to the production of pollution prevention equipment only, and do not include construction cost. Figures refer to fiscal year.

Source: The Japan Association of Industrial Machinery Producers, *The Production of Pollution Prevention Equipment.*

(unit: current prices, million yen)

	Petro-leum & coal	Petro chemi-cal	Chemi-cal	Cera-mics	Iron & steel	Non-ferrous metals	Machi-nery	Other manufact
	27	(26)	(26)	30	31	32	34&	38&&
A.								
1.	471	244	2133	3388	10351	4940	2952	2315
2.	9887	3	0	0	0	0	0	2
3.	724	10	193	4	416	296	0	2
4.	-	-	-	-	-	-	-	-
5.	1052	267	568	16	6129	297	106	141
6.	2199	750	1253	1300	2167	107	224	0
7.	479	115	236	32	932	67	46	189
	14812	1389	4383	4740	19995	5704	3328	2649
B.								
1.	1902	2351	3562	221	4386	1316	3561	2783
2.	0	4	0	0	223	0	384	80
3.	0	0	10	0	0	0	30	15
4.	0	80	5	489	30	27	337	89
5.	-	-	-	-	-	-	-	-
6.	283	25	836	23	1731	81	828	1144
	2185	2460	4413	733	6370	1426	5140	4111
C.								
1.	0	150	164	0	826	0	9	53
2.								
3.	-	-	-	-	-	-	-	-
4.	0	1	0	0	3000	0	0	5
	0	151	164	0	3826	0	9	58
D.								
1.	24	69	128	5	233	3	72	0
2.	2	3	9	3	8	0	5	0
3.	1	2	1	0	0	0	0	0
	27	74	138	8	241	3	77	0
	17024	4074	9098	5381	30432	7133	8554	6818
	(8.8)	(2.1)	(4.7)	(2.8)	(15.6)	(3.7)	(4.4)	(3.5)

	Elect-ricity	Mining	Other non-manu.	Total private sector	Local gov't	Others	Total public sector	Exports
	(70)	10&		IP			IG	E
A.								
1.	8348	904	394	38518	581	399	980	316
2.	0	0	0	9892	0	0	0	0
3.	1533	16	13	3589	29	16	45	0
4.	-	-	-	-	-	-	-	-
5.	0	69	23	8856	61	37	98	345
6.	9202	250	0	17653	0	160	160	318
7.	47	47	70	2350	846	276	1122	0
	19130	1286	500	80858	1517	883	2405	979
B.								
1.	131	178	378	30043	1144	141	1285	266
2.	0	0	3835	4619	14626	2482	17108	150
3.	0	0	345	404	10089	992	11081	7
4.	60	0	87	1512	2672	309	2981	6
5.	-	-	-	-	-	-	-	-
6.	1552	5	1104	8172	8783	2181	10964	0
	1743	183	5749	44750	37314	6105	43419	429
C.								
1.	0	0	3	1239	15260	1261	16521	106
2.								
3.	-	-	-	-	-	-	-	-
4.	0	0	10	3016	211	12	223	0
	0	0	13	4255	15471	1273	16744	106
D.								
1.	9	0	5	574	29	1	30	0
2.	0	2	0	34	(	0	0	0
3.	0	0	6	10	0	0	0	0
	9	2	11	618	29	1	30	0
	20882	1471	6273	130481	54331	8267	62598	1514
	(10.7)	(0.8)	(3.2)	(67.0)	(27.9)	(4.3)	(32.2)	(0.8)

	Total value	Percent of total	Percent of category
	00		
A.			
1.	39814	(20.5)	(47.3)
2.	9892	(5.1)	(11.7)
3.	3634	(1.8)	(4.3)
4.	-	(-)	(-)
5.	9299	(4.8)	(11.0)
6.	13131	(9.3)	(21.5)
7.	3472	(1.8)	(4.6)
	84242	(43.3)	(100.0)
B.			
1.	31594	(16.2)	(36.2)
2.			
3.	21877	(11.2)	(24.7)
4.	11492	(5.9)	(13.0)
5.	4499	(2.3)	(5.1)
6.	19136	(9.8)	(21.6)
	88598	(45.5)	(100.0)
C.			
1.	17865	(9.2)	(84.7)
2			
3.	-	(-)	(-)
4.	3239	(1.7)	(15.4)
	21105	(10.8)	(100.0)
D.			
1.	604	(0.3)	(93.2)
2.	34	(0.0)	(5.3)
3.	10	(0.0)	(1.5)
	648	(0.3)	(100.0)
	194593	(100.0)	-
	(100.0)	-	-

b. 1975

Purchasing sectors:	Food	Textiles	Pulp, paper
Sector code	18	20&	24
A. Air pollution prevention equipment			
1. Dust collection	399	441	2029
2. Heavy oil desulphurization	0	0	0
3. Stack fume desulphurization	1508	4407	2182
4. Stack fume nitrogen oxide removal	0	0	0
5. Exhaust gas treatment	372	205	259
6. High stack	0	0	0
7. Related equipment	33	252	147
Subtotal	2312	5305	4617
B. Water pollution prevention equipment			
1. Industrial water processing	9781	4289	8714
2. Sewerage processing	7	1	0
3. Toilet water processing	62	2	0
4. Sludge processing	1274	213	2098
5. Sea water pollution prevention	12	0	0
6. Related equipment	793	365	577
Subtotal	11929	4870	11389
C. Waste treatment equipment			
1. Household waste processing	0	0	0
2. Industrial waste processing	344	22	64
3. Small-sized incinerators	-	-	-
4. Related equipment	29	1	10
Subtotal	373	23	74
D. Noise and vibration prevention equipment			
1. Noise abatement	8	12	15
2. Vibration prevention	0	0	0
3. Related equipment	0	0	0
Subtotal	8	12	15
Total	14622	10210	16095
Percentage	(2.1)	(1.5)	(2.4)

Note: Samples are 224 firms, of which 203 (90.6 %) responded.

Source: Same as Table 6-1a.

(unit: current prices, million yen)

	Petro-leum& coal	Petro chemi-cal	Chemi-cal	Cera-mics	Iron & steel	Non-ferrous metals	Machi-nery	Other manu-fact.
	27	(26)	(26)	30	31	32	34&	38&&
A.								
1.	1985	1734	3601	9161	32770	3393	5118	4466
2.	79676	0	0	0	0	0	0	0
3.	2738	4632	10855	492	6512	3946	33	1863
4.	3516	4000	800	30	38	0	0	27
5.	2954	675	6941	258	9242	759	1273	1314
6.	1297	540	202	206	1175	0	0	0
7.	2713	752	1651	51	2750	173	388	2418
	94879	12333	24050	10198	52487	8271	6812	10097
B.								
1.	8282	5863	14523	501	15106	3322	8392	5854
2.	1	0	59	0	0	2	635	672
3.	0	0	3	0	20	0	149	281
4.	140	319	623	230	725	232	1006	603
5.	0	22	0	4	0	5	7	220
6.	188	265	2955	21	962	16	1473	2554
	8611	6474	18163	756	16813	3577	11662	10184
C.								
1.	0	0	0	0	0	0	0	124
2.	827	826	1169	40	1519	107	713	1016
3.	-	-	-	-	-	-	-	-
4.	0	24	23	0	34	11	153	68
	827	850	1192	40	1553	118	866	1208
D.								
1.	166	95	20	32	67	2	299	306
2.	0	0	0	0	0	0	2	0
3.	0	0	0	0	0	0	73	0
	166	95	20	32	67	2	374	306
	104483	19752	43425	11026	70920	11968	19714	21795
	(15.3)	(2.9)	(6.4)	(1.6)	(10.4)	(1.7)	(2.9)	(3.2)

	Elect-ricity	Mining	Other non-manu.	Total private sector	Local gov't	Others	Total public sector	Exports
	(70)	10&		IP			IG	E
A.								
1.	3957	171	1516	70741	4197	711	4908	1492
2.	0	0	0	79676	0	0	0	0
3.	54738	0	200	94106	59	1017	1076	22
4.	10	0	50	8471	0	0	0	0
5.	332	0	92	24676	3119	207	3326	1538
6.	3222	0	0	6651	0	0	0	0
7.	1912	229	812	14281	621	875	1496	4
	64171	400	2670	298602	7996	2810	10806	3056
B.								
1.	3601	177	3681	92091	2554	2199	4753	4580
2.	21	0	9698	11096	72177	8846	81023	51
3.	0	0	665	1182	27172	4132	31304	0
4.	280	477	104	8324	22816	2514	25330	200
5.	58	0	220	548	800	77	877	249
6.	1385	86	1109	12749	18487	2522	21009	701
	5343	740	15477	125990	144006	20290	164296	5781
C.								
1.	0	0	0	124	59525	792	60317	0
2.	811	24	1665	9147	1676	679	2355	237
3.	-	-	-	-	-	-	-	-
4.	20	0	62	435	379	16	395	0
	831	24	1727	9706	61580	1487	63067	237
D.								
1.	67	0	232	1321	35	86	121	1
2.	0	2	0	4	11	0	11	0
3.	0	0	8	81	2	0	2	0
	67	2	240	1406	48	86	134	1
	70414	1166	20114	435704	213630	24673	238303	9075
	(10.3)	(0.2)	(2.9)	(63.8)	(31.3)	(3.6)	(34.9)	(1.3)

	Total		Percent of total	Percent of category
	Quantity	Value		
		00		
A.				
1.	7392	77141	(11.3)	(24.7)
2.	20	79676	(11.7)	(25.5)
3.	189	95204	(13.9)	(30.5)
4.	11	8471	(1.3)	(2.7)
5.	526	29540	(4.3)	(9.5)
6.	26	6651	(1.0)	(2.1)
7.	-	15781	(2.3)	(5.0)
	8164	312464	(45.8)	(100.0)
B.				
1.	1702	101424	(14.8)	(34.3)
2.	1640	92170	(13.5)	(31.1)
3.	1946	32486	(4.8)	(11.0)
4.	950	33854	(5.0)	(11.4)
5.	56	1674	(0.2)	(0.6)
6.	-	34459	(5.0)	(11.6)
	6294	296067	(43.3)	(100.0)
C.				
1.	345	60441	(8.9)	(82.8)
2.	5856	11739	(1.7)	(16.1)
3.	-	-	(-)	(-)
4.	-	830	(0.1)	(1.1)
	6201	73010	(10.7)	(100.0)
D.				
1.	139	1443	(0.2)	(93.6)
2.	4	15	(0.0)	(1.0)
3.	-	83	(0.0)	(5.4)
	143	1541	(0.2)	(100.0)
	20802	683082	(100.0)	-
	-	(100.0)	-	-

c. 1980

Purchasing sectors:	Food	Textiles	Pulp, paper
Sector code	18	20&	24
A. Air pollution prevention equipment			
1. Dust collection	839	290	723
2. Heavy oil desulphurization	0	0	0
3. Stack fume desulphurization	298	49	1055
4. Stack fume nitrogen oxide removal	150	25	0
5. Exhaust gas treatment	121	30	117
6. High stack	26	0	0
7. Related equipment	69	34	303
Subtotal	1503	428	2198
B. Water pollution prevention equipment			
1. Industrial water processing	6732	1098	1502
2. Sewerage processing	226	2	0
3. Toilet water processing	16	0	0
4. Sludge processing	1174	51	391
5. Sea water pollution prevention	8	0	0
6. Related equipment	314	226	314
Subtotal	8470	1377	2207
C. Waste treatment equipment			
1. Household waste processing	0	0	0
2. Industrial waste processing	228	21	1204
3. Small-sized incinerators	122	28	14
4. Related equipment	36	9	0
Subtotal	386	58	1218
D. Noise and vibration prevention equipment			
1. Noise abatement	21	0	3
2. Vibration prevention	0	0	0
3. Related equipment	1	0	1
Subtotal	22	0	4
Total	10381	1863	5627
Percentage	(1.6)	(0.3)	(0.9)

Note: Samples are 192 firms, of which 173 (90.1 %) responded.

Source: Same as Table 6-1a.

(unit: current prices, million yen)

	Petro-leum & coal	Petro chemi-cal	Chemi-cal	Cera-mics	Iron & steel	Non-ferrous metals	Machi-nery	Other manu-fact.
	27	(26)	(26)	30	31	32	34&	38&&
A.								
1.	83	750	3078	4318	9272	2325	7437	2910
2.	9604	0	0	0	0	0	0	0
3.	2285	33	196	219	2	556	10	35
4.	857	19	227	0	0	0	0	0
5.	7	0	968	279	1880	285	1902	655
6.	100	0	249	0	176	0	103	0
7.	26	56	497	607	1112	204	735	3152
	12962	858	5215	5418	12442	3370	10187	6752
B.								
1.	503	388	3590	879	5066	1078	11010	3859
2.	0	0	49	2	3	0	4517	721
3.	0	0	290	66	0	6	195	164
4.	195	237	886	0	654	55	2041	390
5.	56	6	0	0	13	0	63	2815
6.	52	224	620	71	153	61	466	832
	806	855	5435	1018	5889	1200	18292	8781
C.								
1.	0	418	0	0	71	0	282	1
2.	1	55	774	33	337	0	351	1821
3.	6	26	355	14	111	90	162	555
4.	0	38	26	28	0	2	44	308
	7	537	1155	75	519	92	839	2685
D.								
1.	1	0	45	19	176	10	776	777
2.	0	0	0	0	4	0	310	9
3.	16	1	3	5	5	7	123	26
	17	1	48	24	185	17	1209	812
	13792	2251	11853	6535	19035	4679	30527	19030
	(2.1)	(0.3)	(1.8)	(1.0)	(2.9)	(0.7)	(4.7)	(2.9)

	Elect-ricity	Mining	Other non-manu.	Total private sector	Local gov't	Others	Total public sector	Exports
	(70)	10&		IP			IG	E
A.								
1.	11701	363	1760	45849	7685	1304	8989	6984
2.	0	0	0	9604	0	0	0	155
3.	20637	0	0	25375	0	0	0	1053
4.	20047	0	0	21325	454	500	954	491
5.	122	0	322	6688	5546	444	5990	10432
6.	5295	0	130	6079	90	0	90	46
7.	905	30	87	7812	1172	21	1193	1000
	58707	393	2299	122732	14947	2269	17216	20161
B.								
1.	6831	89	3131	45756	4864	2517	7381	15003
2.	33	19	8246	13818	139829	17836	157665	2789
3.	7	0	684	1428	48439	8185	56624	82
4.	135	509	1321	8039	21946	1845	23791	934
5.	571	0	41	3573	129	8	137	493
6.	74	0	3060	6467	4434	3143	7577	575
	7651	617	16483	79081	219641	33543	253175	19876
C.								
1.	0	0	0	772	102101	7690	109791	0
2.	1521	0	4037	10383	4203	1250	5453	128
3.	23	0	1592	3098	890	266	1156	0
4.	1	0	1805	2297	3222	17	3239	67
	1545	0	7434	16550	110416	9223	119639	195
D.								
1.	183	3	232	2246	271	2858	3129	61
2.	0	0	153	476	0	317	317	0
3.	5	10	22	225	30	0	30	0
	188	13	407	2947	301	3175	3476	61
	68091	1023	26623	221310	345305	48201	393506	40293
	(10.4)	(0.2)	(4.1)	(33.8)	(52.7)	(7.4)	(60.1)	(6.2)

	Total		Percent of total	Percent of category
	Quantity	Value		
		00		
A.				
1.	12702	61822	(9.4)	(38.6)
2.	8	9759	(1.5)	(6.1)
3.	106	26428	(4.0)	(16.5)
4.	63	22770	(3.5)	(14.2)
5.	472	23110	(3.5)	(14.4)
6.	12	6215	(0.9)	(3.9)
7.	-	10005	(1.5)	(6.2)
	13363	160109	(24.4)	(100.0)
B.				
1.	1569	68140	(10.4)	(19.4)
2.	2804	174272	(26.6)	(49.5)
3.	621	58134	(8.9)	(16.5)
4.	750	32764	(5.0)	(9.3)
5.	486	4203	(0.6)	(1.2)
6.	-	14619	(2.2)	(4.2)
	6230	352132	(53.8)	(100.0)
C.				
1.	231	110563	(16.9)	(81.1)
2.	2197	15964	(2.4)	(11.7)
3.	4778	4254	(0.6)	(3.1)
4.	-	5603	(0.9)	(4.1)
	7206	136384	(20.8)	(100.0)
D.				
1.	2269	5436	(0.8)	(83.8)
2.	24	793	(0.1)	(12.2)
3.	-	255	(0.0)	(3.9)
	2293	6484	(1.0)	(100.0)
	29092	655109	(100.0)	-
	-	(100.0)	-	-

d. 1985

Purchasing sectors:	Food	Textiles	Pulp, paper
Sector code	18	20&	24
A. Air pollution prevention equipment			
1. Dust collection	850	999	1612
2. Heavy oil desulphurization	0	0	0
3. Stack fume desulphurization	13	294	479
4. Stack fume nitrogen oxide removal	28	0	0
5. Exhaust gas treatment	108	141	67
6. High stack	0	0	0
7. Related equipment	55	22	295
Subtotal	1054	1456	2453
B. Water pollution prevention equipment			
1. Industrial water processing	3021	762	1652
2. Sewerage processing	2	12	0
3. Toilet water processing	0	0	0
4. Sludge processing	123	10	66
5. Sea water pollution prevention	0	0	0
6. Related equipment	135	27	99
Subtotal	3281	811	1817
C. Waste treatment equipment			
1. Household waste processing	0	0	0
2. Industrial waste processing	8	2	5
3. Small-sized incinerators	221	83	113
4. Related equipment	1	0	0
Subtotal	230	85	118
D. Noise and vibration prevention equipment			
1. Noise abatement	0	0	6
2. Vibration prevention	0	3	0
3. Related equipment	0	0	0
Subtotal	0	3	6
Total	4565	2355	4394
Percentage	(0.7)	(0.4)	(0.7)

Note: Samples are 180 firms, of which 165 (81.7 %) responded.

Source: Same as Table 6-1a.

(unit: current prices, million yen)

	Petro-leum & coal	Petro chemi-cal	Chemi-cal	Cera-mics	Iron & steel	Non-ferrous metals	Machi-nery	Other manu-fact.
	27	(26)	(26)	30	31	32	34&	38&&
A.								
1.	1025	414	3481	4022	8598	1717	6526	3323
2.	433	0	0	0	0	0	0	0
3.	774	1805	453	313	371	0	75	0
4.	0	1211	263	0	0	0	75	0
5.	195	453	2665	640	3416	654	801	4559
6.	0	0	251	0	0	0	0	1021
7.	217	76	327	131	1051	21	210	2668
	2644	3959	7440	5106	13435	2392	7687	11571
B.								
1.	2029	142	3276	987	3822	1058	16306	4420
2.	0	2	40	18	86	41	361	1036
3.	0	0	0	0	0	23	42	3
4.	231	9	24	56	343	0	93	543
5.	59	0	0	0	20	0	4	1889
6.	53	81	353	24	90	30	102	758
	2372	234	3693	1085	4361	1152	16908	8449
C.								
1.	0	0	2	0	0	0	0	1
2.	0	0	78	0	69	7	371	549
3.	61	113	106	35	78	39	301	417
4.	0	0	13	0	0	0	10	36
	61	113	199	35	147	46	682	1003
D.								
1.	3	9	51	5	728	1	885	486
2.	0	16	78	0	0	0	20	43
3.	0	0	65	12	2	0	89	6
	3	25	194	17	730	1	994	535
	5080	4331	11526	6243	18673	3591	26271	21558
	(0.8)	(0.7)	(1.8)	(0.9)	(2.9)	(0.6)	(4.0)	(3.3)

	Elect-ricity	Mining	Other non-manu.	Total private sector	Local gov't	Others	Total public sector	Exports
	(70)	10&		IP			IG	E
A.								
1.	5673	427	1360	40027	5299	1201	6500	5160
2.	0	0	0	433	0	0	0	101
3.	33603	0	0	38180	0	0	0	672
4.	13576	0	14	15167	0	116	116	7380
5.	36	3	920	14658	5897	466	6363	1227
6.	2486	0	195	3953	101	0	101	75
7.	499	0	312	5883	338	401	739	946
	55873	430	2801	118301	11635	2184	13819	15561
B.								
1.	5604	404	1612	44895	18483	3674	22157	5593
2.	58	0	3438	5094	119557	9533	129090	1389
3.	300	0	112	480	22216	10280	32496	74
4.	19	0	14	1531	45135	1308	46443	1175
5.	30	0	51	2053	212	47	259	321
6.	672	66	1874	4364	17630	7211	24841	203
	6683	470	7101	58417	223233	32053	255286	8755
C.								
1.	0	0	0	3	126723	14910	141633	536
2.	18006	10	193	19298	338	3504	3842	34
3.	124	8	2501	4200	893	567	1460	253
4.	0	0	1	61	7348	2	7350	197
	18130	18	2695	23532	135302	18983	154285	1020
D.								
1.	20	5	29	2228	611	37	648	74
2.	0	0	0	160	0	0	0	0
3.	0	0	21	195	219	92	311	205
	20	5	50	2583	830	129	959	279
	80706	923	12647	202863	371000	53349	424349	25615
	(12.4)	(0.1)	(1.9)	(31.0)	(56.8)	(8.2)	(65.0)	(4.0)

	Total			
	Quantity	Value	Percent of total	Percent of category
		00		
A.				
1.	17342	51687	(7.9)	(35.0)
2.	2	534	(0.0)	(0.4)
3.	149	38852	(6.0)	(26.3)
4.	131	22663	(3.5)	(15.3)
5.	1445	22248	(3.4)	(15.1)
6.	29	4129	(0.6)	(2.8)
7.	-	7568	(1.2)	(5.1)
	19098	147681	(22.6)	(100.0)
B.				
1.	1880	72645	(11.1)	(22.6)
2.	6908	135573	(20.8)	(42.1)
3.	342	33050	(5.1)	(10.2)
4.	592	49149	(7.5)	(15.2)
5.	210	2633	(0.4)	(0.8)
6.	-	29408	(4.5)	(9.1)
	9932	322458	(49.4)	(100.0)
C.				
1.	395	142172	(21.8)	(79.5)
2.	275	23174	(3.5)	(13.0)
3.	6469	5913	(0.9)	(3.3)
4.	-	7608	(1.2)	(4.2)
	7139	178867	(27.4)	(100.0)
D.				
1.	1936	2950	(0.5)	(77.2)
2.	200	160	(0.0)	(4.2)
3.	-	711	(0.1)	(18.6)
	2136	3821	(0.6)	(100.0)
	38305	652827	(100.0)	-
	-	(100.0)	-	-

e. 1990

Purchasing sectors:	Food	Textiles	Pulp, paper
Sector code	18	20&	24
A. Air pollution prevention equipment			
1. Dust collection	269	502	1729
2. Heavy oil desulphurization	0	0	0
3. Stack fume desulphurization	0	117	219
4. Stack fume nitrogen oxide removal	0	76	0
5. Exhaust gas treatment	193	65	255
6. High stack	0	100	0
7. Related equipment	112	89	693
Subtotal	574	949	2896
B. Water pollution prevention equipment			
1. Industrial water processing	11238	521	3517
2. Sewerage processing	23	0	0
3. Toilet water processing	0	0	0
4. Sludge processing	314	1	306
5. Sea water pollution prevention	0	0	0
6. Related equipment	113	5	164
Subtotal	11688	527	3987
C. Waste treatment equipment			
1. Household waste processing	31	0	0
2. Industrial waste processing	12	0	412
3. Small-sized incinerators	126	111	69
4. Related equipment	0	0	71
Subtotal	169	111	552
D. Noise and vibration prevention equipment			
1. Noise abatement	4	201	76
2. Vibration prevention	0	0	5
3. Related equipment	0	0	2
Subtotal	4	201	83
Total	12435	1788	7518
Percentage	(1.6)	(0.2)	(1.0)

Note: Samples are 180 firms, of which 157 (87.2 %) responded.

Source: Same as Table 6-1a.

(unit: current prices, million yen)

	Petro-leum & coal	Petro chemi-cal	Chemi-cal	Cera-mics	Iron & steel	Non-ferrous metals	Machi-nery	Other manu-fact.
	27	(26)	(26)	30	31	32	34&	38&&
A.								
1.	115	519	1665	2657	9906	2213	7522	2083
2.	656	0	0	0	0	0	0	0
3.	136	8	2	1363	0	4	98	0
4.	286	14	178	57	148	21	208	520
5.	0	630	1289	90	3146	2684	3820	2943
6.	0	437	0	0	0	0	0	2704
7.	288	13	1063	109	1361	90	1557	139
	1481	1621	4197	4276	14561	5012	13205	8389
B.								
1.	535	531	4267	1039	4631	1028	11586	8817
2.	0	8	52	4	72	29	215	1722
3.	0	0	0	0	0	0	0	0
4.	0	0	110	0	2	0	33	79
5.	0	0	1	0	0	0	0	473
6.	12	5	128	14	115	13	210	173
	547	544	4558	1057	4820	1070	12044	11264
C.								
1.	0	0	0	0	0	0	412	1110
2.	0	37	5	0	0	0	353	747
3.	28	106	88	56	38	45	310	426
4.	0	0	0	0	0	0	0	10
	28	143	93	56	38	45	1075	2293
D.								
1.	17	46	82	8	384	31	801	229
2.	1	7	10	0	125	0	34	27
3.	22	0	15	0	68	0	14	4
	40	53	107	8	577	31	849	260
	2096	2361	8955	5397	19996	6158	27173	22206
	(0.3)	(0.3)	(1.1)	(0.7)	(2.5)	(0.8)	(3.5)	(2.8)

	Elect-ricity	Mining	Other non-manu.	Total private sector	Local gov't	Others	Total public sector	Exports
	(70)	10&		IP			IG	E
A.								
1.	12605	173	2423	44381	8038	1327	9365	4343
2.	0	0	0	656	0	0	0	10298
3.	9045	494	0	11486	0	0	0	1833
4.	21597	0	58	23163	0	2026	2026	3425
5.	61	37	854	16067	5723	477	6200	435
6.	3371	0	0	6612	288	0	288	0
7.	1069	40	277	6900	979	5028	6007	746
	47748	744	3612	109265	15028	8858	23886	21080
B.								
1.	6127	511	1992	56340	8786	2567	11353	2333
2.	137	58	12405	14725	118785	12846	131631	1498
3.	0	12	44	56	30176	24158	54334	0
4.	308	102	301	1556	75515	5858	81373	1481
5.	0	46	0	520	213	0	213	0
6.	2920	310	136	4318	23386	6900	30286	58
	9492	1039	14878	77515	256861	52329	309190	5370
C.								
1.	0	0	61	1614	183168	25990	209158	1729
2.	5580	978	692	8816	595	153	748	840
3.	93	25	2461	3982	1288	456	1744	3
4.	82	0	158	321	3173	78	3251	0
	5755	1003	3372	14733	188224	26677	214901	2572
D.								
1.	874	975	617	4345	582	696	1278	73
2.	2	0	25	236	0	6	6	13
3.	70	0	33	228	277	40	317	0
	946	975	675	4809	859	742	1601	86
	63941	3761	22537	206322	460972	88606	549578	29108
	(8.1)	(0.5)	(2.9)	(26.3)	(58.7)	(11.3)	(70.0)	(3.7)

	Total		Percent of total	Percent of category
	Quantity	Value		
		00		
A.				
1.	21133	58089	(7.4)	(37.7)
2.	4	10954	(1.4)	(7.1)
3.	87	13319	(1.7)	(8.6)
4.	167	28614	(3.6)	(18.5)
5.	873	22702	(2.9)	(14.7)
6.	12	6900	(0.9)	(4.5)
7.	-	13653	(1.7)	(8.9)
	22276	154231	(19.6)	(100.0)
B.				
1.	918	70026	(8.9)	(17.9)
2.	6169	147854	(18.9)	(37.7)
3.	673	54390	(6.9)	(13.9)
4.	773	84410	(10.8)	(21.5)
5.	1	733	(0.1)	(0.2)
6.	-	34662	(4.4)	(8.8)
	8534	392075	(50.0)	(100.0)
C.				
1.	283	212501	(27.1)	(91.5)
2.	86	10404	(1.3)	(4.5)
3.	3868	5729	(0.7)	(2.5)
4.	-	3572	(0.5)	(1.5)
	4237	232206	(29.6)	(100.0)
D.				
1.	1979	5696	(0.7)	(87.7)
2.	147	255	(0.0)	(3.9)
3.	-	545	(0.1)	(8.4)
	2126	6496	(0.8)	(100.0)
	37173	785008	(100.0)	-
	-	(100.0)	-	-

Table 6-3 Production of pollution prevention equipment, time series

a. Purchased by private sector (unit: current prices, million yen)

	1969	1970	1971
A. Air pollution prevention equipment			
1. Dust collection	23464	38517	47318
2. Heavy oil desulphurization	23860	9892	14810
3. Stack fume desulphurization	1255	3589	8157
4. Stack fume nitrogen oxide removal	-	-	-
5. Exhaust gas treatment	8424	8856	11242
6. High stack	11179	17653	29042
7. Related equipment	2575	2350	8026
Subtotal	70778	80858	118595
B. Water pollution prevention equipment			
1. Industrial water processing	9696	28946	52313
2. Sewerage processing	1547	4619	5008
3. Toilet water processing	1046	404	280
4. Sludge processing	59	1512	1869
5. Sea water pollution prevention	-	-	-
6. Related equipment	2703	8172	9521
Subtotal	19889	44750	70626
C. Waste treatment equipment			
1. Household waste processing	523	1239	0
2. Industrial waste processing	-	-	3195
3. Small-sized incinerators	-	-	-
4. Related equipment	14	3016	390
Subtotal	537	4255	3585
D. Noise and vibration prevention equipment			
1. Noise abatement	-	574	713
2. Vibration prevention	-	34	28
3. Related equipment	-	10	6
Subtotal	-	618	747
Total	91205	130481	193553
Share of private sector (%)	63.9	67.1	64.0

Note: The data here refer to production of pollution prevention equipment only and do not include construction cost. Figures refer to fiscal year.

Source: Same as Table 6-2a.

	1972	1973	1974	1975	1976	1977	1978
A.							
1.	57796	68605	91971	70741	65059	45544	33134
2.	27296	37209	28792	79676	47562	4941	954
3.	14705	36135	145896	94106	97859	44853	20410
4.	-	-	-	8471	5588	2910	4337
5.	9262	29227	24503	24676	23011	14983	13013
6.	8767	8958	7684	6651	9610	3781	5442
7.	7001	14015	19625	14281	13445	10256	6429
	121827	194149	318471	298602	262134	127268	83719
B.							
1.	64364	75640	79462	92091	68638	45662	41819
2.	4588	6647	14698	11096	12227	12628	8142
3.	2812	1201	956	1182	573	899	795
4.	5795	7819	10899	8324	7546	5680	5913
5.	992	876	428	548	2019	1512	1205
6.	5462	7989	9978	12749	10170	7409	8052
	86642	102198	117517	25990	101173	73790	65926
C.							
1.	215	48	323	124	89	48	0
2.	8499	14126	12655	9147	11981	12967	19374
3.	-	-	-	-	-	-	-
4.	174	610	418	435	1016	2142	248
	8888	14784	13396	9706	13086	15157	19622
D.							
1.	863	883	1174	1321	1496	2916	1913
2.	0	6	11	4	2	38	30
3.	101	118	54	81	100	682	624
	964	1007	1239	1406	1598	3636	2567
	218321	212138	450623	435704	377991	219851	171834
	58.3	63.9	66.5	63.8	54.5	37.8	28.0

	1979	1980	1981	1982	1983	1984	1985
A.							
1.	36546	45849	42351	36218	28542	44683	40027
2.	1089	9604	3900	1844	6592	2767	433
3.	10781	25375	27725	35737	38714	37472	38180
4.	13041	21325	32941	10638	20487	20777	15167
5.	8249	6688	9992	18649	9200	15050	14658
6.	4289	6079	7867	6111	12666	3671	3953
7.	8412	7812	5299	4863	3647	3925	5883
	82407	122732	130075	114060	119848	128345	118301
B.							
1.	39979	45756	35768	44033	39219	55008	44895
2.	8723	13818	19238	8917	11296	7376	5094
3.	454	1428	11400	1924	662	248	480
4.	6770	8039	6787	2743	5196	3185	1531
5.	2446	3573	3840	3451	3734	3156	2053
6.	7189	6467	10365	7850	8684	4367	4364
	65561	79081	87398	68918	68791	73340	58417
C.							
1.	67	772	8633	479	0	450	3
2.	10327	10383	11298	3217	8237	1913	19298
3.	-	3098	4030	3966	4100	3914	4200
4.	182	2297	372	432	187	5	61
	10576	16550	24333	8094	13024	6282	23562
D.							
1.	2362	2246	1035	3754	1596	2962	2228
2.	0	476	40	122	8	173	160
3.	218	225	294	524	628	852	195
	2580	2947	1369	4400	2232	3987	2583
	161124	221310	243175	195472	203895	211954	202863
	25.0	33.8	35.9	32.0	31.3	35.6	31.0

	1986	1987	1988	1989	1990
A.					
1.	44753	35159	25161	41476	44381
2.	249	1969	840	214	656
3.	29307	34979	16395	14283	11486
4.	16573	10954	13904	10914	23163
5.	16990	10374	11967	14759	16067
6.	6206	6152	1080	97150	6612
7.	4655	3593	3063	6278	6900
	118733	103180	82139	95074	109265
B.					
1.	41876	33326	40608	55561	56340
2.	5767	9053	9369	11943	14725
3.	68	236	149	354	56
4.	1977	1355	1159	2107	1556
5.	864	1711	406	3115	520
6.	3625	5611	2536	4176	4318
	54177	51292	54227	77256	77515
C.					
1.	10	104	43	2995	1614
2.	12019	10610	17637	11510	8816
3.	3833	3577	2496	3078	3982
4.	411	54	177	1211	321
	16273	14345	20353	18794	14733
D.					
1.	2946	3811	2171	3973	4345
2.	160	311	0	7	236
3.	107	305	278	149	228
	3213	4427	2449	4129	4809
	192396	173244	159168	195253	206322
	28.8	27.8	24.7	28.6	26.3

b. Purchased by public sector

	1969	1970	1971
A. Air pollution prevention equipment			
1. Dust collection	1311	980	2223
2. Heavy oil desulphurization	-	0	0
3. Stack fume desulphurization	3	45	80
4. Stack fume nitrogen oxide removal	-	-	-
5. Exhaust gas treatment	420	98	375
6. High stack	19	160	463
7. Related equipment	773	1122	365
Subtotal	2527	2405	3506
B. Water pollution prevention equipment			
1. Industrial water processing	142	590	2176
2. Sewerage processing	12175	17108	32269
3. Toilet water processing	7550	11081	12683
4. Sludge processing	2659	2981	7353
5. Sea water pollution prevention	-	-	-
6. Related equipment	8938	10964	13131
Subtotal	31682	43419	67927
C. Waste treatment equipment			
1. Household waste processing	12352	16520	28797
2. Industrial waste processing	-	-	1611
3. Small-sized incinerators	-	-	-
4. Related equipment	1501	223	494
Subtotal	13853	16744	30902
D. Noise and vibration prevention equipment			
1. Noise abatement	-	30	58
2. Vibration prevention	-	0	0
3. Related equipment	-	0	0
Subtotal	-	30	58
Total	48062	62598	102393
Share of public sector (%)	33.7	32.2	34.1

Source: Same as Table 6-1a.

(unit: current prices, million yen)

	1972	1973	1974	1975	1976	1977	1978
A.							
1.	5597	4715	6484	4908	8065	6164	7518
2.	0	0	0	0	0	0	0
3.	82	1589	817	1076	101	48	0
4.	-	-	-	-	135	56	0
5.	422	1121	1881	3326	3640	3762	7202
6.	359	428	0	0	745	32	110
7.	670	1212	638	1496	1269	2615	691
	7130	9065	9820	10806	13955	12677	15521
B.							
1.	6013	2401	4275	4753	6988	6813	12489
2.	47057	58255	70670	81023	106268	142136	168265
3.	18599	18092	30104	31304	32196	38044	48394
4.	8599	11508	23110	25330	34858	30165	32486
5.	1368	144	250	877	86	760	1758
6.	16193	15125	15089	21009	23966	26104	40026
	98888	105689	143967	164296	204362	244022	303418
C.							
1.	42886	48418	54923	60317	79573	84041	83703
2.	1887	2445	6614	2355	979	2503	1736
3.	-	-	-	-	-	-	0
4.	702	377	1511	395	1248	4040	12575
	45475	51240	63028	63067	81799	90584	98014
D.							
1.	45	20	5	121	515	1022	1122
2.	0	0	47	11	40	0	0
3.	0	20	0	2	9	3	79
	45	40	52	134	564	1025	1201
	151538	166034	216867	238303	300680	348308	418154
	40.4	34.0	52.0	34.9	43.3	60.0	68.2

	1979	1980	1981	1982	1983	1984	1985
A.							
1.	7916	8989	5519	6435	9532	6874	6500
2.	0	0	0	0	0	0	-
3.	0	0	0	3119	35	4	-
4.	530	954	175	867	3	2041	116
5.	8028	5990	4189	6654	5807	10426	6363
6.	367	90	341	585	1239	100	101
7.	1429	1193	1217	1974	795	1124	739
	18270	17216	11441	19634	17411	20569	13819
B.							
1.	17865	7381	7076	6749	7224	11347	22157
2.	172614	157665	150153	113690	127287	121301	129090
3.	49475	56624	57092	50153	48555	33948	38394
4.	38488	23791	36825	41416	39229	40134	46443
5.	625	137	1231	2088	25	143	259
6.	44410	7577	16228	30516	33683	27733	24841
	323477	253175	268605	244612	256003	234606	255286
C.							
1.	102600	109791	103036	118943	137796	89727	141633
2.	4969	5453	7304	3563	2470	1774	3842
3.	-	1156	1574	1629	1470	2751	1460
4.	1566	3239	2693	6884	7240	13178	7350
	109135	119639	114607	131019	148976	107430	154285
D.							
1.	878	3129	3978	1151	617	1481	648
2.	0	317	0	0	0	1	-
3.	435	30	284	1062	1438	1135	311
	1313	3476	4262	2213	2055	2617	959
	452195	393506	398915	397478	424445	365222	424349
	70.2	60.1	58.9	65.1	65.2	61.4	65.0

	1986	1987	1988	1989	1990
A.					
1.	6052	5113	2738	38311	9365
2.	-	-	0	0	0
3.	-	14	21	0	0
4.	150	3	0	0	2026
5.	4013	4528	3714	6636	6200
6.	71	215	385	229	288
7.	1670	3407	3518	243	6007
	11956	13280	10376	45419	23886
B.					
1.	9279	8364	11218	8037	11353
2.	121749	134320	147056	116264	131631
3.	49208	49142	40195	44889	54334
4.	53702	62038	68658	70865	81373
5.	457	33	0	82	213
6.	23200	23377	16299	22637	30286
	257595	277274	283426	262774	309190
C.					
1.	131355	119496	161791	150398	209158
2.	4616	1363	1787	3121	748
3.	1596	1231	1369	1792	1744
4.	10592	899	4856	3867	3251
	148159	122989	169803	159178	214901
D.					
1.	211	370	910	934	1278
2.	-	-	0	14	6
3.	352	-	252	360	317
	563	370	1162	1308	1601
	418273	413913	464767	468679	549578
	62.6	66.4	72.1	68.6	70.0

Table 6-4 Pollution prevention investment and selected SNA series

a. private sector (unit: billion yen)

Year	Production of equipment, for private sector	Pollution prevention investment, industry (%)	SNA fixed capital formation, plant and equipment, private sector
1965		29.7(3.1)	5151.4
1966		26.8(2.9)	6038.6
1967		46.2(3.5)	7948.4
1968		62.4(3.7)	9926.6
1969	91.2	106.7(5.0)	12569.1
1970	130.5	163.7(5.3)	15406.3
1971	193.6	270.6(6.5)	15358.5
1972	218.3	323.2(8.3)	16200.8
1973	312.1	514.7(10.6)	20814.4
1974	450.6	917.0(15.6)	24660.1
1975	435.7	964.5(17.7)	24290.5
1976	378.0	781.9(13.5)	25224.0
1977	219.9	405.5(7.2)	26260.6
1978	171.8	326.5(5.5)	28102.6
1979	161.1	290.1(4.5)	32937.6
1980	221.3	312.8(3.9)	37615.9
1981	243.2	403.7(4.8)	39744.9
1982	195.5	451.6(5.1)	40572.5
1983	203.9	454.0(6.4)	41163.3
1984	212.0	347.5(4.5)	46047.7
1985	202.9	366.5(4.9)	51757.9
1986	192.4	267.2(3.6)	53336.5
1987	173.2	242.8(3.6)	55917.3
1988	159.2	281.5(4.1)	63901.5
1989	195.3	276.6(3.2)	73801.8
1990	206.3	340.9(3.3)	83410.0

Source: The Japan Association of Industrial Machinery Producers (Table 6-3a above) (sample of about 200 machinery producers), the Ministry of International Trade and Industry *Plant and Equipment Investment Plans in Major Industries* (sample of about 1,000 corporation with capital of 100 million yen and over under MITI's jurisdiction), and the Economic Planning Agency, *Annual Report on National Accounts*.

investment, but this is for their internal use only.

The data we have used in this paper relate to machinery only and do not include civil engineering and construction cost, whereas capital formation data compiled by the Ministry of International Trade and Industry (pollution

b. public sector (unit: billion yen)

Year	Production of equipment, for public sector	Administrative investment			SNA fixed capital formation, public sector
		Sani-tation	Sewerage	Total	
1965		44	72	2699	2782.9
1966					3440.1
1967		47	109	3527	3704.6
1968		56	142	4104	4374.5
1969	48.1	68	165	4847	4904.7
1970	62.6	92	224	5911	5890.5
1971	102.4	140	371	7621	7199.5
1972	151.5	199	504	9321	8873.4
1973	166.0	250	546	10692	10830.2
1974	216.9	333	625	14204	12081.0
1975	238.3	377	816	16514	13417.6
1976	300.7	383	876	17598	14556.7
1977	348.3	429	1164	20868	16838.3
1978	418.2	507	1447	24373	20099.1
1979	452.2	521	1624	26110	22005.3
1980	393.5	548	1848	27876	22888.3
1981	398.9	578	2011	28793	24278.8
1982	397.5	562	2021	28762	24121.4
1983	424.5	568	1896	27987	23579.9
1984	365.2	539	1872	27640	23105.4
1985	424.3	558	1989	26506	21648.2
1986	418.3	565	2206	27861	22270.6
1987	413.9	605	2646	30412	23729.8
1988	464.8	642	2647	31679	25046.8
1989	468.7	700	2751	33828	25881.6
1990	549.5				28104.9

Source: The Japan Association of Industrial Machinery Producers (Table 6-3b above), the Ministry of Home Affairs, *Administrative Investment* and the Economic Planning Agency. *Annual Report on National Accounts.*

prevention investment), the Ministry of Home Affairs (administrative investment), and the Economic Planning Agency (SNA) include both machinery and construction.

As for details of supplying sectors of fixed capital formation, one should re-

Table 6-5 Heavy oil desulphurization

Year	Number of units (units)	Capacity (million kl)	Cumulative investment (billion yen)
1966			0.1
1967	1(0, 1)	2,231	12.2
1968			31.9
1969	13(11, 2)	16,990	56.6
1970			66.2
1971	20(16, 4)	29,555	80.8
1972			107.8
1973	19(14, 5)	49,997	141.0
1974	32(26, 6)	55,707	161.0
1975	39(30, 9)	73,957	214.4
1976	42(31,11)	78,367	247.0
1977	42(31,11)	80,117	250.4
1978	43(32,11)	81,290	251.0
1979	43(32,11)	81,696	251.7
1980	56(44,12)	83,611	257.8
1981	56(44,12)	83,611	260.2
1982	56(44,12)	83,611	261.4
1983			265.5
1984			267.3
1985			267.6
1986			267.6
1987			269.0
1988	41(29,12)	74,460	269.6
1989			269.8
1990	38(26,12)	70,810	270.2

Note: Desulphurization capacity data are compiled by the Petroleum Association and refer to annual capacity which is derived from BPSD (barrels per stream day) figure assuming that the facility will be utilized 365 days a year. Figures in parantheses refer to the number of units utilizing indirect and direct desulphurization process, respectively. Cumulative investment series has been derived by the present author from production data on 'heavy oil desulphurization' for the private sector compiled by the Japan Association of Industrial Machinery Producers, converted to 1970 prices. Gross output deflator of the general machinery sector (P34) has been used in the conversion.

Sources: *Environment White Paper*various issues. The Japan Association of Industrial Machinery Producers, *The Production of Pollution Prevention Equipment*, various issues.

fer to the fixed capital formation matrix compiled every five years, coordinated

Table 6-6 Flue gas desulphurization

Year	Number of units (units)	Capacity (million Nm^3/h)	Cumulative investment (billion yen)
1966			0.3
1967			1.1
1968			3.2
1969			4.5
1970	102	5.4	8.1
1971	183	9.3	16.2
1972	323	18.0	30.7
1973	543	28.8	63.0
1974	768	42.7	164.4
1975	994	79.5	228.6
1976	1,134	103.8	295.8
1977	1,192	110.5	326.1
1978	1,227	114.8	339.9
1979	1,266	117.5	347.0
1980	1,329	122.0	363.1
1981	1,362	126.5	380.3
1982	1,366	127.2	402.5
1983	1,405	129.1	426.9
1984	1,583	133.4	450.6
1985	1,741	154.5	474.8
1986	1,758	155.0	493.9
1987	1,789	169.2	517.4
1988	1,810	176.3	528.6
1989	1,843	178.8	538.4
1990	1,914	193.8	545.7

Note: Data on capacity are based on a series of surveys conducted by the Environment Agency. The figures up to fiscal 1982 are as of January 1 of the pertinent fiscal year and those from 1983 on are as of March 31 (end of the fiscal year). Cumulative investment series has been derived by the present author from production data on 'stack fume desulphurization' for the private sector compiled by the Japan Association of Industrial Machinery Producers, converted to 1970 prices. Gross output deflator of the general machinery sector (P34) has been used in the conversion.

Sources: Environment Agency *Environment White Paper*, various issues. The Japan Association of Industrial Machinery Producers, *The Production of Pollution Prevention Equipment*, various issues.

by the Administrative Management Agency.[8] This source provides such data for individual sectors corresponding to the classification scheme of the input-

Table 6-7 Flue gas nitrogen oxide removal

Year	Number of units (units)	Capacity (million Nm^3/h)	Cumulative investment (billion yen)
1972	5	0.1	—
1973	10	0.4	—
1974	20	1.2	—
1975	45	4.3	5.8
1976	71	8.2	9.6
1977	93	13.7	12.5
1978	109	22.2	15.4
1979	122	28.4	24.1
1980	140	39.1	37.6
1981	175	63.6	58.1
1982	188	71.7	64.7
1983	231	95.1	77.6
1984	253	103.5	90.8
1985	305	109.9	100.4
1986	323	125.9	111.2
1987	348	138.1	118.6
1988	379	142.1	128.0
1989	430	159.1	135.5
1990	538	200.0	150.2

Note: Data based on a series of surveys conducted by the Environment Agency. The figures up to fiscal 1982 are as of January 1 of the pertinent fiscal year and those from 1983 on are as of March 31 (end of the fiscal year). Cumulative investment series has been derived by the present author from production data on 'stack fume nitrogen oxide removal' for the private sector compiled by the Japan Association of Industrial Machinery Producers, converted to 1970 prices. Gross output deflator of the general machinery sector (P34) has been used in the conversion.

Sources: *Environment White Paper*, 1993 edition, Detailed Chapter Volume, p.73. The Japan Association of Industrial Machinery Producers, *The Production of Pollution Prevention Equipment.*

output tables. One can also refer to converters for fixed capital formation in the private and public sectors in the annual input-output tables in order to know the supplying sectors.

If we can assume that pollution prevention investment is of the same con-

Table 6-8 Recycling of industrial water

(unit: 10,000 m^3/day)

	Industrial water total	Plain water	Recycled (%)	Cumulative investment (bil. yen)
1964	66,916	47,755	15,036(31.5)	—
1965	70,524	49,162	17,826(36.3)	—
1966	76,755	53,107	21,093(39.7)	1.8
1967	83,829	57,698	24,180(41.9)	5.9
1968	93,306	64,738	28,907(44.5)	11.9
1969	106,557	74,406	35,790(48.1)	21.9
1970	123,181	85,042	43,986(51.7)	50.8
1971	137,502	95,247	53,310(56.0)	102.7
1972	143,738	101,458	58,889(58.0)	166.2
1973	159,669	113,915	70,658(62.0)	233.8
1974	166,454	120,040	77,790(64.8)	289.0
1975	166,732	121,625	81,432(67.0)	351.8
1976	172,748	127,863	88,030(68.8)	398.9
1977	176,882	131,707	92,947(70.4)	429.8
1978	176,720	132,990	95,430(71.8)	458.0
1979	181,310	137,820	100,790(73.1)	484.5
1980	180,190	138,850	102,225(73.6)	513.5
1981	176,448	137,300	101,517(73.9)	535.7
1982	172,254	133,914	98,775(73.9)	563.1
1983	171,178	133,868	98,786(73.7)	587.8
1984	174,976	136,884	101,938(74.4)	622.6
1985	175,591	137,309	102,381(74.5)	651.1
1986	173,137	136,290	101,870(74.5)	678.4
1987	173,738	137,303	102,869(74.9)	700.8
1988	176,795	139,525	105,856(75.8)	728.3
1989	181,823	143,796	108,856(75.7)	766.5
1990	186,300	146,763	111,461(75.9)	802.5

Note: The data on industrial water refer to establishments with 30 or more employees. The difference between total and plain water is attributable to sea water. Cumulative investment series has been derived by the present author from production data on 'industrial water processing' for the private sector compiled by the Japan Association of Industrial Machinery Producers, converted to 1970 prices. Gross output deflator of the general machinery sector (P34) has been used in the conversion.

Sources: Ministry of International Trade and Industry, *Census of Manufactures*. The Japan Association of Industrial Machinery Producers, *The Production of Pollution Prevention Equipment*.

Table 6-9 Sewerage treatment

Year	Number of treatment terminals (number)	Area under sewerage treatment (hectares)(%)	Cumulative investment (billion yen)
1965	n.a.	49,686(4.0)	—
1966	n.a.	58,324(4.7)	8.0
1967	n.a.	66,962(5.4)	15.6
1968	n.a.	77,518(6.2)	23.7
1969	n.a.	88,466(7.1)	36.3
1970	238	99,242(8.0)	53.4
1971	n.a.	116,752(9.4)	85.3
1972	n.a.	134,619(10.8)	131.8
1973	n.a.	154,206(12.4)	183.8
1974	n.a.	170,996(13.7)	233.0
1975	341	183,966(14.8)	288.2
1976	n.a.	198,211(15.9)	361.2
1977	n.a.	214,371(17.2)	457.1
1978	n.a.	235,887(19.0)	570.9
1979	n.a.	260,677(20.9)	685.2
1980	439	283,646(22.8)	785.0
1981	463	306,636(24.6)	878.5
1982	479	332,433(26.7)	949.0
1983	499	360,543(29.0)	1029.3
1984	533	390,567(31.4)	1106.1
1985	568	426,627(34.3)	1187.9
1986	596	461,696(37.0)	1266.2
1987	617	500,011(40.2)	1356.5
1988	648	543,889(43.0)	1456.3
1989	693	592,556(47.6)	1536.1
1990	744	644,957(51.8)	1620.3

Note: The percentage of area under sewerage treatment is calculated as the ratio to planned area as of 1989 (1,244,772 hectares). Cumulative investment series has been derived by the present author from production data on 'sewerage processing' for the public sector compiled by the Japan Association of Industrial Machinery Producers, converted to 1970 prices. Gross output deflator of the general machinery sector (P34) has been used in the conversion.

Source: Ministry of Home Affairs, *Administrative Investment*. The Japan Association of Industrial Machinery Producers, *The Production of Pollution Prevention Equipment*.

struct as the fixed capital formation in general, we have the clues for knowing

Table 6-10 Municipal waste treatment

Year	Total volume (ton/day)	Incinerated (ton/day)(%)	Land fill (ton/day)(%)	Cumulative investment (billion yen)
1965	44,522	16,896(37.9)	17,659(39.6)	n.a.
1966	48,340	21,899(45.2)	16,594(34.3)	n.a.
1967	53,825	25,459(47.3)	20,292(37.7)	n.a.
1968	62,005	29,959(48.3)	22,470(36.2)	n.a.
1969	70,115	35,758(51.0)	24,688(35.3)	12.8
1970	76,998	45,559(55.3)	25,715(33.5)	29.8
1971	83,328	37,717(45.3)	27,543(33.1)	57.8
1972	91,757	42,604(46.5)	30,587(33.3)	100.2
1973	95,052	45,170(47.5)	32,003(33.7)	143.4
1974	84,205	45,983(54.6)	25,430(30.2)	181.6
1975	87,167	50,380(57.8)	24,461(28.1)	222.7
1976	87,406	52,915(60.6)	23,529(26.9)	277.3
1977	90,285	57,140(63.3)	23,726(26.3)	334.0
1977	113,775	62,014(58.6)	42,216(39.9)	334.0
1978	118,335	65,394(58.9)	43,642(39.3)	390.7
1979	121,904	67,887(59.0)	44,509(38.7)	458.6
1980	120,371	68,739(60.4)	42,139(37.1)	578.1
1981	116,818	71,102(64.5)	35,651(32.3)	592.2
1982	121,857	75,264(65.3)	37,261(32.3)	666.1
1983	116,364	75,022(67.6)	32,841(29.6)	753.0
1984	117,916	77,841(69.1)	31,535(28.0)	809.8
1985	119,041	80,370(70.6)	30,007(26.4)	899.5
1986	122,599	84,548(71.9)	29,008(24.6)	984.0
1987	126,956	89,116(72.6)	28,773(23.6)	1064.4
1988	132,582	93,552(72.8)	29,613(23.0)	1174.2
1989	136,912	98,424(73.9)	28,772(21.6)	
1990	138,196	100,482(74.4)	27,519(20.4)	

Note: Data compiled by the Environment Agency, *Environment White Paper*, various issues. Cumulative investment series has been derived by the present author from production data on 'household waste' in the public sector in Table 6-3, converted to 1970 prices. Percentages are as against total volume processed excluding the portion which is self-disposed.

the total magnitude, from the portion related to machinery sectors.

Table 6-11 Industrial waste treatment

	Total volume (1000 tons/year)			
	1975	1980	1985	1990
Cinder	1,203	1,797	2,409	2,678
Sludge	37,660	88,190	112,821	171,450
Oils	2,289	2,419	3,672	3,471
Acids	9,872	10,219	4,320	2,674
Alkalis	14,435	6,090	923	1,547
Plastics	1,480	2,232	2,816	4,334
Paper	991	1,624	1,472	1,193
Wood	7,890	6,628	8,058	6,573
Textiles	204	101	98	99
Organic wastes	2,596	4,323	2,207	3,543
Rubber	597	92	78	94
Metals	9,985	13,111	8,877	8,533
Glass and ceramics	2,870	2,297	3,910	5,295
Smelter residuals	60,950	60,561	41,649	42,507
Building materials	34,144	30,007	48,948	54,798
Dusts	8,101	11,731	6,224	7,491
Domestic animal residues	41,222	49,691	62,558	77,236
Others		1,199	1,230	1,218
Total	236,489	292,312	312,271	394,736
Cumulative investment (bil. yen)	39.2	82.6	110.3	150.8

Note: Data on industrial waste are compiled by the Ministry of Health and Welfare. Cumulative investment series has been derived by the present author from production data on 'industrial waste processing' for the private sector compiled by the Japan Association of Industrial Machinery Producers, converted to 1970 prices. Gross output deflator of the general machinery sector (P34) has been used in the conversion.

6.3 Pollution prevention capacity

Our next task is to examine the build-up of pollution prevention capacity. We focus our attention on the following.

Air:

(a) Heavy oil desulphurization (Table 6-5)

(b) Flue gas desulphurization (Table 6-6)

	Percentage recycled (percent) 1979	1983	1987	1990
Cinder	23.3	42.9	34.2	17.5
Sludge	10.5	24.8	13.6	11.5
Oils	24.8	24.6	28.0	24.5
Acids	19.4	33.6	16.1	26.5
Alkalis	19.5	28.8	24.1	8.0
Plastics	25.8	24.4	26.5	31.2
Paper	35.6	43.8	32.4	63.2
Wood	22.3	95.1	78.6	43.3
Textiles	40.2	50.6	33.3	51.1
Organic wastes	31.9	83.3	74.5	31.3
Rubber	19.3	25.6	14.5	12.5
Metals	98.7	97.5	99.5	93.0
Glass and ceramics	36.5	37.9	47.9	50.8
Smelter residuals	81.4	75.1	87.9	84.7
Building materials	1.6	10.1	15.1	16.0
Dusts	64.7	58.2	78.0	
Domestic animal residues				
Others				
Total	54.6	58.5	55.4	n.a.
Cumulative investment (bil. yen)	76.1	96.9	125.3	150.8

Sources: *Environment White Paper*, various issues. The Japan Association of Industrial Machinery Producers, *The Prodction of Pollution Prevention Equipment.*

(c) Flue gas nitrogen oxides removal (Table 6-7)

Water:

(e) Industrial waste water processing (Table 6-8)
(f) Sewerage treatment (Table 6-9)

Waste:

(g) Municipal waste treatment (Table 6-10)
(h) Industrial waste treatment (Table 6-11)

We omit detailed discussion for lack of space, and interested readers are referred to the tables.

Table 6-12 Trends in air quality

Year	Sulphur dioxide (ppm)	Carbon monoxide (ppm)	Floating particles (mg/m^3)
1965	0.057	3.3	
1966	0.057	3.2	
1967	0.059	4.1	
1968	0.055	4.4	
1969	0.050	5.1	
1970	0.043	4.5	
1971	0.037	6.0	
1972	0.031	5.4	
1973	0.030	5.3	
1974	0.024	5.0	0.059
1975	0.021	4.7	0.051
1976	0.020	4.2	0.050
1977	0.018	3.7	0.047
1978	0.017	3.3	0.048
1979	0.016	3.2	0.045
1980	0.016	2.9	0.043
1981	0.014	2.9	0.043
1982	0.013	2.6	0.041
1983	0.012	2.4	0.036
1984	0.012	2.5	0.041
1985	0.011	2.4	0.037
1986	0.010	2.4	0.041
1987	0.010	2.4	0.041
1988	0.010	2.4	0.039
1989	0.011	2.4	0.039
1990	0.011	2.3	0.042

Note: SO_2 levels are the average of 15 general observation stations from which time-series data are available. CO levels relate to the average of 15 auto exhaust gas observation stations. Floating particles relate to the average of 40 general observation stations.

6.4 Improvement in environmental quality

The next question is how to establish a relationship between pollution prevention effort and actual level of environmental quality. This is a difficult question to tackle in our framework. Environmental quality is a concept which inherently focuses on individual regions, and the span of the region is best

	Nitrogen monoxide		*Nitrogen dioxide*	
Year	General stations (ppm)	Auto exhaust stations (ppm)	General stations (ppm)	Auto exhaust stations (ppm)
1965		0.047	0.026	0.026
1966		0.037	0.030	0.030
1967		0.036	0.025	0.025
1968		0.060	0.023	0.031
1969		0.133	0.023	0.061
1970		0.022		
1971	0.034		0.021	0.032
1972	0.032		0.020	0.034
1973	0.030	0.103	0.025	0.037
1974	0.029	0.096	0.027	0.040
1975	0.027	0.089	0.026	0.040
1976	0.027	0.093	0.027	0.042
1977	0.023	0.083	0.026	0.042
1978	0.024	0.080	0.028	0.043
1979	0.023	0.078	0.028	0.042
1980	0.022	0.075	0.027	0.043
1981	0.022	0.074	0.026	0.042
1982	0.023	0.069	0.025	0.042
1983	0.021	0.070	0.025	0.040
1984	0.021	0.069	0.025	0.038
1985	0.021	0.073	0.024	0.037
1986	0.023	0.073	0.026	0.039
1987	0.024	0.071	0.028	0.041
1988	0.023	0.072	0.028	0.042
1989	0.023	0.068	0.028	0.042
1990	0.023	0.068	0.028	0.041

For automobile exhaust gas, the data for 1965-1969 refer to Kasumigaseki, Tokyo. *Environment White Paper, 1971.* For general air, the data for 1968 and 1969 refer to the average of 7 observation stations. *Environment White Paper, 1976.*

Source: Environment Agency, *Environment White Paper.*

defined with reference to people's daily lives. The national average pollution level, for instance, does not mean very much. However, the regional dimension is not explicitly recognized in the framework we are discussing here. With this

Table 6-13 Trends in water quality

Year	Water containing hazardous materials (%)	BOD level: 8 main urban rivers (mg/l)	BOD level: Sumida river (mg/l)	COD level, Tokyo Bay (mg/l)
1965				
1966		23.5	16.0	3.5
1967				
1968				
1969				2.1
1970	1.40			3.8
1971	0.63	16.6	13.0	5.2
1972	0.28	13.6	7.5	4.7
1973	0.23	13.1	7.5	5.8
1974	0.20	10.5	5.6	4.9
1975	0.17	11.1	4.9	2.7
1976	0.09	10.4	6.0	3.3
1977	0.08	11.2	3.9	3.4
1978	0.07	8.8	4.7	3.2
1979	0.06	8.6	4.8	3.0
1980	0.05	8.8	5.1	2.7
1981	0.05	6.8	3.8	2.7
1982	0.03	6.8	3.7	2.9
1983	0.04	6.8	3.7	2.7
1984	0.03	6.8	4.6	2.6
1985	0.02	6.8	3.6	3.4
1986	0.02	6.7	3.8	3.1
1987	0.02	7.2	3.1	3.6
1988	0.02	5.9	2.8	3.3
1989	0.01	6.0	2.6	3.3
1990	0.01	5.6	2.7	3.3

Note: Figures given for hazardous materials refer to the percentage of samples. BOD level refers to the average of 8 main urban rivers.

Source: Environment Agency, *Environment White Paper*.

limit in mind, we have to rely in this paper on observations from selected points.

There are data sets which enables us to handle a detailed regional analysis, such as grid data encompassing the natural environment, land use, population density, production facilities and so forth. In between the aggregate data and grid data, we have a whole set of data from numerous observation points which were put into place as environmental policy developed. One can conceivably come up with some aggregate value based on them, but at the cost of losing some of the observations from earlier years.

As for the air pollution level, we chose to show SO_2, NO_2, CO, and floating particles. The sulphur dioxide data are the average of 15 observation stations engaged in general air quality from which the longest observation period is available. The nitrogen monoxide data are from two sources: one is the average of 22 stations dealing with automobile exhaust gas and the other the average of 29 stations dealing with general air quality. The nitrogen dioxide data are also from two sources: one is the average of the same 15 observation stations as above which deal with the general air quality, and the other is the average of 22 observation stations, which deal with automobile exhaust gas. Carbon monoxide data are from 15 of the stations engaged in the observation of automobile exhaust gas from which the longest time series data are available. Small floating particles data are from 40 stations engaged in general air quality observation.[10]

About one-half of the NO_X comes from mobile sources, which is not shown in the tables. Passenger vehicles, small-sized and large-sized trucks used to have nearly the same share in NO_X output in the mid-1970s. Since then, a series of regulations have been implemented. In the case of passenger vehicles (gasoline and LPG), taking the 1973 unregulated level as 100%, NO_X exhaust has been reduced to 8% in 1978. Likewise, NO_X exhaust from small-sized trucks (gasoline and LPG) has been reduced to 16% and from large-sized trucks (diesel) to 26%. As a result, while total NO_X from mobile sources still constitutes about one-half of the total, nearly half of it is attributable to large trucks.[*Environment White Paper*, 1990, General Discussion volume p.129, Detailed volume p.88]

6.5 From environmental accounting to modeling

I have suggested earlier that, in dealing with resource allocation by social purpose including environmental protection, it is both practical and useful to distinguish six phases which include the following.[Uno 1989a, pp.181-208]

(1) Financing account by social purpose
(2) Expenditure account by social purpose

(3) Interindustry relation account
(4) Capital stock account
(5) Performance account
(6) Social status account (which is linked to social indicators and subjective evaluation)

The first and the second are usually obtained by rearranging various statistics on the government budget, business investment, production cost, etc. One advantage of rearranging statistics by social purpose is that, compared to the original data where activities are recorded by institutional units, it reveals the activities of various institutions in a particular sphere of social concern. It is useful in distinguishing the role of the public sector, for instance. The finance account reveals who is paying for the particular social objectives, including lending and providing of subsidy. The expenditure account represents spending by business, household, and the government in the form of investment, consumption, or as intermediate input.

The third phase is what we commonly see in national income statistics and input-output tables, and deals with the interindustry repercussion of the expenditure, employment pattern with due consideration given to interindustry relations, etc. We opt not to alter the existing SNA framework which is aimed at recording, as a general rule, market transactions. Imputation of various aspects which do not take place in the market mechanism, such as the value of leisure time, 'true' value of resources consumption, environmental damage, etc. usually swamps market transactions in their magnitude and probably obscures the validity of established core accounts. Rather, the new dimensions should be constructed in the form of a satellite account, as we see below, and used in conjunction with the core account to arrive at some adjusted figures such as the value added adjusted for environmental damage, the measures of economic well-being, GNP including the production of intangibles, etc.

Energy requirement, generation of pollutants (not emission) in the production process, and the like are also closely related to the third phase described here, although this aspect is often expressed in physical terms.

The fourth phase deals with capital stock, both private and public, which is related to a particular social concern. When one shifts attention from current production (economic point of view) to well-being of the population, stock variables become more relevant.

The fifth phase is what I call performance account. This deals with the 'service' or the 'function' rendered by the capital stock for various social purposes. In our case, a pertinent variable would be pollutants removed or waste treated. In this regard, whereas the capital stock account is in terms of pecuniary unit, the performance account should be in terms of physical unit to describe the capacity of such facilities. Correspondingly, the linkages between variables expressed in money terms, such as value of production, and those espressed in physical units, such as input of resources and pollutants generated, needs to be established. Since these characteristics vary according to the sectors in question, we need proper disaggregation.

The sixth phase is the social status account, which describes the conditions prevailing in various aspects of social life. In our case, this would be environmental quality.

Then comes the problem of constructing a policy model based on the accounts. This is commonly done in the case of phase three, which deals with the production sphere. It is also useful to extend the methodology to other spheres. One advantage of econometric modeling is that it enables us to examine the complex interrelationship among various variables which is not apparent from casual and intuitive observation. It also enables us to experiment with simulation. When policy variables are embedded in the model specification, the effectiveness of policy intervention can also be examined.

It is interesting to see how modeling is done in the environmental field. Let us take a look at some of the modeling attempts.

A. *Pollution Subsystem in COSMO (Comprehensive System Model) by the Economic Planning Agency* [1973, pp.26-27, 67, 264-372]

COSMO is a large-sized, multisectoral, econometric model constructed for the purpose of dealing with various social concerns which emerged after a long and rather successful industrialization drive. One of the most serious problems the model considered was environmental disruption.

Here we look at the relation between the economy-industry subsystem and the pollution subsystem of the model. The economy-industry subsystem has 14 sectors, whereas the pollution subsystem distinguishes 18 sectors for SO_X, 20 for BOD, and 29 for industrial wastes. The value of production, the number of workers, and the energy consumption are determined for each sector. The volume of pollutants generated is calculated based on these variables, multiplied by the unit pollutant output coefficients for each pollutant category.

The environmental standards are determined as follows. For SO_X, the output level is expressed in terms of volume (ton) per year. For BOD, the output per day (ton) is used. Treatment rate (%) is used for industrial wastes. In view of the fact that pollution prevention depends very much on the technical development of pollution prevention equipment, the possible dates for the introduction of new technologies to actual plants have been examined. The stipulation in environmental regulation, in conjunction with new technology, determines the volume of pollutants that has to be removed. This, when multiplied by the unit investment requirement (i.e., investment requirement/unit of pollutant) and the unit pollutant removal cost (i.e., cost/unit of pollutant), determines the amount of investment and the pollution prevention cost. These variables are linked to the economy-industry subsystem with a time lag of one period.

In the economy-industry block, pollution prevention investment becomes a component of investment demand. The pollution prevention cost, on the other hand, affects the real economy via two channels. (1) Based on polluter-pays principle, the cost is internalized through price increases. In this case, the effect will be a sequence of price rises (producer prices—wholesale prices—export and import prices), the final result of which would be the reduction in exports and the expansion of imports. (2) The cost is internalized through decreased business profits. The larger the cost, the greater the negative effect on "business mind" in investment decisions. If "business mind" is weakened and the incentive for capital formation is negatively affected, production will then be adversely affected.

B. *Environment block in the Environment Agency's econometric model* [Environment Agency 1977, pp.76-104]

The model aims at describing the process of environmental pollution and its removal as exactly as possible; at the same time, it intends to reveal the interrelationship between environmental preservation and economic activity. To this end, the model explicitly considers the following: (1) The relationship between economic activity and the output of pollutants. (2) The relationship between the pollution emission targets and the quantity of pollutants to be removed. (3) The relationship between pollutant removal and the required investment (and capital stock) for pollution prevention. (4) The induced final demand by pollution prevention investment. (5) The effects of pollution prevention cost on prices and the repercussion of the price hike. (6) The demand curtailment effects of the pollution prevention cost.

The pollutants included in the model are sulphur oxides, nitrogen oxides, water pollutants (COD), industrial wastes, and general wastes.

The structure of the model to satisfy such requirements needs be as follows. (1) The pollutant emission varies by industry; thus, the use of an interindustry model is indispensable in order to describe the economic activities by industry. (2) A price determination mechanism needs to be introduced in the model explicitly so that the effects of pollution prevention cost on prices, which vary from one industry to another, can be analyzed.

The model consists of three main blocks. (1) The supply block determines the supply quantities and supply prices by industry under the constraints of capital stock at the beginning of the period and the available labor force. (2) The demand block determines the demand levels for individual industry by linking the Keynesian-type final demand model and the input structure of the input-output tables. The supply and demand blocks are solved simultaneously in order to obtain the supply quantities and prices. (3) The environment block determines, for the given level of production, the output and emission of pollutants by category. In addition, the required level of pollution prevention investment is calculated. This in turn is fed back to the demand structure through the expansion of effective demand; at the same time, the price change induced by the pollution prevention cost affects the relative prices and the demand structure.

C. *The framework of NNW (net national welfare)*[Economic Council 1973; Uno 1989b]

Environmental accounting has emerged as an important tool in describing the interaction between human activities and the environment. Rather than focusing on a particular aspect of the environment, this approach attempts to fit various aspects into a statistical accounting framework in a consistent manner, taking the national accounts as a starting point. In Japan there have been attempts to extend the concept of national accounts to an aggregate welfare measure (Net National Welfare, NNW) in line with the framework developed by James Tobin and William Nordhaus [1972] when they suggested MEW (Measures of Economic Welfare).

In an effort to obtain a more direct measure of the economic well-being of the population, the NNW framework incorporates the following items (in many cases, they are obtained by appropriately adjusting the concept of SNA): (1) NNW government consumption, (2) NNW personal consumption, (3) flow of services attributable to household-related social overhead capital, (4) flow of services attributable to household durable goods, (5) leisure time, (6) non-market activities such as housewive's domestic services, etc., (7) damage due to environmental pollution (deduction), (8) loss due to urbanization (deduction). Investment in plant and equipment, which is an important component of GNP, is excluded from the NNW as a measure of economic well-being on the grounds that it is "intermediate" from the point of view of quality of life. Investment in social overhead, housing, and, for that matter, purchases of consumer durables are also excluded. Instead, flow of services from social capital, housing capital, and stock of consumer durables is imputed and included.

The NNW Measurement Committee write in their report that two methods are conceivable in estimating the social cost of environmental pollution. (a) The direct estimation in money terms of damage caused by the above-stated discharge of polluting factors, such as to health, human life, animals and plants, and properties. (b) The method of estimating expenses necessary for the proper treatment of polluting factors overflowing into the present environment to an assumed "normal physical environmental level." It is a method to estimate expenses necessary to make the external diseconomies internalized.

For lack of sufficient data for the former, which they believed to be the preferred method, they have adopted the alternative. First, they set a base year in which environmental quality is imagined to have been acceptable. The base year taken is generally 1955. Excess emission is estimated, and the removal cost is then calculated.

(1) Air pollution: We can distinguish stationary sources such as factories and power plants and mobile sources, namely automobiles. For the former, the quantity of emission of sulphur oxides by each industry is calculated from the quantities of consumption of crude oil and coal and their sulphur content ratios. Taking 1955 as the base year, the expenses necessary to remove excess sulphur oxides discharge is calculated. In the case of soots and dusts, it is assumed that the total discharge is causing environmental pollution, and the total removal cost is correspondingly equated to damage. For exhaust gas from

automobiles, the discharge quantity of carbon monoxide (CO), nitrogen oxide (NO_X), and hydrocarbons (HC) is calculated. The base year in this case is 1961, and the cost involved in reducing the actual pollution level to that of 1961 is assumed to represent damage.

(2) Water pollution: The NNW framework takes BOD (biological oxygen demand) as an indicator of water pollution. Taking 1955 as the base year, the cost necessary to remove excess BOD is regarded as the damage attributable to water pollution. The calculation is conducted separately for discharge from industry and from household. The BOD discharge load units are obtained from a special survey. The treatment costs for foods, textiles, pulp, and chemical industries are derived from the tertiary sewerage treatment cost, while the costs for other, less polluting sectors are derived from the secondary sewerage treatment cost. The treatment cost for household discharge refers to secondary treatment of drainage treatment facilities.

(3) Wastes: The environmental damage cause by industrial wastes is estimated based on the unit waste discharge coefficients for 13 items in each industrial sector, which is multiplied by the annual value of production. The figures thus obtained are adjusted by the discharge ratios of industrial wastes into the environment. The ratio is assumed to be as high as 89% in 1965 and 74% in 1970, which has subsequently dropped to 20% in the case of manufacturing sectors and 50% in others. Finally, imputation of the damage is based on the cost of proper treatment for the 13 items and the quantity of improperly discharged wastes.

In the case of household wastes, the total discharge quantity is estimated. Depending on the physical nature of the wastes, the portion being disposed of without adequate treatment is then distinguished. The cost of damage caused by household wastes is then obtained by multiplying it by the unit cost of proper treatment.

It has now become possible to link the framework explained here to the variables generated within a multisectoral industry model, COMPASS (comprehensive model for policy assessment). The model has 25 industrial sectors, with potential output block, demand block, energy consumption block, prices and wages block, and others. The framework presented here forms a part of the COMPASS and an aggregate indicator of the economic well-being of the population. Pollution prevention investment is not 'productive' in the conventional sense in that it does not contribute to the production of 'goods and services' but merely prevents 'bads' from happening. It is particularly costly at the initial stage when the pollution is recognized as a problem and social consensus is reached to improve the situation by enforcing strict environmental standards. At this stage, pollution prevention equipment has to be installed to the existing capital stock in addition to the new facilities. After a while, this stock effect will be eased considerably. At the same time, technological change is induced in the production process or to the final products, and pollution prevention becomes a part of the more or less normal production activity, taking such forms as heat recovery, waste recovery, production of by-products, etc. Sometimes the production process itself is altered into a new method. The entire process of legislative action, technological change, diffusion of new technology, increase in production cost and substitution effect, the stock effect, etc. is so complex that it is difficult to describe in a model framework. Our model treats this aspect in a block where the ratio of pollution prevention investment for the individual industrial sector is treated as an exogenously determined variable. However, pollution prevention effort is not simply a drain of economic resources because it contributes to improving the quality of life. If we can agree that quality of life, which includes flow of consumption as well as benefits derived from capital stock (housing and social overhead capital), as the final objective of economic activity, then pollution prevention effort has direct bearings. This brings us to the environmental accounting which includes both the cost side and benefit side of pollution prevention.

6.6 Concluding remarks

In this chapter, we focused our attention on pollution prevention investment and demonstrated the feasibility of constructing an accounting system which can be used in assessing the effectiveness of environmental policy. This framework is also amenable to the established SNA. The framework leaves the SNA concept intact while providing greater details on the pollution prevention activities, both in monetary and physical terms, within individual sectors to cope with various environmental hazards. The effectiveness of such activities can then be expressed in terms of the reduced damage done to the environment, together with positive aspects of economic activities, and included in an accounting framework which describes various attributes of economic well-being.

I may add that a similar framework can be envisaged for energy conservation and recycling of resources.

FOOTNOTES

[1] See Uno [1985a, b] for an earlier effort to compile a satellite account on various social concerns including environment. This was preceded by Uno [1978] in which a system of social indicators and simple accounting was introduced, encompassing economic, sociodemographic, and environmental spheres. The expenditures related to environmental aspect and not enumerated here include the following:

a. operating cost of pollution prevention facilities (except where such expenditures are included in the current input for production).

b. the cost of shifting to low-pollution raw materials (such as low-sulphur crude or natural gas)

c. introduction of new (possibly more costly) production methods

d. the cost of research and development (for R&D expenditures by purpose, see Uno [1991e, p.96]

e. the cost of energy conservation, and

f. the benefits of recycling and by-products (such as scrap metals, heat recovery, electric power generation, sulphur, etc.) (this is discussed in Chapter7)

g. the cost of additional equipment for pollution prevention, such as catalyst for automobiles.

[2] Data compiled by the Japan Association of Industrial Machinery Producers (Nihon Sangyo Kikai Koryokai). Since 1966, the Association has published "The Production of Pollution Prevention Equipment" (Kogai Boshi Sochino Seisan Jissekini Tsuite) based on a survey of the member companies. The 1970 survey covered 187 firms, of which 155 responded; the 1989 survey covered 179 firms, of which 156 responded.

[3] However, we need to bridge the production figures to capital formation statistics which include construction.

[4] Empirical data on energy conservation investment, classified by industrial sector, is available in Chapter8. See Ministry of International Trade and Industry, *Plant and Equipment Investment Plans in Major Industries.*

[5] The survey, which is reported annually in MITI's *Plant and Equipment Investment Plans in Major Industries* for recent years while in earlier years the data were published separately, started in 1965. Sectoral figures are available based on a scheme unique for the survey up to 1985, which are now based on the standard industrial classification. The survey covers around 1500 establishments. An establishment engaged in various economic activities is listed separately for the respective industrial sectors.

[6] This point can be supplemented by the *Summary of the Trends of Investment in Plant and Equipment by Small Manufacturing Firms* by the Small Business Finance Corporation. The ratios of pollution prevention investment to total in individual industrial sectors do not seem to diverge too widely between the two sources, and the use of MITI's sectoral data as the representative one seems to be justified on these grounds.

[7] The latter is available from the Economic Planning Agency. Formally, this is not a part of the SNA, but consistency with the aggregate data is maintained.

[8] The fixed capital formation matrices for the past 20 years are introduced in Uno [1991e].

[9] See Uno [1989a][1990c], pp.195-205, for time-series compilation of final demand converters in standardized 36 sector tables.

[10] OECD [1977] provides air quality objectives (SO_2, particulates, and NO_2) and automobile emission standards (CO, HC, and NO_X) for Japan and selected countries around 1975. OECD suggests that the tables suggest that "policy objectives are more ambitious in Japan than in many other countries."

Chapter 7

Produce-Consume-and-Recycle

7.1 The end of produce-consume-and-forget regime

"Sustainable development" is the key word in our effort to restructure our society to make it amenable to global environment. In reality, however, the self-perpetuating expansion of our economy does not seem to automatically induce changes in industrial technology and lifestyle leading to sustainable development.

The Brundtland report wrote in 1987 that "Suitainable development is development that meets the needs of the present without compromising the ability of future generations to meet their own needs. It contains within it two key concepts: the concept of 'needs', in particular the essential needs of the world's poor, to which overriding priority should be given; and the idea of limitations imposed by the state of technology and social organization on the environment's ability to meet present and future needs."[World Commission on Environment and Development, 1987, p.43] Explicitly referring to sustainable development, the report writes, "Humanity has the ability to make development sustainable—to ensure that it meets the needs of the present without compromising the ability of future generations to meet their own needs. The concept of sustainable development does imply limits—not absolute limits but limitations imposed by the present state of technology and social organization on environmental resources and by the ability of the biosphere to absorb the effects of human activities. But technology and social organization can be both managed and improved to make way for a new era of economic growth."[WCED, 1987, p.8] "But physical sustainability cannot be secured unless development policies pay attention to such considerations as changes in access to resources and in the distribution of costs and benefits."[WCED, 1987, p.43]

This new concept is not without ambiguity, however, when it comes to the assessment of the divergence of our societies from sustainability and the implementation of policies aimed at correcting it. The need to clarify this new development concept in operational terms has to be stressed. First of all, an accounting system has to be developed which reveals interaction between the environment and the economy. This points to the need for resource accounting that encompasses resource stocks, flow of resources among countries, resource flows within an economic system, and the generation and removal (or emission) of wastes and pollutants, leading to modeling and empirical analysis covering the entire process. Once proper policy tools have been identified, this methodology then can lend itself to the formulation of policy. Second, a new concept to gauge the needs of the population has to be devised. The concept should go beyond the conventional concept of consumption of goods and services and should include the changing quality of the environment and the benefits derived from consumer durables and housing. By explicitly considering the stock, rather than simply focusing on the annual flow of consumption, one can measure the quality of life (or lack of it) in its full span.

The first point corresponds to sustainability of the global eco-system and resource base. The second point relates to the sustainability of the quality of life. The two formulations of sustainable development can be linked to each other within an extended system of input-output framework and policy modeling based on it.

7.2 Sustainability reconsidered

As we have seen in chapter 1, the environmental burden can be broken down into four factors: (1) population size, (2) income per capita, (3) resource use per unit of income, and (4) environmental damage per unit of resource use.

If we allow the population to grow and if we allow people to attain higher incomes while assuming the remaining two factors to be constant, we inevitably reach the conclusion that the global community will soon overload the global commons with resource depletion and environmental degradation.

Thus, attempts by economists and statisticians to define the concept of sustainability in an operational term can be interpreted within this context.

Our purpose in this chapter is to look into the third factor, namely, resource use per unit of income. In particular, we will look into recycling which, if successful, can substantially reduce the environmental burden of our economic and social activities. The economic system up to now has operated on the implicit assumption that an unlimited amount of resources can be made available as soon as the demand for them arises, and that production of disadvantages such as deterioration of air and water quality can be accommodated by Nature (Figure 7-1a). It is recognized that resources are exhaustible and that the uncontrolled emission of hazardous materials, gases, and waste heat has caused environmental problems of global scope. One way of rectifying the situation is to close the resource flow by recycling (Figure 7-1b).

Other logical solutions are apparent from the identity presented above. Arresting the growth of the population is an obvious option, and an important one, although how much can be done based on the conventional tool box of economic policy is rather uncertain, especially when the income level is low and children are important factors in household production. Hopefully, demographers will have a better answer in this regard.

I have looked into the second factor which relates to income levels in Chapter 4. The argument there is that rather than focusing on consumption, which is the current flow of goods and services, we should be concerned with the levels of economic well-being, the latter being defined in terms of services derived from the stock of consumer durables, housing, quality-of-life related social capital, and current consumption net of environmental degradation and urban congestion. In essence, this is a suggestion to focus not on current income and consumption but on quality of life. We return to this point in Chapter 11 below.

Regarding the fourth factor which relates to environmental damage caused

Figure 7-1 Establishing a recycling economy

a. Produce-consume-and-forget

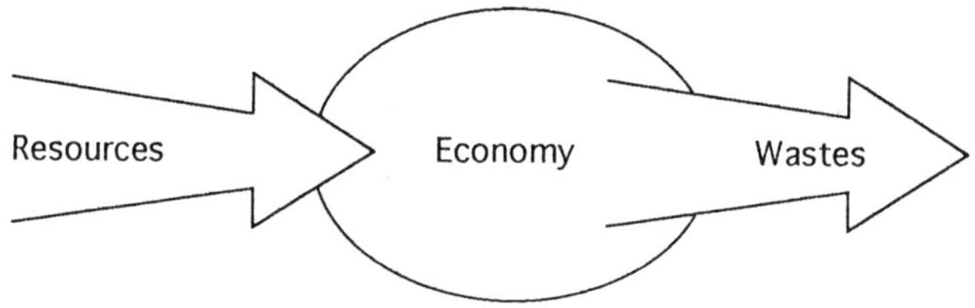

b. Produce-consume-and-recycle

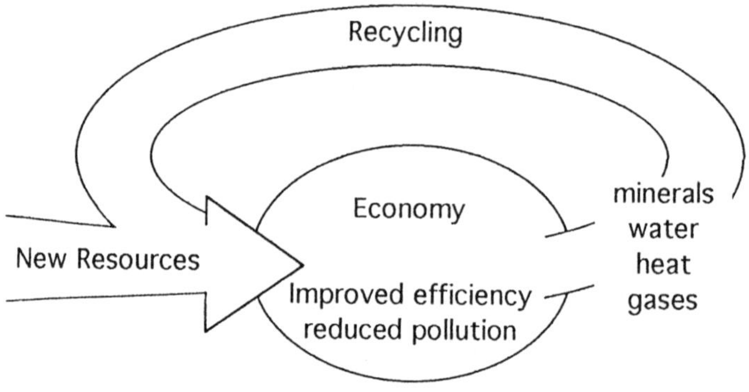

by man, I have shown previously that an increased provision of pollution prevention facilities can actually diminish such damage (Chapter 6). In addition, we should be concerned with technological progress which results in pollution prevention and increased efficiency (Chapter 8). Lifestyle change, often induced by higher income levels and urbanization, is also a factor (Chapter 4). In this chapter, we focus on the sustainability of our socioeconomic system from the point of view of resource flow.

Today's environmental concern extends beyond pollution. We aim at explicitly considering the recycling of various resources. As is depicted in Figure 7-1, we are accustomed to live in a produce-consume-and-forget system. We exploit natural resources without paying attention to whether they are depletable or constitute part of the natural environment on which we depend so heavily, and emit pollutants into the environment without due consideration to the hazards they present not only for health but also to the global ecosystem. In view of the fact that we are now concerned with conservation of the global eco-system, we have to put an end to this way of living. To this end, we should look into the third factor mentioned above and seek ways of reducing our dependence on forest, energy, ferrous and nonferrous metals. We need to develop a conceptual framework which will enable us to keep track of resource use per unit of activity. We need an accounting framework which properly traces the flow of material inputs into our economic system, with particular attention to recycling. If part of waste heat and materials can be recyled and put into effective use, the resource burden can be that much reduced. The ultimate goal would be a produce-consume-and-recycle system.

The analytical framework for the concept of sustainability of the resource base can be formally put forward in an input-output format. I need not elaborate here the input-output analysis itself because it has been widely used among economists and statistical offices the world over. What I find most illuminating in this context is the table which indicates the material flow among industrial sectors in physical terms. Usual input-output tables are formulated in value terms, which is the only common demoninator across various activities. Physical tables correspond to transaction tables in value terms but on the other hand show the transactions in physical terms wherever possible.

The observed changes in input coefficients are attributable to technological change, which is reflected in the industrial composition of inputs as well as factor inputs including capital and labor. Also reflected in shifts in input coefficients over time are the changes in relative prices among industries. A transaction is obtained in value terms as the multiple of price and quantity; however, the price for a particular commodity may differ depending on, for example, whether the transaction is based on large, long-term contracts or small, ad hoc ones. There may not be a uniform price for a particular industrial sector. Transaction tables in physical terms are useful in overcoming this difficulty. For our purpose, which is to establish an analytical framework that enables us to trace material flow within an economic system, the framework can offer a general accounting framework.

However, it is not possible to compile a complete physical table for the entire economy. The reasons are as follows. First, its construction presupposes that the transactions can be broken down into price times quantity, while in reality, service-related activities comprise more than 50% of the total economy where quantitative measurement is not feasible. Second, even in sectors engaged in physical production, the commodities included tend to be divergent within a single sector, making physical aggregation impossible. This is particularly the case for machinery sectors and 'others'. Third, information to identify output sectors is typically lacking.

For those reasons, physical tables are not universally available. However, input-output tables compiled in Japan for recent years include partial physical tables covering major raw materials and their products.

There are several different ways of treating the generation of scraps and by-products in transaction tables. In Japan this is based on the methodology suggested by Richard Stone, which is to record them as negative output. Therefore, where one sector produces products (positive output) and scraps and by-products (negative output) simultaneously, which is usually the case, the recorded output is obtained after netting out. To remedy this defect, a table concerning output (generation) and input (use) of scraps and by-products must be compiled. In fact, Japanese transaction tables identify both positive and negative outputs by attaching special codes after the sector codes (i.e., 2 referring to input of scraps, 3 to output of scraps, 4 to input of by-products, and 5 to output of by-products). Physical tables also follow this convention, enabling us to identify the generation of scraps and by-products and their disposal.

This will serve as the central accounting framework for tracing the material flow within an economic system. Referring to Figure 7-1 which depicts the produce-consume-and-forget and produce-consume-and-recycle systems, the recycling process can be described in this framework. One advantage of this framework is that it is perfectly general in providing an overall physical material flow consistent with the transaction in value terms, allowing aggregation and disaggregation as needed and whenever empirically feasible. Thus, we have a statistical framework which enables us to trace the input of resources, throughput within an economic system, output of waste and pollutants, and recycling into input again to the degree they are actually done.

If this is more or less a picture from the point of view of resource-consuming countries, the logical extension would be to cover the point of view of resource-producing countries. This can be done by attaching the framework discussed above to international trade matrices concerning resource flow among countries. Within individual resource-producing countries, one can envisage statistics on current production and existing stock (the latter including, for example, distinction such as proven reserves, etc.). New discoveries can also be included in the picture. Another linkage between production and consumption of resources is provided by changing prices, which are endogenously determined within the economic system. One should not forget technological change in resource exploration and extraction as well as that at the consuming end. These factors should be enumerated within the framework of resource accounting.

7.3 Sustainability of resource base: accounting for recycling

In this section, we focus our attention on operationalizing the concept of sustainability of our resource base. In particular, we aim at establishing a statistical framework suitable for describing a produce-consume-and-recycle system. We take up the cases of pulp and paper, iron and steel, and energy. The actual state-of-the-art of recycling is described, and the conceptual framework is then applied, taking numerical examples from Japan.

(a) Paper

According to an estimate [Morisawa 1990], 47.9% of paper consumed is

Figure 7-2 Flow of pulp and resources

(1,000 cubic meters)

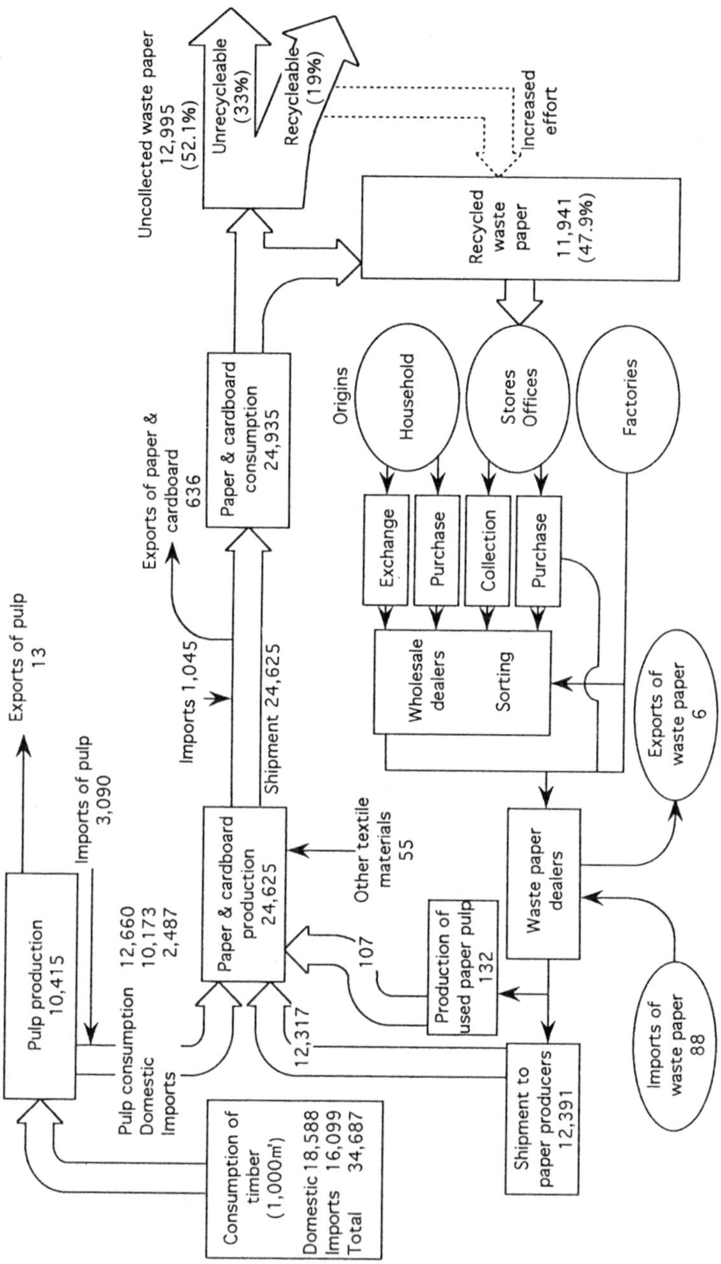

Note: The data refer to 1988.

Source: Used Paper Recycling Promotion Center, *Annual Statistics on Used Paper.*

Table 7-1 Recycling of used paper, 1988

	Total consumption (1,000 tons)	Recycled (1,000tons)(%)
Plain white paper, art paper, and magazines	11325.0	3185.1(28.1)
News paper	3262.4	2948.7(90.4)
Brown paper	541.6	354.3(55.3)
Corrugated cardboard	7414.9	5110.0(68.9)
Paper mat	3093.8	588.5(19.0)
Total	25637.7	12186.4(47.5)

Source: Used Paper Recycling Promotion Center, *Used Paper Supply and Demand Statistics.*

recycled in Japan (1988 figure). The ratio varies according to the kind of paper as follows (Table 7-1).[Used Paper Recycling Center, 1980]

Figure 7-2 describes the flow chart of paper production and recycling of used paper. In Japan, paper and cardboard production in 1988 stood at 24,936 thousand tons, of which 11,941 thousand tons (47.9%) were collected after consumption. Of the remaining 52.1%, about 19% can be recycled with extra effort, and about 33% is not collectible. Waste paper is classified into industrial waste and household waste. The former refers to waste paper from those establishments which handle large amounts of paper, such as newspaper companies, printing shops, and cardboard box producers, and the latter refers to used paper from households, supermarkets, department stores, etc. Industrial waste paper is collected by contracted specialized dealers who deliver the collected paper to used paper dealers. Household waste paper is collected through "group exchange" (where schoolchildren and local residents engage in collection voluntarily often to financially supplement their activities), tissue paper exchangers (where housewives accumulate waste paper and exchange for tissue paper, toilet paper, paper towels, etc.), or purchasers. The waste paper thus collected is then sorted and delivered to the paper producers.

Time series data are provided in Table 7-2. In 1988, raw material for paper and cardboard production consisted of 50.4% of pulp, 49.4% of waste paper, and 0.2% of other textile materials. In other words, waste paper provides about one-half of the raw material requirement of the paper industry. As shown in Table 7-3, waste paper usage stood at 24.2% for paper production and 82.6% for cardboard production.

Table 7-2 Consumption of waste paper for paper production

(unit: 1,000 tons)

Year	Paper production Pulp	Waste paper	Total	Cardboard production Pulp	Waste paper	Total
1983	8431	2831	11263	1867	6350	8217
1984	8723	3029	11752	1937	6752	8689
1985	8853	3061	11914	1919	7467	9386
1986	9227	3145	12372	1794	7673	9466
1987	9725	3203	12928	1888	8481	10369
1988	10759	3427	14186	1901	8997	10898

Source: Japan Paper and Pulp Co., *Paper and Pulp Statistics Illustrated.*

The use of recycled paper expanded with the development of DIP (de-inking pulp) technology. The DIP processing consists of melting of the sheet paper, dust removal, de-inking, washing, and bleaching.

Compared with the usual production of pulp from wood, pulp production from recycled paper consumes only 1/3 of the energy such as heavy oil and electricity, in addition to conserving forest resources. The increased recycling of paper will thus conserve energy resources and help protect natural forests.[Japan Paper and Pulp Co., 1990]

Table 7-3 shows the recycling of waste paper in an input-output framework. The data refer to 1985. The recycling of waste paper in Table 7-2 amounted to 10,528 thousand tons in 1985. The corresponding figure in Table 7-3 stands at 10,320 thousand tons in quantity terms (205.8 billion yen in value terms). We may say that the correspondence is fairly accurate between the two. The major activity generating waste paper is private consumption, followed by manufacturing activity such as paper production, printing and publishing.

(b) Iron and steel—the case of automobiles [*Environment White Paper*, 1991]

The automobile industry is one of the leading sectors of the economy and turns out new cars every day. The production of new cars prompts the scrapping of old ones. This means that car disassembling is needed at about the same speed as car assembling. However, whereas new cars are produced by large firms, scrapping is conducted by small firms, and the process is some-

Table 7-3 Recycling of waste paper in an input-output framework

Output sector	Value (bil. yen)	Quantity (1000 ton)
Manufacturing total	(-) 44.341	(-) 1,748
(24) Pulp and paper	(-) 22.995	(-) 906
(25) Printing and publishing	(-) 21.346	(-) 841
Private consumption	(-) 150.968	(-) 5,951
Imports	(-) 10.488	(-) 2,621
Total	(-) 205.797	(-) 10,320
Input sector		
Manufacturing total	205.349	
(24) Pulp and paper	205.349	
Exports	.448	
Total	205.797	

Note: The group of by-products and scrap corresponds to "pulp" (code 1811-011).
Source: Management and Coordination Agency, *1985 Input-Output Tables.*

times disorderly. Disassembly technology is undeveloped, and recycling is faced with a wall of cost (Figure 7-3).

The disassembly process relies heavily on manual work. First, batteries and fuel tanks are removed to prevent any danger of explosion, as well as used tires. Doors, fenders, and the like are theoretically removed for resale in the used parts market, but this is seldom done in reality. Then a machine called a nibbler removes the tires, wheel cover, engine, and radiator. The remaining body is pressed and sent to a shredder, which cuts it into mixed pieces of steel, wiring, plastics, and seat cloth. An electric magnet is used to sort only steel pieces from this mix. Copper, aluminum and other valuables are then selected, usually by manual operation. This leaves a mountain of what is called shredder dust. With the increased use of plastic parts in an automobile, the percentage of shredder dust will tend to increase. Steel pieces are sent to a steel mill where they are melted by electric furnace. Coupled with scrap iron from other sources, about 32% of steel production was attributable to recycled steel in 1990. Through the continuous casting process, melted steel is finally turned into structural materials and steel bars.

How many resources can be obtained when an automobile is scrapped? The case here refers to a 1986 vintage 4-door sedan with displacement capacity of

Figure 7-3 Flow of used automobiles

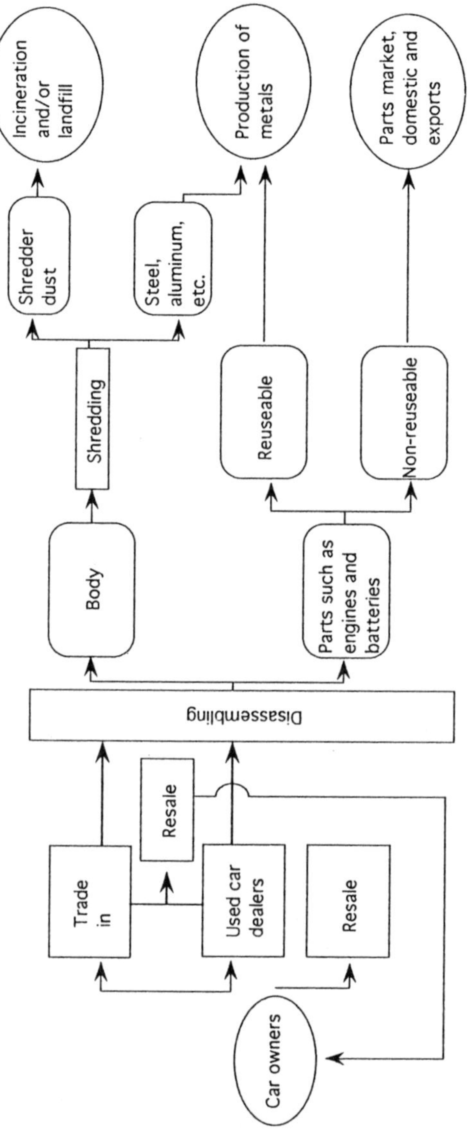

Source: Japan Automobile Federation data.

1,500 cc. Steel of course is the main structural material of the body, suspension, and engine and amounts to about 450 kg. This will never be used again to make steel for car production, but it is good enough to produce various steel products such as those for building construction. From the aluminum wheel cap, about 20 kg of aluminum can be recovered. In addition, the engine and transmission contain retrievable amounts of aluminum. Platinum and other rare metals can be recovered from the catalyst converter. The present converters use platinum, rhodium, and sometimes palladium as catalysts to convert the exhaust emission of carbon monoxide and hydrocarbons into less harmful carbon dioxide and water. According to an estimate, catalytic converters consume about 40% of the platinum and 80% of the rhodium produced in the West. Although the amount contained in a car is minimal (about 0.8 g), it is recovered fully because of its high price. Tires are used as an energy source. Rubber produces a heat value close to heavy oil, prompting its use as fuel in industry. Used parts such as engine, doors, alternator, etc., can also be recovered, and sold to used parts dealers. However, many usable parts are disposed of as industrial waste because of difficulty in recovering them and lack of sales channels. The steering wheel, seats, dashboard, window glass, and plastic bumpers, which make up nearly 30% of the total weight of the car, also have to be disposed of as waste. Most of these materials are not suitable for incineration, and land fill is usually the last resort.

The case of automobiles is typical example of the disposal of consumer durables made of steel, plastics, and the like. Although the production process is highly automated, disassembly technology has yet to be developed. At present, they produce various socials cost at the disposal stage, a cost not borne by either the consumers or the producers. And they create a considerable environmental burden, too. It is imperative to develop technology and a logistic system to extend the years of service of these products. It is also highly desirable to establish a recovery channel as well as a disassembly and disposal process.

Table 7-4 lists time series data on the generation of scrap iron obtainable from Japanese input-output tables. Included in the table are final demand items such as private consumption, private investment, and government investment. Scrap iron is also generated in the intermediate production processes, but its output is netted out with input, resulting in a net input. Table 7-5

Table 7-4 Output of scrap iron, time series

(unit: billion yen)

Year	Private consumption	Private investment	Government investment
1965	3.852	60.000	13.350
1970	13.000	42.817	15.600
1975	27.552	70.030	16.964
1976	13.346	45.544	13.403
1977	32.943	84.866	20.760
1978	31.946	81.072	20.434
1979	38.920	99.617	23.959
1980	42.469	109.093	26.426
1981	31.183	28.425	6.886
1982	23.548	60.490	14.653
1983	20.927	147.542	15.466
1984	34.534	125.372	15.200
1985	42.609	121.999	29.675
1986	30.597	109.226	21.309
1987	29.101	129.557	20.267
1988	36.273	127.698	25.289
1989	33.404	132.082	22.470
1990	32.850	130.716	21.991

Sources: Administrative Management Agency (Management and Coordination Agency since 1985) and the Ministry of International Trade and Industry, *Input-Output Tables*, various issues.

refers to 1985 data and gives a detailed picture of recycling of scrap iron in Japanese input-output tables.[Management and Coordination Agency, *1985 Input-Output Tables*, exposition volume, p.220]

(c) Energy [*Environmental White Paper*, 1991]

Increased energy consumption is often regarded as occuring concomitant with the expansion of household and industrial activities, causing a depletion of fossil fuel reserves, air pollution, and global warming. There are only two obvious solutions to the problem, namely, to reduce the loss in energy use and/or to reduce the energy use itself.

At present, only 35% of the total energy supply is used effectively, while the remaining 65% is being directly emitted into the environment as waste heat. For example, in the case of thermal power stations, about 40% of the energy input is transformed into electricity; in the case of a gas oven, about

Table 7-5 Recycling of scrap iron in an input-output framework

Output sector	Value (bil. yen)
Manufacturing total	(-) 446.541
(31) Iron and steel	(-) 301.654
(33) Fabricated metals	(-) 44.990
(34) General machinery	(-) 38.434
(35) Electrical machinery	(-) 17.709
(36M) Transport equipment, auto	(-) 27.385
(36S) Transport equipment, other	(-) 13.458
(37) Precision instruments	(-) 2.235
(38&)Other manufacturing	(-) 0.676
Construction	(-) 11.884
Private consumption	(-) 42.609
Fixed capital formation, public	(-) 29.675
Fixed capital formation, private	(-) 121.999
Imports	(-) 102.252
Total	(-) 754.960
Input sector	
Manufacturing total	748.729
(26) Chemicals	0.327
(31) Iron and steel	748.402
Business and personal services	2
Exports	6.229
Total	754.960

Source: Management and Coordination Agency, *1985 Input-Output Tables*

50% of the energy input is effectively turned into hot water; and in the case of automobiles, only about 20% of the energy input is actually used in driving, while the rest becomes waste heat. In steel plants about 36% of the energy becomes waste heat, and in cement factories about 20%, while of the energy input for power generation, which comprises 37% of the total energy input of the whole economy, 24% becomes waste. In the case of household use, 7% out of 17% becomes waste. For transportation and industrial use, the ratios are 12% out of 16% and 14% out of 35%, respectively.

Given the compelling concern for preserving the global environment, it is imperative to use all the means possible to reduce such a wasteful energy use. It may be necessary to go beyond the conventional cost consideration and look at the issue from a broader perspective. Possible options include the following.

First, energy conservation has to be promoted alongside the production

Table 7-6 Recycling of waste heat, the case of power generation

(unit: 10**10 Kcal)

Year	New source of energy	Refuse power gemeration proportion	Primary energy, domestic source	Primary energy, total
1960	4640	0 (-)	57064	100810
1965	2604	5 (20)	56193	168910
1970	3260	10 (45)	50782	319708
1971	3357	11 (51)	48761	324790
1972	3448	11 (51)	46137	324790
1973	3651	20 (89)	40846	385409
1974	3523	29 (131)	45071	384679
1975	3372	34 (153)	44473	366224
1976	3643	56 (248)	46970	387333
1977	3662	72 (322)	43324	387270
1978	3655	74 (328)	48486	386453
1979	3994	76 (337)	53965	411134
1980	3918	107 (477)	59267	397198
1981	3913	116 (518)	59040	382168
1982	4172	137 (608)	60532	364296
1983	4438	168 (747)	63909	383540
1984	4772	210 (931)	65893	403067
1985	4875	227 (1008)	74195	405323
1986	5022	274 (1218)	73307	402217
1987	5303	274 (1218)	76241	422377
1988	5573	325 (1443)	77108	445366
1989	5839	342 (1522)	78004	461729
1990	6098	407 (1807)	81426	486161

Note: Year refers to fiscal year (from April to March next year). Figures in parentheses are electric power generating capacity in terms of million WH.

Source: Compiled from Agency of Natural Resources and Energy, *Comprehensive Energy Statistics, and the Institute of Energy Economics,* Statistical Material on Energy.

and delivery of energy as a whole. This is not limited to the introduction of energy-saving technology and equipment. Energy supply and demand have to become more efficient, through the introduction of co-generation, combined cycle power generation, and heatpump, among others.[1] Recovery of waste heat from factories and urban facilities is also a possibility. A heat pump is a useful device to recover heat from sewerage, river water, subways, and underground power lines where the temperature is not too different from atmospheric temperature. Rationalization of the urban transportation system and redesign of city structures may also become necessary. Second, promotion of energy-efficient automobiles and consumer durables continues to be an important element, as well as of energy-saving production processes. Insulation of office buildings and private houses needs to be improved, and energy use within offices and homes has to be rationalized. Third, apart from improvements in the energy efficiency of office and household equipment, efforts should be made to use them in an energy-efficient manner. For this purpose, it is necessary to promote the energy consciousness of the population.

We take up a particular case below where heat is recovered from the incineration of refuse and used for power generation. This enables energy to be transmitted to distant places, whereas the direct use of heat is limited to within a particular district. Power generation combined with a refuse incinerator not only enables self-sufficiency in energy of the facility but also the selling of surplus electricity to utility companies. This case is now spreading rapidly among local governments. Starting with large-scale facilities, it has become economically feasible even for small-scale facilities such as those with incineration capacity of 200 tons/day or less. In the case of the Tokyo metropolitan area, for example, electric power generated by this method amounted to 370 million KWH, of which 190 million KWH (52%) were sold to the utility company.

The input-output framework is not suitable for tracing the energy flow throughout the economic system. More useful information, including the input of primary energy and its transformation to electricity, etc. before final consumption, can be found in energy matrices. This framework distinguishes various sources of energy as well as consuming sectors.

The picture for the nation as a whole is shown in Table 7-6. Column 2 refers to refuse power generation, which is expanding fairly rapidly. Although it still constitutes a small fraction of the total primary energy supply, it is increasing at a considerable pace (an increase of about 300% in the past 10 years). According to Table 6-10 in the previous chapter, the output of household waste is increasing steadily, and the portion which is incinerated is increasing at an even faster pace. This substantially reduces the volume of wastes to be disposed of finally. Presumably, only a fraction of this total is used for power generation and other means of heat recovery. The production of electricity using the heat generated at the time of incineration, coupled with the use of that energy for regional heat supply, serves many purposes. This amounts to adding a nonconventional source of energy, which is incineration of wastes. It certainly is a viable way of achieving a produce-consume-and-recycle system.

7.4 Sustainability of the quality of life

Production and current income are not final goals of our economic activities. Recognizing the fact that our well-being is derived from a stock of durable goods, housing, and social capital in addition to food and clothing, we should take into account the benefit derived from them. In other words, we should pay more attention to the stock variables. Thus, if these items are built to last longer or designed to be energy-efficient and pollution-free, we can keep the levels of living intact while increasing the chance for achieving sustainability of our resource base. Damage done by environmental deterioration is obviously a negative factor and would have to be deducted.

This proposal can be implemented based on the existing framework of the SNA (System of National Accounts), without replacing it but rather supplementing it. A major addition would be statistics on various stocks and their characteristics as to resource use and output of pollutants and wastes. Pollution prevention facilities are also an indispensable segment of the additional accounts. It is noted that statistical data in physical terms are needed in addition to the ones in monetary terms.

The report of the World Commission on Environment and Development further stimulated discussion.

The message of this chapter is rather straightforward including the following. First, the need to operationalize the concept of sustainable development is pointed out. My suggestion is to distinguish sustainability of the resource base and sustainability of the quality of life.

Second, to take account of the former, one can envisage an accounting framework which encompasses stock of resources, current production of resources, international trade, input into an economic system, interindustry flow of resources and their products, and output of resources and pollutants. For this purpose, statistical data in physical terms are as important as those in monetary terms.

Third, on the part of the resource-consuming countries, the need to close the resource flow by recycling is suggested. This is termed a shift from today's produce-consume-and-forget system to a produce-consume-and-recycle system which makes our resource base sustainable and thereby helps leave the global ecosystem intact.

Fourth, in order to accommodate the need for economic development in a world plagued by wide income differentials, a new concept of development has to become the policy goal. This is referred to as the sustainability of the quality of life. A statistical framework is suggested to explicitly record the performance of our society in this regard.

Fifth, the two accounting frameworks, one related to sustainability of the resource base and the other to sustainability of the quality of life, can be linked by an input-output framework. In other words, being a general description of interindustry transactions of resource flow and production of goods and services, the input-output framework would provide linkage between the two.

Finally, since the input-output framework provides a snapshot at one point in time, my suggestion is to incorporate structural information obtainable from a time series of input-output tables into a multisectoral econometric model.[2] This way, other supplementary information can also be incorporated into the integrated, inherently dynamic framework. Examples are prices of goods and services (including resources, exchange rate, interest rate, and wage rate), final demand (including consumption, capital formation, and others), resource importation, research and development (which affect not only technology in narrow sense but also the demand for resources and pollution removal), among others. This also paves the way for policy simulation for the various options open to us.

FOOTNOTES

[1] Energy loss can be reduced by utilizing the heat generated at different temperatures. Co-generation is a plant design which allows the repeated use of heat obtained by fossil fuel burning, by first utilizing high-temperature gas for power generation and recovering heat from the exhaust from fuel burning at the same time. When the demand for electricity

and heat is properly combined, total energy effciency can reach as high as 70% to 80%. Mobile power is obtained by the combustion method which drives a piston by high-temperature (1500 degrees C or above), high-pressure gas generated at the time of explosive burning of fuel, the gas turbine method which drives a turbine in high-temperature gas (about 1100 degrees C), or the steam turbine method which drives turbines by steam (about 500 degrees C). Compared with the steam turbine method which is generally seen in industrial use today, the former two enable us to utilize high temperature gas. Of the total electric power generation in Japan of 185.1 MW, co-generation stands at 1.135 MW for industrial use and 0.148 MW for commercial and household use. Combined cycle power generation is another method to achieve a higher energy efficiency of about 44%. This compares favorably with the case of conventional thermal power generation, where energy efficiency averages about 37%. This method uses high-temperature gas of about 1100 degrees C to drive a gas turbine for power generation. Exhaust gas then can be utilized for steam which drives another turbine. Heatpump is a technology which recovers waste heat for effective use. It utilizes low-temperature waste heat such as exhaust gas, atmosphere, and sewerage and river water as the heat source, and electricity or gas for driving the pump, thereby obtaining a higher energy output than input. Under ideal condtions, the heat obtained reaches 3.5 times the energy used for driving the pump. It makes use of coolant, which obtains heat under low pressure and releases it under high pressure.

[2] Such a framework already exists. See the multisectoral model named COMPASS (comprehensive model for policy assessment) described in Uno [1987]. Full incorporation of the environmental block and quality of life block remains a task for the future.

Chapter 8

Environment-Related R&D

8.1 Limits to growth and technology

We have seen in Chapter 2 that environmental issue, which in general terms can be expressed as the output of various pollutants per geographic area, can be formulated as the multiple of four factors, namely:

(1) population density,
(2) per capita income,
(3) resource intensity (i.e., resource requirement per unit output), and
(4) efficiency of resource use (i.e., emission of pollutants per resource use).

One scenario for the global community would be that, as the population continues to grow at a rapid pace and some of the less developed countries take off in their industrialization drive and start enjoying higher per capita income, the environmental disruption will inevitably worsen. Human beings will be suffocated or fall ill or be buried under environmental waste. Considering the historical fact that industrialization requires energy, steel, etc., human beings will be without resources.[1]

This eventuality is a logical conclusion if we leave out factors (3) and (4) or assume that they would tend to worsen with industrialization, or remain constant at best. In other words, human civilization is doomed if we preclude technological change.

On the other hand, we have a totally different picture if we assume that factors (3) and (4) tend to improve over time, thus counteracting the increased environmental burden stemming from factors (1) and (2). Resource intensity and efficiency of resource use relate to technology. Research and development are precisely directed to this point. The technological frontier is constantly expanding through R&D, which is an organized endeavor with specific purposes in mind, rather than sporadic inventions by individual persons. Price structure is a factor determining the actual selection of technology from various options, as is the mechanism of financing investment and entrepreneurship itself. Diffusion of technology within an economy or across national boundaries is an important topic in itself.

We focus our attention in this chapter on the R&D which has a direct bearing on the environment. In section 2, we examine the data concerning R&D on the environment and energy. In section 3, we look at the research effort in pollution prevention and energy fields based on the Japanese experience. In section 4, we examine the development and diffusion of new technology aimed at pollution abatement. Section 5 is devoted to the discussion on alternative energy, also relying on the Japanese experience. Related to this, energy conservation technology is discussed in Chapter 3 where development in the industrial sector, transportation sector, household and commercial sector, and energy transformation sector is examined. Finally, section 6 takes up some policy issues.

8.2 Environment-related R&D

Time series data on R&D expenditures are shown in Table 8-1. Total R&D expenditures in 1990 amounted to 12.089 trillion yen, which is about 2.8% of GNE. (The GNE figures are available in Table 11-1.) The table also provides a breakdown into specific purposes of research including defense, space development, ocean development, information processing, enrivonmental protection, and energy.[2] In 1990, the research expenditures on environment protection stood at 235 billion yen and energy-related research at 913 billion yen, which is 1.9% and 7.6% of total R&D expenditures, respectively.

The total number of researchers numbered 484300 persons in 1990, of which 19800 were engaged in energy-related research. The number of researchers in the environment field is not readily available.

It is shown that R&D for environmental protection purposes increased very rapidly in the early 1970s. In fact, it is calculated that it has jumped from 16.9 billion yen in 1970 to 56.6 billion yen in 1975 in real terms (1970 prices). R&D deflators are estimated based on the price indices for research expenditures including wages and salaries, raw materials, land, buildings, machinery and equipment, and others.[Science and Tehnology White Paper, various issues] The increase was modest between 1975 and around 1977 when the expenditures hovered around 52-54 billion yen. It is increasing rapidly again in recent years, reaching 75 billion yen in 1990. We have shown in chapter 6 that pollution abatement investment also started to take off in the early 1970s in contrast to experiences in other industrialized countries, where it started to increase in the mid-1980s. We may say that, in Japan, both the search for new technology for pollution abatement and diffusion of that new technology in the form of capital formation proceeded simultaneously around 1970.

Energy is also very closely related to the environment as a major source of pollutant emission and as a typical case of depletable resource. Both of these concerns surfaced in the early 1970s and were prompted by deteriorating air pollution and the oil crisis. Statistical information on energy-related R&D, entitled *Survey of Research and Development on Energy*, was first compiled for 1976 as a supplement to the *Survey of Research and Development* by the Bureau of Statistics. The number of units enumerated in the survey was about 7100 (5300 companies with capital of one billion yen or more, 950 research institutions, and 860 universities and colleges conducting R&D in the natural sciences.) Survey items included research themes, persons engaged in R&D on energy, and R&D expenditures on energy. Sectoral enumeration is available.

In 1976, energy-related R&D stood at 77 billion yen (1970 prices), already surpassing environmental protection research by nearly 50%. In 1990, it has jumped to 294 billion yen (1970 prices), or an increase of 3.8 times in real terms. During the same period, R&D for environmental protection showed an increase of 1.5 times. It is also noted from Table 8-1 that a bulk of the energy-related research was nuclear (81% in 1976), and the ratio declined to about 44% in 1990.

This, of course, is due to the expansion of research activities in non-nuclear fields. The details of energy-related R&D are shown in Table 8-2. In 1990,

Table 8-1 R&D expenditures by purpose

	Total R&D expenditures	(R&D deflator)	Defense	Space development
1970	1195328	(=100.00)	11065	16217
1971	1345919	(107.06)	12305	23153
1972	1586708	(117.36)	14096	31456
1973	1980896	(142.74)	15575	31687
1974	2421367	(176.52)	16156	48501
1975	2621827	(190.83)	16949	73132
1976	2941373	(208.01)	18825	79500
1977	3233543	(219.65)	21826	69400
1978	3569953	(226.71)	24272	83900
1979	4063627	(244.27)	27649	96300
1980	4683768	(263.74)	29599	92900
1981	5363986	(272.06)	32573	107500
1982	5881539	(278.95)	36487	119300
1983	6503737	(280.67)	39452	116100
1984	7176511	(286.41)	44607	121300
1985	8116399	(286.98)	58677	152000
1986	8414993	(280.38)	66133	158700
1987	9016186	(282.10)	74135	171000
1988	9775165	(288.42)	82700	177900
1989	10909335	(300.76)	93068	195200
1990	12089593	(310.51)	104268	195100

Note: The survey covers business firms (with capital of 100 million yen and over), research institutions, and universities.

total R&D expenditures concerning energy amounted to 914 billion yen. This may be compared with 160 billion yen in 1976. Using the deflator provided in Table 8-1, it is calculated that this is an increase of 3.8 times during the 15-year period.

The effort in earlier years was concentrated on nuclear fields, especially nuclear power plants, nuclear fuel cycle, and nuclear fusion. In the mid-1970s, the picture was very different, with nuclear research dominating energy-related R&D, with a share of 81%. Energy conservation at that time shared a mere 5%. In 1990, the largest item listed in the table was nuclear energy, which shares about 44% of the total R&D expenditures in energy, followed closely by energy conservation with a share of about 40%. In value terms, they stood at 401 billion yen and 370 billion yen, respectively. Of the latter, transportation

(unit: million yen)

	Ocean delopment	Information processing	Environmental protection	Energy	of which nuclear
70	5391	31251	16900		
71	7117	41259	32732		
72	12627	51033	51317		
73	11725	82236	84883		
74	15751	88411	94358		
75	15547	99907	108077		
76	18100	116400	107700	160201	129954
77	19200	116200	81400	231213	170189
78	22400	130200	135800	277321	182335
79	27100	155900	142500	364052	207525
80	29400	162700	143600	501021	235826
81	39300	213700	150800	636200	265718
82	40700	249800	137400	677200	284300
83	44500	289600	144100	681416	314511
84	37600	388200	139500	712017	317378
85	40600	449400	149900	758324	348574
86	50700	492800	151600	822590	399852
87	61200	604700	154500	852440	421165
88	58400	744700	170400	888280	449687
89	54400	1012600	194900	906216	425957
90	52000	1121200	234800	913970	401974

Source: Bureau of Statistics, *Report on the Survey of Research and Development*. R&D for defense purpose is from Science and Technology Agency, *Science and Technology White Paper*. Energy-related R&D is from Table 4-4 below.

was the focal point starting in the early 1980s. This was not so in the 1970s when energy conservation in the industrial field was dominant. Thus, the last 15 years have seen much effort being directed toward energy-efficient technology rather than seeking the practical applicability of nuclear energy.

Research concerning fossil energy has been kept at around 80 billion yen a year throughout the 1980s, after taking off rapidly in the late 1970s. Utilization of natural energy such as solar, bio-mass, and geothermal came to focus in the early 1980s. According to the table, research interest in these fields seems to have receded somewhat in the late 1980s.

Before leaving this section, it should be noted that the larger percentage of environment-related R&D is borne by private industry rather than the

Table 8- 2 R&D expenditures concerning energy

Energy sources	1976	1977	1978
A. Fossil energy			
1. Oil	10791	7972	10889
2. Gas			
3. Coal	1746	2827	7085
4. Others	—	156	1479
Subtotal	12537	10955	19453
B. Natural energy			
1. Geothermal	1725	2550	1584
2. Solar	4244	4426	6485
3. Marine	—	314	690
4. Wind	85	190	482
5. Biomass	—	1113	3002
6. Others	2052	158	203
Subtotal	8106	8751	12445
C. Nuclear energy			
1. Power plant technology	58669	52299	46416
2. Multipurpose utilization	8718	8069	6551
3. Nuclear fuel cycle	9832	36257	46628
4. Nuclear fusion	14185	15944	26418
5. Other nuclear	26697	27555	34166
6. Nuclear ships	433	278	229
7. Utilization of radiation	5546	16907	9370
8. Radiation protection	5844	12881	12556
Subtotal	129954	170189	182335
D. Energy conservation			
1 Industrial use		17169	23371
2. Office and household		4397	7446
3. Transportation		8141	14638
4. Conversion and storage		6000	10401
5. Hydrogen energy		1680	1956
6. Others		2712	2181
Subtotal	8028	40099	59992
E. Other energy	1576	1218	3096
Total	160201	231213	277321

Note: Conversion and storage refers to that of electric power.

Source: Bureau of Statistics, *Report on the Survey of Research and Development on Energy.*

(unit: million yen, current prices)

	1979	1980	1981	1982	1983	1984
A						
1	18866	27551	38782	47245	47456	44192
2		6333	3712	3695	2948	4010
3	12116	45859	39751	35716	32703	31979
4	886	1326	1335	1133	1742	1097
	31869	81069	83580	87790	84848	81278
B						
1	2789	5668	9107	7901	7145	5653
2	11661	20237	20087	20936	19120	19918
3	2063	1472	2062	1889	1622	1577
4	757	2007	1659	1713	1008	1077
5	4611	5580	9286	7204	6598	7879
6	368	587	765	700	620	1069
	22250	35551	42965	40343	36113	37174
C						
1	67469	85296	88919	93296	107291	107140
2	6735	7047	9891	9619	7217	8418
3	34634	39325	47878	48976	62505	65673
4	39816	54349	60986	75822	75902	80450
5	35439	18486	21738	20462	22263	24401
6	220	4689	6766	7128	7350	473
7	15223	18204	17510	18764	21930	22709
8	7989	8429	12030	10251	10053	8114
	207525	235826	265718	284318	314511	317378
D						
1	30247	41879	44961	51964	48904	45351
2	24566	31185	39926	21807	16887	17359
3	25398	41197	111298	141815	134198	158422
4	9861	15042	24317	26360	22327	19591
5	2258	2989	3825	3215	3229	3358
6	1435	2279	1840	2509	1330	6324
	93765	134571	226167	247669	226875	250404
E	8643	14004	17733	17115	19068	25783
	364052	501021	636164	677235	681416	712017

	1985	1986	1987	1988	1989	1990
A						
1	40369	45991	31626	35071	42685	43313
2	3805	3746	3776	4186	5600	6438
3	33824	37060	27232	25823	24797	36389
4	5642	3467	3104	3252	3402	3888
	83640	90265	65740	68330	76484	89939
B						
1	6317	5543	5829	4415	4171	3701
2	20063	16373	16138	14249	14870	15906
3	1266	1137	1187	947	847	744
4	1019	908	987	1076	1513	1267
5	7863	8316	8006	7137	5629	5902
6	897	838	1412	837	899	691
	37425	33115	33559	28661	27929	28210
C						
1	124265	163621	162230	172669	150026	139057
2	10287	13853	14177	16793	17923	17183
3	75089	105279	122768	133098	123483	107506
4	75248	42282	42699	43454	46040	50873
5	22333	32589	33989	36334	38109	35508
6	11106	1979	2180	2152	2221	2083
7	23631	35077	22271	24804	26615	31114
8	6617	5172	20850	20381	21542	18648
	348574	399852	421165	449687	425957	401974
D						
1	47436	42304	38640	40393	67892	46286
2	17950	22998	23471	24535	29940	31215
3	171048	181693	214729	210867	230336	243775
4	21312	22519	25411	28618	35664	31873
5	5222	5675	5779	6786	7627	6823
6	2312	2614	4709	7173	6680	10001
	265277	277802	312740	318372	352556	369974
E	23408	21557	19236	23230	23291	23875
	758324	822590	852440	888280	906216	913970

government. Concerning environment protection, R&D expenditures in 1990 were allocated among private and government institutions as follows [Bureau of Statistics, *Report on the Survey of Research and Development*]:

Private industry	171 billion yen
Private research institutes	14
Universities and colleges	18
Government research institutes	31
Total	235

An OECD publication [OECD 1992, p.38] states concerning the situation in some selected European countries: "The enterprise sector is already the main source of finance and main performer for environmental R&D, and this role is likely to increase. Yet, there has traditionally been ambivalance both in public attitudes to industry as a source of pollution and in industry's attitude to the environment." Although the situation in Japan is fundamentally similar, the role of the private sector is probably more established and dominant.

Concerning R&D in energy fields, the picture in 1990 was as follows[Bureau of Statistics, *Report on the Survey of Research and Development*, 1992, pp.36-37]:

Private industry	349 billion yen
Private research institutes	261
Universities and colleges	42
Government research institutes	262
Total	914

In terms of research personnel in energy fields, the government contributed 3100 out of a total of 19800 in the same year. The dominance of effort by the private sector in this field is merely a reflection of the total R&D scene in Japan. International comparison of R&D in environment-related fields is hampered by the unavailability of comprehensive environmental R&D data. What is collected by the OECD, for example, concerns government R&D, which is included in many OECD publications [OECD 1992, 1993]. But its usefulness is limited by the reason just mentioned.

From an organizational point of view, it is important to recognize the technological interrelation among various lines of activities within a business firm. Diversification of products and technology have occurred around a 'core'

Table 8-3 Energy conservation investment

		1985		1986	
Textiles	[20&]	15711	(26.2)	9628	(19.5)
Pulp & paper	[24]	48452	(19.3)	25866	(13.5)
Petrochemicals	[26]	12615	(11.6)	9781	(6.2)
Petroleum refinery	[27-]	9890	(5.0)	13874	(6.2)
Nonmetallic minerals	[30]	14504	(25.4)	611	(5.6)
Iron & steel products	[31]	92628	(15.5)	115781	(20.2)
Nonferrous metals	[32]	691	(1.5)	11587	(25.3)
General machinery	[34]	—	(—)	—	(—)
Electronics	[35-]	—	(—)	—	(—)
Electrical machinery	[35-]	3855	(0.4)	3452	(1.1)
Motor vehicles	[36M]	13248	(2.3)	8740	(1.2)
Other manufacturing	[38&&]	51237	(6.7)	27821	(6.0)
Manufacturing total	[0M]	262831	(7.4)	227143	(8.3)
Mining	[15]	—	(—)	—	(—)
Electric power	[70-]	22664	(1.3)	30176	(1.4)
Gas	[70-]	1944	(0.9)	1070	(0.5)
Wholesale and retail	[40]	913	(1.0)	984	(5.2)
Leasing		—	(—)	—	(—)
Services	[75]	—	(—)	—	(—)
Nonmanufacturing total		34401	(1.4)	38820	(1.4)
All industry	[00]	297232	(4.9)	265963	(4.9)

Note: Total investment in (a) refers to responding firms. Figures in parenthesis are percentage shares in total investment.

according to a survey based on a bibliographic database by the Japan Information Center of Science and Technology [JICST 1993]. In this sense, it was simply not a matter of money to promote R&D in different fields including pollution abatement and energy conservation. Business experience in pollution abatement and energy conservation is introduced in Schmidheiny [1992], based on cases from all over the world. A detailed exposition on this point, however, is outside the scope of this paper.

8.3 Diffusion of new technology

New technology has to be implemented in new capital stock in order to actually achieve the desired objectives. In this section, we examine the plant and equipment investment for energy conservation, followed by an overview of the introduction of alternative energy sources.

	1987		1988		1989		1990	
Te	9612	(13.4)	7463	(7.7)	3090	(6.7)	13065	(12.6)
Pu	12696	(5.3)	19223	(9.0)	22748	(10.9)	58894	(13.7)
Ch	23331	(8.3)	17952	(6.0)	17272	(6.5)	11055	(3.1)
Pe	11223	(4.1)	4817	(1.5)	12874	(3.3)	8957	(1.8)
No	8917	(6.7)	7481	(6.2)	9213	(5.6)	12580	(8.1)
Ir	108303	(23.6)	42860	(9.0)	37552	(5.9)	65269	(10.0)
No	2884	(3.6)	3044	(3.8)	3158	(5.5)	3859	(4.4)
Ge	1365	(2.4)	1892	(2.2)	1765	(1.9)	5393	(7.0)
El	714	(0.3)	1065	(0.9)	59	(0.3)	926	(0.8)
El	2982	(1.2)	2722	(1.2)	1136	(0.5)	4444	(0.9)
Tr	6849	(0.9)	13548	(1.9)	13664	(1.4)	10138	(1.0)
Ot	6872	(3.7)	7688	(3.0)	5365	(2.9)	10301	(2.7)
Ma	195748	(6.5)	129755	(4.3)	127896	(3.9)	204911	(4.7)
Mi	283	(1.4)	51	(0.3)	312	(1.6)	0	(0.0)
El	24088	(1.0)	18673	(0.7)	15228	(0.7)	48505	(1.8)
Ga	1224	(0.6)	1370	(0.6)	1234	(6.2)	336	(0.1)
Wh	207	(1.3)	1232	(2.0)	0	(0.0)	1576	(1.9)
Le	7820	(0.8)	5885	(0.9)	7652	(1.0)	1154	(0.8)
Bu	26	(0.0)	2419	(1.5)	4085	(2.9)	1743	(2.4)
No	33648	(0.9)	29630	(0.8)	28511	(0.9)	53314	(1.6)
To	229396	(3.4)	159385	(2.4)	156407	(2.5)	258225	(3.4)

Source: MITI, *Plant and Equipment Investment Plans in Major Industries.*

Empirical figures are available only for recent years when the second oil crisis prompted interest in energy conservation. Starting from less than 200 billion yen in 1979 and about 300 billion yen in 1980, plant and equipment investment for energy conservation reached nearly 450 billion yen in 1981 and 1982. This was precisely when the crude oil price (CIF) approached 55000 yen per kl. The energy conservation investment actually started to decline in the early 1980s as the crude oil price declined to 10000 yen per kl or less. This is a reflection of lower oil prices and, in Japan, appreciation of the yen currency.

Detailed data were made public starting 1985, which are reported in Table 8-3. The data comprise a part of the annual survey conducted by the MITI for the Council of Industrial Structure, with which it is affiliated. In its 1990 edition, the sample was reported to comprise about 2000 major enterprises under the jurisdiction of the MITI, of which some 1120 responded. Of this, 325 enterprises were engaged in investment in energy-related fields.

Where an enterprise was engaged in various activities, individual activities

Table 8-4 Alternative energy investment

a. Breakdown by industrial sector

Industry	Sector code	1986
Food processing	[18]	—
Textiles	[21&]	1046
Pulp & paper	[24]	26855
Petrochemicals	[26]	1046
Petroleum refinery	[27-]	16066
Nonmetallic mineral products	[30]	373
Iron and steel products	[31]	8177
Nonferrous metal products	[32]	0
General machinery	[34]	—
Electronics	[35-]	—
Electrical machinery	[35-]	897
Motor vehicles	[36M]	7158
Other manufacturing	[38&]	14124
Mining		—
Leasing		—
Total		75196

b. Breakdown by energy source

	1986	
Coal	63760	(84.9)
LNG	3362	(4.5)
Hydro	—	(—)
Geothermal	3079	(4.1)
Solar energy	5	(0.0)
Wastes	756	(1.0)
Others	4234	(5.6)
Total	75196	(100.)

Note: Figures in parenthesis are % in total alternatative energy.

Source: Same as Table 8-3.

were classified to relevant industrial sectors. The investment figures are enumerated on the basis of progress of construction rather than payments. Energy conservation equipment in the survey refers to the following:

(1) equipment designed to recover waste energy;

(2) equipment which is retrofitted to the main facility in order to improve energy efficiency;

(unit: million yen, %)

	1987	1988	1989	1990
Fo	4	17	0	0
Te	2957	1102	4331	379
Pu	15605	9318	21248	20157
Ch	11319	215	6247	1956
Pe	1337	0	4100	7400
No	129	223	2148	4644
Ir	3069	15458	17587	14373
No	190	2404	627	357
Ge	92	102	207	272
El	301	0	115	145
El	—	820	450	991
Tr	5265	360	3565	3858
Ot	163	301	885	6169
Mi	—	86	414	0
Le	—	—	1256	—
Total	40431	30406	63180	60701

	1987		1988		1989		1990	
Co	25605	(63.3)	16138	(53.1)	40080	(63.4)	30450	(50.2)
LN	2854	(7.1)	6104	(20.1)	3960	(6.3)	8587	(14.1)
Hy	—	(—)	0	(0.0)	1889	(3.0)	860	(1.4)
Ge	0	(0.0)	0	(0.0)	440	(0.7)	0	(0.0)
So	28	(0.1)	3	(0.0)	142	(0.2)	142	(0.2)
Wa	5939	(14.7)	8128	(26.7)	15825	(25.0)	20577	(33.9)
Ot	6003	(14.8)	33	(0.1)	844	(1.3)	85	(0.1)
To	40431	(100.)	30406	(100.)	63180	(100.)	60701	(100.)

(3) equipment which is considerably superior compared with the existing facility in its energy efficiency; and

(4) improvement of production processes for the purpose of improving energy efficiency.

Those for the production of equipment for energy conservation and facilities to produce energy-efficient products are not included.

According to Table 8-3, 1990 seems to have witnessed a reversal in the decline of energy conservation investment. In the manufacturing industry, it jumped nearly 47% from 128 billion yen in 1989 to 205 billion yen. The ratio to total investment of the reporting firms stood at 4.7%. The survey from more recent years indicates that the higher level continued in 1991 and 1992. A similar trend was observed for the nonmanufacturing sector (the bulk of which is electric power generation) where it increased from 29 billion yen in 1989 to 53 billion yen in 1990.

The ratio to total investment was 4.7% for manufacturing, 1.6% for nonmanufacturing, and 3.4% for all sectors. The ratio, however, fluctuated rather widely, as can be observed in the table.

We turn now to Table 8-4 which shows the recent trend in investment in order to shift to alternative energy sources. The data are based on a survey entitled "Survey on Plant and Equipment Investment on Petroleum Alternatives" conducted by the MITI, which is also a part of the annual investment survey reported above. The samples are similar to the above survey. Of the 895 enterprises which responded, 84 were engaged in investment in order to shift to nonpetroleum energy.

Petroleum alternatives in the survey refer to furnaces themselves and accompanying pollution prevention facilities, etc. which are needed in the introduction of coal, natural gas, and other alternatives to petroleum. The cases include where an existing heavy oil boiler was converted into using natural gas, etc., and also where new facilities were installed which use nonpetroleum energy sources.[3]

According to the table, the investment for this purpose has fluctuated between 30 to 75 billion yen during the five-year period listed, with no clear trend. The 1990 figure of 61 billion yen is about 1/4 of the energy conservation investment reported in Table 8-3.

Panel (b) of Table 8-4 is interesting in that it shows which energy is being contemplated as an alternative to petroleum in Japanese industry. In 1986, the earliest year for which the data collection is available, coal was practically the only alternative. The share of oil has declined from 85% in 1986 to 50% in 1990. It should be noted that the use of wastes as alternative energy has come to a practical stage, as can be witnessed by the fact that around 30% of investment for alternative energy use is directed to this category in the 1989-1990 period, compared with a mere 1% in 1986. Another major source is LNG.

8.4 Pollution prevention technology, some cases

R&D adds to the stock of scientific and engineering knowledge commonly available to the global community. There are economic, institutional, and technological obstacles in the diffusion process. Public acceptance is also an issue, especially concerning nuclear energy and urban waste disposal plants. In this section, we take up some of the key technologies which are in their infancy and at their stages of implementation. We examine pollution prevention technology, energy conservation technology, and the development of new sources of energy.

The performance of an economy used to be, and still is to a considerable degree, measured by the increase in annual flow of value added. GDP is, of course, a concise indicator of the levels of economic activities. Increasing attention is being paid to before and after such a flow, namely, the effect on the stock of resources and the environmental impact. Greater care must be taken to control pollution, minimize waste, and recycle and reuse energy and materials.

Sulphur oxide pollution: In order to meet the emission standards set for each facility as a result of consultation involving local residents, local governments, and the production unit concerned, as well as for the area as a whole as stipulated in the Pollution Control Programs, Japan has relied on heavy oil desulphurization and flue gas desulfurizers.[4]

(1) For the desulphurization of heavy oil, direct and indirect desulphurizers were installed starting in 1967. Time series data on the installation of desulphurization units are available in Table 6-4. The capacity expansion was complete by 1980. With the desulphurization of heavy oil, coupled with the importation of low sulphur crude oil (with S content less than 1%), the average sulphur contents in heavy oil have been reduced from 2.5% in 1967 to 1.43% in 1974. The 1990 figure stood at 1.02%.

(2) R&D on flue gas desulphurization (FGD) was initiated by the Ministry of International Trade and Industry in 1966 as a part of its Large Scale Industrial Technology Development Plan, and the operation of practical units started in 1970. Time series data on this account are shown in Table 6-5.

Coal liquefaction and gasification can work as pre-combustion control methods to contain sulphur emission by removing pollutants in the process. In order to remove pollutants during combustion, new types of boilers are used such as atmospheric and pressurized fluidized bed combustion (FBC) systems.

Supplies of by-product gypsum from the FGD process continue to increase, posing a difficult disposal problem.

Nitrogen oxide pollution from stationary sources: The emission standards have been strengthened since 1973. In 1985, controls for small-sized boilers were added. Starting in 1987, gas turbines and diesel engines were also regulated, with implementation starting in 1988 for new facilities and 1990 for existing ones. In 1990, the regulation was extended to cover gas engines and gasoline engines.

(1) The research effort at reducing nitrogen oxide emission was started in 1975. Existing methods for controlling NO_X emission during combustion include two-stage combustion, low NO_X burner, and flue gas recirculation (FGR).

(2) The post-combustion removal of nitrogen oxides represents the newest technology made available through extensive R&D and has been commercialized in countries such as Japan and Germany for their power plants. A selective catalytic reduction (SCR) system removes 80% to 90% of NO_X when used in conjunction with combustion modification methods [OECD 1987, p.37]. The diffusion process of flue gas denitrification units in Japan is described in Table 6-6.

NOx reduction from mobile sources: The regulation on NO_X emission from automobiles has been strengthened, as shown in Table 5-6. NO_X emission from mobile sources proved to be a very stubborn problem. In Japan, only 6.9 observation stations built to monitor automobile exhaust gas cleared the environmental standards. Nitrogen oxide emission attributed to automobiles amounted to 67% and 47% in Tokyo and Osaka, respectively. The Central Council for Environmental Pollution Control recommended in 1989 short-term (less than 5 years) targets and long-term (less than 10 years) targets for emission reduction, primarily for diesel vehicles. This regulation also applies to the reduction of particulates from diesel engines.

Preferential tax measures are being utilized in order to encourage a shift to vehicles which satisfy the latest regulation in the introduction of low-pollution vehicles such as electric cars.

Photochemical air pollution: The Environmental Agency and the Meteorological Agency both monitor the weather data, issuing warnings as needed. Factories are directed to reduce their emission level and automobile users are asked not to make unnecessary trips. The control of volatile organic compounds (*VOC*), which include hydrocarbons (*HC*), is particularly important since these substances are involved in the formation of photochemical oxidants (ozone). *VOC* emission often results from the evaporation of hydrocarbons in a variety of locations, requiring vapor recovery methods.

Hydrocarbon: The control technology of hydrocarbon from automobiles has been discussed in chapter 5.

Chlorofluorocarbons (CFCs): In order to prevent depletion of the ozone layer, the global community agreed in the Vienna Convention for the Protection of the Ozone Layer (adopted in 1985 and put into effect in 1988) and the Montreal Protocol on Substances that Deplete the Ozone Layer (adopted in 1987 and put into effect in 1989). This was followed up domestically by the Ozone Layer Protection Law in 1988.[5] The production of five kinds of *CFC*s and three kind of halons were to be reduced according to the schedule specified in the Montreal Protocol. Covered by the Protocol are *CFC* 11, 12, 113, 114, and 115 (hereafter referred to as specified *CFC*s) and halon 1211, 1301, and 2402. The development of technology for the destruction of specified *CFC*s is underway. *CFC*s have been widely used for cleaning purposes in the semiconductor industry and as a refrigerant in air conditioners. Attempts have been made to encourage the rapid diffusion of emission control and recovery facilities through tax incentives and low-interest loans. In the meantime, participating parties to the Montreal Protocol met in Helsinki, and agreement was reached on a total phaseout of specified *CFC*s by the end of the century (Helsinki Declaration).[6]

Stipulation in the Montreal Protocol on specified *CFC*s and halons are cited below.

Specified CFCs:	After 1989, 100% or less of 1986 level
	After 1994, 25% or less 〃
	After 1996, to be abolished.
Specified halons:	After 1992, 100% or less of 1986 level
	After 1994, to be abolished.

In Japan, the production and consumption of specified *CFC*s have been reduced to about 61% and 55% of the reference point in 1986, respectively.[7]

Water pollution: There has been considerable achievement in improving the water quality in Japan in the past 20 years. We have noted the environmental quality as measured by *BOD* and *COD* in Table 6-12 which we need not repeat here. These indicators relate to organic matters contained in the water. As to cadmium and other substances hazardous to human health, the situation has also improved. The ratio of water samples exceeding environmental quality standards has been reduced from 1.4% in 1970 to 0.02% in 1990.[8] Total mercury should not be found in water according to the environmental standards. The ratio of samples containing mercury has been reduced from 1.0% in 1970 to 0% in 1990. In fact, improvement has been achieved within several years after implementing the stringent standards.

According to the Japanese experience, the abatement of water pollution was mainly achieved through purification and recycling within industrial plants and increased provision of sewerage for household and urban water consumption. Stricter environmental standards and their forceful implementation through the regional Pollution Control Programs were also contributing factors. Relevant data are found in Tables 6-7 and 6-8. In both cases, the results have been achieved by implementing the existing technology and/or by incremental improvement thereof rather than waiting for technological breakthrough, as far as the description of the *Environment White Paper* reflects reality. The practice of recycling industrial water is widespread in Japan, as shown in Table 6-7, and this was achieved by shifting to a closed water circulation system.

Wastes: We have examined the disposal of municipal wastes in Table 6-9 and that of industrial wastes in Table 6-10. In the former, the percentage incinerated has been on the increase. We noted in Table 3-3 that the incineration of waste is now contributing a new source of energy. The trend is recapitulated here.

1975	34 10**10 kcal
1980	107
1985	227
1990	407

This now represents 0.5% of the domestic energy supply or 0.08% of the total energy supply including imports.

New technology is involved also in the pipeline facilities for the transport of wastes.

As for industrial waste, a larger amount is now being recycled, as shown in Table 6-10. The total volume recycled has increased, but the percentage recycled has not improved greatly. The responsibility for the treatment of industrial waste rests with the factories involved, but the role played by enterprises specializing in the treatment of industrial waste has become increasingly important.

Recycling technology, information systems to control the flow of industrial wastes, and technology to repair the environmental damage done by the illegal disposition of wastes are some of the examples of technological development in this field. The Ministry of International Trade and Industry has subsidized the Clean Japan Center in order to design a pilot plant for the recycling of industrial waste. The MITI also has a research institute dedicated to this purpose.

In order to facilitate the recycling of raw materials, product design and material are being examined for certain items, including automobiles, air conditioners, television sets, washing machines, and refrigerators. In order to ease the collection of aluminum and steel cans, specific marking is being considered. In the construction industry, the reuse of iron slag, coal ashes, asphalt, concrete, and wood is being promoted.

8.5 New sources of energy, some cases

In a review of government policy among its member countries, IEA [1992a, p.185] summarized the Japanese energy policy as follows: "Japan's energy policy has proved successful in limiting the risks inherent in Japan's vulnerable energy situation and in providing the basis for rapid economic growth. It has been particularly successful in diversifying the sources of energy supply, developing nuclear energy and promoting efficiency in energy use. Some of this success has been bought at a cost - for example the reliance on high cost nuclear energy and LNG in electricity generation - but this has been the result of deliberate choice arising from Japan's energy situation."

Apart from the endowment of energy resources, which is practically none (excluding nuclear) as can be seen in Table 3-3, Japan is aiming at stabilizing CO_2 emission on a per capita basis in year 2000 at about the 1990 level. The energy strategy to this end includes the promotion of nuclear power, introduction of the renewable energies, promotion of CHP, greater use of natural gas, and promotion of energy efficiency. The effort to arrest global warming tends to re-emphasize nuclear energy in Japan.

Nuclear energy: In 1990, over 80% of government energy R&D was devoted to nuclear energy. Nuclear-related technology has dominated the Japanese research effort, as can be witnessed in Table 8-2. Nuclear power stations are operated by nine privately owned regional utilities, but the Japan Atomic Power Company is responsible for the construction and operation of nuclear plants. This company was established by the utilities and the Electric Power Development Co., which is 70% government-owned.

France is another country in which nuclear power plays a major role in the energy supply. "For economic-political reasons and because of the availability of its own energy resources, France has undertaken a very goal-oriented energy conservation program since the oil crisis of 1973 and has been concentrating on reducing its consumption of oil and coal, and especially on the possibility of developing its nuclear energy"[Balaceanu, Bertrand, and Lacour 1989]. The French energy balance in 2000 will be 8% coal, 33% oil, and 41% nuclear.

Indeed, based on technological and economic considerations (i.e. assuming that current constraints of public acceptability are totally removed), experts from the OECD Nuclear Energy Agency argue that there would be an opportunity for greater penetration of the electricity generation sector by nuclear plants [Gehrisch et al. 1989]. "In the absence of reprocessing, in addition to electricity a significant amount of spent fuel is produced. The unit of measurement is the tonnage of heavy metal contained; by 2000, it is expected that some 150,000 HM will have been produced in the OECD. According to the nuclear capacity curve ... a further 650,000 ton HM would have been added by 2030. ... This would require a storage volume of the order of 260,000 to

130,000 cubic meters and there is no reason to suppose that this amount of storage could not be possible. However this level of unreprocessed spent fuel storage appears to be incompatible with the basic scenario, as the spent fuel will be wanted for reprocessing." "A major factor in deciding on the timing and rate of introduction of fast breeder reactors (FBRs) will be the availability of plutonium to fuel them. By, say, 2020, there will be of the order of 650,000 ton of spent fuel containing about 6,500 tonnes of plutonium in various forms of storage or repository. Depending on how much of the spent fuel is in retrievable storage (and as yet there are no operating final repositories for spent fuel) the majority of this plutonium would be available, after reprocessing, for FBR fuel fabrication."

"The main impediment to FBR introduction would be the availability of freshly reprocessed plutonium. That depends on decisions having been taken well in advance on setting up the necessary plants, including facilities for handling the highly radioactive wastes that are produced. Experts in the relevant subjects have confidence that appropriate technology is already available so from that point of view ... there should be no inherent technical constraint on a fairly widespread introduction of FBRs as from 2020."

"It is recognized, however, that this development would not be undertaken without some hard thinking on matters related to policy on non-proliferation of nuclear weapons, such as trade-in and frequent movements of plutonium between fuel cycle facilities."

At present Japan is probably the only country in the global community which actively maintains the nuclear energy policy along the course described above, although some others, including the USA, former Soviet Union, UK, and France, had extensively sought feasibility along this line earlier. The Japanese government and the nuclear industry selected the village of Rokkasho on the northern tip of mainland Japan for the site of reprocessing, together with radioactive waste storage. The logic for resorting to nuclear in Japan is as follows [MITI 1992, p.118]:

(1) Compared with crude oil which depends heavily on the Middle East, uranium is available from various suppliers in the world, most notably from politically stable developed countries.
(2) Nuclear fuels can be stored and transported easily because they are capable of generating a large amount of energy per physical quantity. Stock-

piling is not needed, as in the case of crude oil, because nuclear fuels continue to burn for about 3 years. Coupled with nuclear fuels in the production process, several years worth of fuels can be stored domestically.

(3) By establishing a nuclear fuel cycle which utilizes plutonium and recovered uranium, Japan will be able to obtain a semi-domestic source of energy.

(4) It is free of CO_2 emission, which is the cause of global warming. It is also free of SO_X, NO_X, and other air pollutants.

(5) The cost of electricity for the duration of the operation of the power plant is the lowest (about 9 yen per kW as of 1989) among alternative energy sources.

The Rokkasho plant thus constitutes the backbone in the Japanese nuclear energy policy. The government has defined it as a national project. The plant consists of a uranium enrichment facility, reprocessing facility for irradiated fuel, and low-level waste burial facility. Reprocessing also yields radioisotopes created by splitting uranium atoms, including cesium, iodine, strontium, and technetium, which are referred to as high-level waste. Storage of high-level waste at Rokkasho is planned to be temporary, while its final burial site has not been determined due to public opposition. The Japan Nuclear Fuel Industry Co., jointly owned by electric utilities, is responsible for establishing a nuclear fuel cycle. The construction of the reprocessing facility, which relies on French technology, was started in April 1993 for completion in the year 2000, with a planned budget of more than 840 billion yen (1985 prices).[*The Japan Economic Journal* (J), Apr. 28, 1993] When complete, Japan will realize a nuclear fuel cycle in which used fuels are reprocessed through chemical procedures that extract fission by-product plutonium to be used in FBRs, thus establishing everlasting cycles. The reprocessing also recovers uranium not consumed in the reactor. The Rokkasho plant will be the largest of its kind in the world, producing 5 tons of plutonium per year from 800 tons of spent fuel. This is equal to spent fuel from 30 one-mega kW-class nuclear power plants. Presently, Japan possesses only a small facility at Tokaimura and is obliged to resort to reprocessing facilities in France and the UK.

Japan constructed an experimental FBR in 1987, which is named "Joyo". A prototype FBR, named "Monju" after a Buddhist goddess, was constructed on Tsuruga Peninsula facing the Japan Sea. Its operation is slated for October 1993. Japan is contemplating the construction of a second prototype FBR starting in the second half of the 1990s, establishing the FBR technology for commercial use. While the Rokkasho plant is capable of reprocessing about 70% of spent fuel, the second plant will enable Japan to treat all the spent fuel on Japanese soil.

There are various scenarios as to the future Japanese nuclear policy. [*The Japan Economic Journal* (J), Jan.11, 1993] The most optimistic scenario foresees operation of the first commercial FBR around 2030, according to the long-term plan. Replacing obsolete reactors with FBRs, Japan will complete its nuclear fuel cycle. The second scenario foresees a fate similar to that of other leading countries. The USA has stopped development of FBR together with reprocessing on the ground of nuclear nonproliferation. Germany abandoned a nearly complete FBR in 1991 on economic grounds. The UK is also closing down a FBR in 1994 on economic grounds. France, which has been leading FBR development, decided in the summer of 1992 to suspend the reopening of an experimental reactor indefinitely after experiencing some technical difficulties. Japan will have to resort to sea transportation of irradiated fuel and plutonium to and from French and British reprocessing facilities, triggering opposition on an international scale. Ultimately, economic factors, coupled with the fear of nuclear proliferation, will oblige Japan to follow suit, according to the second scenario. The third scenario foresees some delays in establishing the nuclear fuel cycle, although it is unlikely that Japan would abandon the plan. The construction of a commercial FBR will more likely start in the 21st century. It was reported that Japan's Atomic Energy Commission was contemplating to postpone the construction of next FRB till after 2005, delaying the current plan by five years. Commercial use will not start 5 to 10 years after the current target of the year 2030.[*The Japan Economic Journal* (J), Jan. 6, 1993] In Europe, France and the UK will pursue the production of uranium-plutonium mixed oxide (MOX) for plutonium thermal use in widely used light water reactors rather than the production of plutonium for FBRs.

According to an estimate by IAEA as of December 1992, the world inventory of plutonium is as follows [*The Japan Economic Journal* (J), Jan.6, 1993]:

Civilian:	Current	80 tons
	Year 2000	140 tons
Military:	USA	90 - 100 tons
	Russia	95 - 120 tons

With the end of the cold war, nuclear warheads are being disassembled, leaving behind a huge amount of plutonium, causing a "plutonium glut" in the global community. This has given rise to an argument which calls for reconsideration of achieving the nuclear fuel cycle in Japan, while the USA is leading an effort to develop a new type of reactor fueled with surplus plutonium from the Russian arsenal. Joint plans are already under way to use Russian surplus uranium in commercial reactors.

Plutonium thermal use is widely recognized as technology intermediate to FBRs and the complete nuclear fuel cycle. Japan has developed an advanced thermal converter reactor (ATR) for this purpose. The pilot plant, named "Fugen" after another Buddhist goddess, started operation in 1988. ATRs are capable of burning uranium, enriched uranium, or plutonium.

At any rate, Japan would have to address itself to two serious issues in pursuing its nuclear policy. One is the problem of nuclear waste. Final disposal of high-level waste, consisting of irradiated fuel and/or refuse from reprocessing, reactors from decommissioned nuclear power plants, etc., poses a particularly difficult problem because it remains dangerous for hundreds of thousand, or even millions of years. Second is the international diffusion of nuclear power generation. Japanese experts seem confident that nuclear is now a safe technology for commercial use in Japan. If so, it would be extremely difficult to refuse other countries from having access to nuclear technology, whether on technological or political grounds. The problem is compounded by the first issue. We discussed the policy option on this point in chapter 2.

Renewable energies: The "Sunshine" project of the Agency of Industrial Science and Technology, which was started in 1974, is aimed at new energy technology such as coal liquefaction, solar, wind, and geothermal energy, and the "Moonlight" project, which was started in 1978, aimed at energy efficiency technology. The government hopes to increase the new sources of energy (e.g. solar and wind) from 1.3% of the total energy supply to 5.3% in 2010. The New Energy and Industrial Technology Development Organization (NEDO), which is a joint venture between the government and private sector, is promoting renewable energy research and large scale pilot projects in photovoltaics and geothermal power generation.

MITI [1992] lists the following as new technologies available for practical use:

(1) Clean coal technology
(2) Solar energy
(3) Wind and tidal power generation
(4) Geothermal
(5) Methanol
(6) Fuel cells
(7) High-efficiency gas turbine

The same source mentions the following as future technology potentials of which are under serious discussion:

(1) Hydrogen energy
(2) Biomass
(3) Superconductivity
(4) Energy storage

Clean coal technology: Coal is abundant, and its deposits are fairly dispersed around the world. However, emission of pollutants such as CO_2, SO_X, and NO_X is an obstacle which has to be overcome for extensive use in the future. Difficulty in handling solid fuel is another problem. One notable development in clean coal technology is a fluidized bed combustion boiler which burns crushed coal (about 4 to 6 mm in diameter) in a fluidized bed of preheated lime and sand. The process removes sulphur contained in the coal by reaction with lime, generating gypsum. NO_X emission is also contained. Small- to medium-sized boilers of this type are in practical use, while large-sized ones may be tried for power generation and other industrial purposes. A pressurized type is also under study which will make the design more compact.

Coal handling technology includes coal-water mixture (CWM) fuel, coal cartridge system (CCS), etc. CWM uses a slurry consisting of a very fine powder of coal (70%) and water (30%) mixed with surface-active agents. A coal-oil mixture (COM) has been put to commercial use for the first time in the world by the Tokyo Electric Company in 1984 after 70 billion yens worth of effort. It is hoped that CWM can be used for diesel engines and gas turbines. CCS uses fine coal in a container for transportation, combustion, and recovery of waste ashes. Its practical feasibility is being investigated.

Research on coal gasification was activated after the oil crisis in Japan, USA, and Germany. The Japanese effort is a part of the Sunshine Project, and focus is placed on high calorie gas for city gas and low calorie gas for power generation, among others. Coal gasification combined-cycle power generation has a higher energy efficiency than the conventional coal-burning power plant and is slated for commercial application in the early 21st century. A study leading to the construction of a pilot plant was started in 1986. According to the study, the current efficiency of about 38% for a pulverized-coal-fired power plant is expected to be improved by 5%. This means a reduction of CO_2 emission of 10% [Sasaki 1989].

Coal liquefaction technology is aimed at obtaining light oil from coal and is also being investigated within the Sunshine Project.

Solar energy: Solar energy is used as a heat source for air conditioning or for power generation. In Japan, there are about 350000 solar systems and 4 million solar water heaters already in place. Photovoltaic power generation uses solar cells in converting solar light into electricity. More efficient solar cells are being developed under the Sunshine Project. One of the project's aims is to reduce the production cost, which stood at around 2300 yen per watt in the early 1980s, to 100 yen per watt by 1990. The production of photovoltaic cells in Japan has increased rapidly [MITI 1992]:

1976	22 kWp		
1980	291 kWp	(of which power generation	67 kWp)
1985	9250 kWp	(〃	3431 kWp)
1990	14570 kWp	(〃	5534 kWp)

In 1990, about 38% of total production (in terms of capacity) was for power generation and research purposes and the remainder for consumer appliances.

It is reported that MITI's Agency of Industrial Science and Technology is aiming at reducing the cost of photovoltaic power generation to 20 to 30 yen per kWH by the year 2000, equivalent to the cost of electricity at the consuming end today. By 2010 it is intended to reduce the cost further to around 15 yen per kWH, allowing it to be incorporated in the utility service. The efficiency of energy conversion technology remains around 15%; even through mass production current technology does not promise an electricity cost of less than 40 yen per kWH. It is deemed necessary in the medium-range time frame (say, year 2000) to develop the production technology of three types of photovoltaic cells, polycrystal silicon, amorphous silicon, and cadmium-tellurium. At the same time, the integration of photovoltaic cells in building materials will be sought in order to facilitate their installation. In the long-run (year 2010), an energy conversion efficiency of 20 to 40% is sought based on single crystal silicon cells and compound semiconductor cells.

Solar thermal electric power generation uses a large number of reflectors, generating steam, which then turns turbines to produce electricity. A pilot plant having a capacity of 1000 kW was constructed in Japan as part of the Sunshine Project but proved uneconomical under the Japanese climatic conditions.

Windmill power energy: The cost of electricity using wind power is currently around 5 yen per kWH, assuming an average wind speed of 6 m/s. This is competitive, compared with 11 yen in the case of thermal plants and 9 yen in the case of nuclear plants. Electric utilities are purchasing at 17 to 19 yen per kWH under the new scheme which allows transactions between the private power suppliers and the utilities. With this price level, it is calculated that 100 million kW capacity generating 75 billion kWH is economically feasible. This amounts to about 10% of the total electricity generated by 10 utilities in Japan. The existing capacity is only 4000 kW as compared with 1.9 million kW in the USA and 1.1 million kW in Europe. The Sunshine Project aims at developing a 500-kW-type generating system.

Geothermal: There are 10 existing sites with 270000 kW capacity. Geothermal energy is covered by the Sunshine Project. Refer to Table 3-3 for the diffusion of geothermal power generation.

Methanol: Methanol can be produced from a variety of raw materials, including natural gas, tar sands, coal, wood, and others, and can be used for power generation and for vehicle fuel. A methanol power generation plant utilizes methanol as a substitute for coal, petroleum, or LNG. It promises to be clean, with no SO_X emission and with very little NO_X emission compared with fossil fuels. However, it is a different story if one looks at the total fuel cycle emissions from the initial resource extraction to end use. One calculation shows the emission of greenhouse gases from alternative fuels as follows [DeLuchi et al. reported in IEA 1990c, p.74]:

	CO_2-equivalent total emissions (bil. tons/year)	Changes in emission (% to petroleum)
Methanol from biomass	0	-100%
Double efficiency vehicles	0.668	-50%
CNG from natural gas	1.081	-19%
LNG from natural gas	1.135	-15%
Methanol from natural gas	1.293	-3%
Gasoline and diesel	1.336	—
Methanol from coal,		
30% more efficient	2.026	+52%
baseline	2.639	+98%

In the calculation, all CH_4 emissions as well as NO_2 emissions from power plants are converted to equivalent CO_2 emissions. The report concludes that "It is clear from the use of coal to make any highway fuel would substantially accelerate CO_2 emissions, relative to the base-case use of petroleum. ... Significant reductions in emissions of greenhouse gases can be attained only by greatly increasing vehicle efficiency or by using biomass or nonfossil electricity as the fuel feedstock. In the case of biomass, the net impact will approach zero if CO_2-absorbing replacement crops are planted to sustain the methanol economy." IEA [1990c, p.9] puts the costs of methanol from gas (US$ per barrel-gasoline energy equivalent) at $30-$67, methanol from coal at $63-$109, and from biomass at $64-$126 as compared with assumed price of crude oil at $18 and conventional gasoline at $27.

Fuel cells: This is a process which generates electricity (and water) from hydrogen and oxygen. Technically, phosphoric acid fuel cells are already at the stage of practical use, while economics is the factor which determines the diffusion of this system. Basic research is being performed under the Moonlight Project for molten carbonate fuel-type and solid oxide-type systems. The Japanese technology in this field is believed to be high.

As for phosphoric acid fuel cells, on-site cogeneration systems are being produced commercially for relatively large heat demand (200 kW) in such places as hotels and hospitals. The 5000-kW class fuel cells are under development for regional power and heat supply. On-site installation eliminates the

6% loss normally incurred during the transmission and distribution of electricity. Based on an electrochemical reaction, the power generation efficiency is as high as 40 to 60%; including the utilization of waste heat, the total energy efficiency can reach 80%. "If a 1 million kW natural gas-fired power plant were to be switched to a phosphoric acid fuel cell, which can supply thermal energy, the natural gas consumption and CO_2 emission would be reduced by about 320,000 t/year and about 800,000 t/year, respectively" [Sasaki 1989, p.190]. NO_X emission will also be eliminated.

High-efficiency gas turbines: This also relates to the energy conservation technology in the industrial sector discussed above. A nine-year R&D project was started in 1988 aimed at improving heat efficiency in gas turbines. The average heat efficiency of Japanese power plants has reached around 40% in recent years. In order to improve this further, the turbine inlet temperature will rise from the current 1300 degrees C to 1500 degrees in the coming decade. The Moonlight Project included construction of a pilot plant of 100000 kW capacity and heat-resistant alloys for turbine blades. High-efficiency gas turbine cogeneration, which uses the high-temperature exhaust gas of the gas turbine to turn a steam turbine for power generation, is expected to attain a heat efficiency of 55%. Heat-resistant ceramics are being studied for gas turbine high-temperature parts. The Moonlight Project aims at developing a 300-kW class ceramic gas turbine for cogeneration by 1996.

8.6 Further remarks

Last but not least, we have to examine the interaction between technology, R&D policy and the economy. We have seen cases where R&D efforts and the diffusion process through investment are affected by economic variables such as prices. Higher oil prices inducing energy-conserving investment form one example. There are also cases where technological breakthroughs have been prompted by the introduction of new regulations. The widespread use of catalyst converters and less polluting engines are cases in question.

To reduce the quantity of greenhouse gases, the selection of appropriate policy measures has to be sought. Interactions between technology, market mechanisms, and government policy need to be studied in more detail. Especially important seem to be the time lags involved in recognizing the problem, responding via the research effort, and actually implementing the new technology.

An OECD report [OECD 1992] noted a shift away from a narrowly technical focus of environmental research. "So long as environmental questions were considered essentially as specific problems requiring direct solutions, the social aspects of environmental problems generally did not enter the mainstream of environmental thinking." The report identified fields of focus for social science research in the environmental area including technology assessment, social impact assessment, environmental decisions, and public attitudes, among others. In view of the fact that new policy tools are being sought that utilize the market mechanism, such as a carbon tax, in addition to command-and-rule type of regulations, more emphasis will have to placed on the integration of the social sciences into environmental research.

What prompts R&D? There are two factors at work. One is the regulation imposed by the central and local governments in order to make the polluters internalize the social cost. This may be the case for environmental regulation or government research programs in the fields where the market does not function due to bulkiness of the endeavor or due to discounting of the future. The other are the market forces such as the damage cost of environmental disruption and higher prices of energy. It is often the case that new regulations are introduced in response to public opinion calling for government intervention in the automatic solution through the market mechanism. The underlying factor here is the market signal, actual or foreseen. Seen in this way, one should not focus on one factor to the neglect of the other. Two views are often different sides of the same coin.

The OECD [1985a, p.34] discusses whether regulation is a source of innovation. "Regulation, and in particular that concerning the environment, may stimulate innovation through direct or indirect, expected or unexpected mechanism." "All regulation imposes a constraint to which industry tries to react in the most efficient fashion, i.e. at the least cost. In many cases, the most economic solution is an innovative one. In other words, 'the expected' is that industry finds an innovative solution to the regulatory constraints. Regulation is thus a stimulus to this particular form of innovation. ... Moreover, this effort to innovate may stimulate a whole series of peripheral innovations which would not have otherwise appeared - this is 'the unexpected'."

The other point of view which stresses market signals in innovative activities can be found in the discussion on the substitutability of energy and capital. In response to a sharp rise in crude oil prices in the 1970s, a number of articles appeared aiming at revealing whether such substitutability existed in reality, and if so, to what extent. Whereas some observers conclude based on their econometric analysis that capital and energy are complementary in the short-run and substitutable in the long-run, others claim that there is no such evidence. Our observation in this chapter on energy substitution investment relates to this point. R&D expenditures for environmental protection and new sources of energy probably can be dealt with along this line, using environmental damage cost and energy saving as a proxy of benefit to be derived from such activities, together with the user cost of capital.

From an analytical point of view, linking technology to economy is not an easy task. One possibility is to focus on individual technology and its diffusion process. In this case, the description tends to be piecemeal and case-oriented. This specialized research is of course very valuable. However, if our purpose is to obtain an overall picture and still not lose sight of industry-specific technology, the use of capital stock vintage model seems to be warranted in view of the fact that the R&D effort is built into capital stock via the flow of investment embodying new technology. This process may be approximated by the average age of capital stock, although this approach lacks specifics concerning technology. A more direct way of incorporating the diffusion process in the analytical framework would be to derive the cumulative sum of such investment flow. We have constructed such a model at the disaggregated level (25 industrial sectors), introduced in [Uno 1987]. The kind of data which we have developed here are well fitted for expanding the model along this line.

FOOTNOTES

[1] Meadows, et al. [1972] is a forerunner in explicitly pointing out the "limits to growth". According to their basic scenario, it is assumed that physical, economic, and social relationship which prevailed so far will prevail into the future. They concluded that economic growth would have to come to an end by resource exhaustion and environmental disruption. Their

analysis is based on a system dynamics, parameters of which are provided by the model builders rather than by econometric estimation, a point which invited criticism for lacking empirical basis. Another point of focus was the possibility of technological change. Their response to this point was that technological change and new discovery of resources only delay the eventuality and will not provide a true alternative. It should also be noted that economists tend to pay more attention on the price mechanism and substitution among various resources whereas engineers tend to rely on fixed coefficients. A fact remains that the global community is still faced with the problem of sustainability after twenty years of *Limits to Growth*.

[2] Defense-related R&D correspond to the portion of government science and technology budget which is allocated to the Defense Agency, as listed in *Science and Technology White Paper*.

[3] MITI, *Plant and Equipment Investment Plans in Major Industries*, 1990 edition, p.373.

[4] For the formation of environmental policy in Japan, see Uno [1987], pp.271-310, and annual *Environment White Paper*, among others. English version of the White Paper is entitled *Quality of the Environment in Japan*.

[5] Formal title is "Law concerning the Protection of the Ozone Layer through Control of Specified Substances and Other Measures."

[6] The Montreal Protocol was further strengthened in 1990 and 1992. According to the revision in 1992, production of specified *CFCs* will be reduced to less than 25% of the 1986 level by 1994 and will be abolished after 1996. Production of halon will have to be kept less than the 1986 level by 1992 and will be abolished after 1994. In addition, the coverage of the Protocol has been extended considerably.

[7] *Environment White Paper*, Detailed Discussion Volume, 1993 edition, p.259.

[8] The items included are cadmium, cyanogen, organic acid, lead, chrome (sexivalent), arsenic, alkyl mercury, and PCB.

Part IV

Natural Environment and the Economy

Chapter 9

Land Use

9.1 Availability of land

Land is not reproducible. In other words, despite rapid economic growth and urbanization, the total amount of land available for human habitation remains constant. In Japan, the environmental impact of socioeconomic development has been compounded by the lack of space. The efforts to overcome the limitation posed by land, such as reclamation of the sea and water surface, aggravated the problem. The density of production and human habitatation in Japan is one of the reasons why this country experienced the worst kind of environmental disruption before the other industrialized countries. It is well-known that the environmental pollution issue was most prominent in Japan toward the end of the 1960s and early 1970s. For this reason, Japan was also the first country to implement the most stringent pollution-prevention measures. Energy efficiency was also attained simultaneously. According to an OECD publication, for example, the annual investment by industry in pollution prevention recorded a peak in 1973 and 1974 in absolute terms whereas, in the United States and most EEC countries this occurred in the mid-1980s and thereafter. Japan also promptly developed technology that reduced the pollution attributable to car exhaust emissions.[1] Land is also a form of asset. This fact is reflected in the System of National Accounts in the sense that the current value of land is listed among other forms of financial assets in addition to tangible ones such as production facilities and housing. In Japan, land has

Table 9-1 Land use in selected countries

(unit: 1000 square km, %)

	Arable and permanent crop land	Grassland	Forest and wooded land	Other areas	Total
Japan	46(12.3)	6(1.7)	253(67.1)	71(18.8)	377
United States	1899(20.7)	2415(26.3)	2939(32.1)	1914(20.9)	9167
EEC	839(35.9)	566(24.2)	579(24.8)	351(15.0)	2335
France	191(34.8)	116(21.1)	152(27.6)	91(16.5)	550
Germany	124(35.5)	57(16.2)	104(29.7)	65(18.6)	350
UK	70(29.0)	116(48.0)	24(9.9)	32(13.1)	241
OECD	4175(13.5)	7786(25.1)	10300(33.2)	8740(28.2)	31000
World	11434(10.9)	26192(25.0)	37235(35.6)	29870(28.5)	104731

Note: Land area figures have been rounded from the original source.

Source: OECD, *Environmental Data*, 1991.

been traded in the market just as any other form of assets to a greater degree compared with other OECD countries, especially European countries. The rapid expansion of the economy and, to a greater degree, rapid concentration of population meant a sharp rise in land prices, inducing household and industry alike to purchase land as the most secure form of asset. The problem of land being treated as a form of asset is an interesting question in itself, but we focus our attention on the physical aspect of land in this chapter. The peculiarity of Japan in the availability of land is illustrated in Table 9-1 which compares the total land area as well as its breakdown into various uses with OECD countries. The total land area of Japan is about 4% of the USA and about 16% of the EEC. In the main four islands constituting the Japanese archipelago, the size of the GNP is as large as 60% of the USA. It is also noted from Table 9-1 that the area of forests stands at around 67% of the total, exceeding the OECD average of 33%. This is contrary to the popular belief that Japan is a country of congested urban and industrial areas. Table 9-2 reveals the reasons why. The table shows the physical configuration of land in Japan. As can be seen in the part (a), which classifies land area by gradient, most of the land consists of steep mountains. Areas with a gradient of 15 degrees or more amount to 48% of the total area; and areas with 8 to 15 degrees constitute an additional 24%. Virtually all of the economic activities such as farming, industrial production,

Table 9-2 Gradient and configuration of land area

(unit: square km, %)

a.	Area by gradient	
	0 ° - 3 °	48458 (14.0)
	3 ° - 8 °	50309 (14.6)
	8 ° - 15 °	81341 (23.5)
	15 ° - 20 °	56386 (16.3)
	20 ° - 30 °	80526 (23.3)
	30 ° and over	28519 (8.3)
	Total	345539 (100.0)
b.	Area by configuration	
	Mountain	230331 (61.0)
	Hill land	44337 (11.8)
	Upland	41471 (11.0)
	Lowland	51963 (13.8)
	Inland water, etc.	9232 (2.4)
	Total	377334 (100.0)

Note: Area by gradient excludes area of lakes, marshes, and rivers. Data are based on the 1982 Research and Management on Land Numerical Information.

Source: National Land Agency, *Summary Statistics on Land.*

and human habitation had to be conducted in the remaining 30% or so.

9.2 Environmental impact of changing land use

A typical shift of land use occurs when natural forest, grassland, and farm land are shifted to various urban-type uses. Table 9-3 captures such dynamics of changing land use over the last 20 years. Among the sources of supply of land which are given in Table 9-3a, farm land, forest, and reclamation are distinguished; among new uses of land, urban-type uses such as residential, industrial, public facilities, and leisure facilities are noted and for rural-type use, agricultural and forest are listed. Thus, over time, the table enables us to trace the shift of land use (see Table 9-3b) and establish linkages with capital formation for various purposes. It can be seen that farm land constituted the bulk of land supply converted to urban-type use, while forest was turned into agricultural land. There are also cases when forest directly turned into urban-type use. Typically, of the total land turned into urban-type use, farm land shares about 2/3 and forest 1/3. The reclamation of sea and inland water also

Table 9-3 Transformation of land use, annual tables

(unit: square km, %)

1970-1972 average

Changed from:	Farm land	Forest	Reclaimed	Total
Changed to:				
A. Urban-type use:				
1. Residential	192.0	100.0	3.0	295.0
2. Industrial	75.0	7.0	18.0	100.0
3. Public facilities	117.0	35.0	4.0	156.0
4. Leisure facilities	—	76.0	—	76.0
5. Other	62.0	—	10.0	72.0
Subtotal	446.0	218.0	35.0	699.0
B. Rural-type use:				
1. Agricultural	—	313.0	23.0	336.0
2. Forest	199.0	—	—	199.0
Subtotal	199.0	313.0	23.0	535.0
C. Other use:	38.0	2.0	—	40.0
Total	683.0	533.0	58.0	1274.0

1975

Changed from:	Farm land	Forest	Reclaimed	Total
Changed to:				
A. Urban-type use:				
1. Residential	113.0	32.0	2.0	147.0
2. Industrial	38.0	5.0	22.0	65.0
3. Public facilities	81.0	61.0	3.0	145.0
4. Leisure facilities	12.0	63.0	—	75.0
5. Other	30.0	—	10.0	40.0
Subtotal	274.0	161.0	37.0	472.0
B. Rural-type use:				
1. Agricultural	—	287.0	39.0	326.0
2. Forest	88.0	—	—	88.0
Subtotal	88.0	287.0	39.0	414.0
C. Other use:	27.0	25.0	—	52.0
Total	389.0	473.0	76.0	938.0

1976

Changed from:	Farm land	Forest	Reclaimed	Total
Changed to:				
A. Urban-type use:				
1. Residential	110.0	21.0	2.0	133.0
2. Industrial	33.0	5.0	21.0	59.0
3. Public facilities	69.0	60.0	3.0	132.0
4. Leisure facilities	9.0	42.0	—	51.0
5. Other	34.0	—	8.0	42.0
Subtotal	255.0	128.0	34.0	417.0
B. Rural-type use:				
1. Agricultural	—	296.0	10.0	306.0
2. Forest	69.0	—	—	69.0
Subtotal	69.0	296.0	10.0	375.0
C. Other use:	28.0	35.0	—	63.0
Total	352.0	459.0	44.0	855.0

1977

Changed from:	Farm land	Forest	Reclaimed	Total
Changed to:				
A. Urban-type use:				
1. Residential	101.0	25.0	1.0	127.0
2. Industrial	32.0	8.0	9.0	49.0
3. Public facilities	75.0	20.0	3.0	98.0
4. Leisure facilities	5.0	19.0	—	24.0
5. Other	31.0	—	9.0	40.0
Subtotal	244.0	72.0	22.0	338.0
B. Rural-type use:				
1. Agricultural	—	186.0	30.0	189.0
2. Forest	53.0	—	—	53.0
Subtotal	53.0	186.0	30.0	242.0
C. Other use:	27.0	42.0	—	69.0
Total	324.0	300.0	25.0	649.0

1978

Changed from:	Farm land	Forest	Reclaimed	Total
Changed to:				
A. Urban-type use:				
1. Residential	97.0	19.0	4.0	120.0
2. Industrial	33.0	9.0	10.0	52.0
3. Public facilities	89.0	38.0	4.0	131.0
4. Leisure facilities	3.0	9.0	—	12.0
5. Other	31.0	—	9.0	40.0
Subtotal	253.0	75.0	27.0	355.0
B. Rural-type use:				
1. Agricultural	—	131.0	1.0	132.0
2. Forest	53.0	—	—	53.0
Subtotal	53.0	131.0	1.0	185.0
C. Other use:	31.0	28.0	—	59.0
Total	337.0	234.0	28.0	599.0

1979

Changed from:	Farm land	Forestland	Reclaimed	Total
Changed to:				
A. Urban-type use:				
1. Residential	102.0	20.0	1.0	123.0
2. Industrial	36.0	8.0	12.0	56.0
3. Public facilities	88.0	25.0	5.0	118.0
4. Leisure facilities	2.0	2.0	—	4.0
5. Other	35.0	—	7.0	43.0
Subtotal	264.0	55.0	26.0	344.0
B. Rural-type use:				
1. Agricultural	—	113.0	2.0	115.0
2. Forest	47.0	—	—	47.0
Subtotal	47.0	113.0	2.0	162.0
C. Other use:	30.0	32.0	—	62.0
Total	341.0	200.0	28.0	568.0

1980

Changed from:	Farm land	Forest	Reclaimed	Total
Changed to:				
A. Urban-type use:				
1. Residential	88.0	17.0	0.0	106.0
2. Industrial	34.0	19.0	5.0	59.0
3. Public facilities	87.0	26.0	3.0	117.0
4. Leisure facilities	2.0	4.0	—	6.0
5. Other	34.0	—	8.0	42.0
Subtotal	246.0	66.0	17.0	329.0
B. Rural-type use:				
1. Agricultural	—	96.0	6.0	102.0
2. Forest	45.0	—	—	45.0
Subtotal	45.0	96.0	6.0	146.0
C. Other use:	30.0	32.0	—	62.0
Total	320.0	194.0	22.0	536.0

1981

Changed from:	Farm land	Forest	Reclaimed	Total
Changed to:				
A. Urban-type use:				
1. Residential	81.0	15.0	1.0	97.0
2. Industrial	39.0	10.0	8.0	57.0
3. Public facilities	85.0	33.0	4.0	122.0
4. Leisure facilities	3.0	7.0	—	10.0
5. Other	31.0	—	7.0	38.0
Subtotal	240.0	65.0	20.0	324.0
B. Rural-type use:				
1. Agricultural	—	97.0	2.0	99.0
2. Forest	41.0	—	—	41.0
Subtotal	41.0	97.0	2.0	140.0
C. Other use:	26.0	34.0	—	60.0
Total	307.0	196.0	22.0	525.0

1982

Changed from:	Farm land	Forest	Reclaimed	Total
Changed to:				
A. Urban-type use:				
1. Residential	103.0	21.0	1.0	125.0
2. Industrial	25.0	9.0	3.0	38.0
3. Public facilities	78.0	42.0	3.0	123.0
4. Leisure facilities	4.0	7.0	—	11.0
5. Other	32.0	—	5.0	37.0
Subtotal	242.0	79.0	12.0	334.0
B. Rural-type use:				
1. Agricultural	—	121.0	5.0	126.0
2. Forest	37.0	—	—	37.0
Subtotal	37.0	121.0	5.0	163.0
C. Other use:	25.0	37.0	—	62.0
Total	304.0	237.0	18.0	559.0

1983

Changed from:	Farm land	Forest	Reclaimed	Total
Changed to:				
A. Urban-type use:				
1. Residential	71.0	14.0	1.0	86.0
2. Industrial	37.0	7.0	6.0	50.0
3. Public facilities	70.0	40.0	4.0	114.0
4. Leisure facilities	5.0	13.0	—	18.0
5. Other	31.0	—	5.0	36.0
Subtotal	214.0	74.0	16.0	304.0
B. Rural-type use:				
1. Agricultural	—	85.0	1.0	86.0
2. Forest	32.0	—	—	32.0
Subtotal	32.0	85.0	1.0	118.0
C. Other use:	10.0	30.0	—	40.0
Total	256.0	189.0	17.0	462.0

1984

Changed from:	Farm land	Forest	Reclaimed	Total
Changed to:				
A. Urban-type use:				
1. Residential	75.0	16.0	1.0	92.0
2. Industrial	40.0	18.0	7.0	65.0
3. Public facilities	65.0	37.0	4.0	106.0
4. Leisure facilities	5.0	22.0	—	27.0
5. Other	34.0	—	6.0	40.0
Subtotal	219.0	93.0	17.0	329.0
B. Rural-type use:				
1. Agricultural	—	122.0	1.0	123.0
2. Forest	29.0	—	—	29.0
Subtotal	29.0	122.0	1.0	152.0
C. Other use:	30.0	30.0	—	60.0
Total	278.0	245.0	18.0	541.0

1985

Changed from:	Farm land	Forest	Reclaimed	Total
Changed to:				
A. Urban-type use:				
1. Residential	73.0	21.0	1.0	95.0
2. Industrial	40.0	21.0	5.0	66.0
3. Public facilities	64.0	43.0	5.0	112.0
4. Leisure facilities	5.0	28.0	—	33.0
5. Other	33.0	—	5.0	38.0
Subtotal	215.0	113.0	16.0	344.0
B. Rural-type use:				
1. Agricultural	—	82.0	2.0	84.0
2. Forest	29.0	—	—	29.0
Subtotal	29.0	82.0	2.0	113.0
C. Other use:	28.0	31.0	—	59.0
Total	272.0	226.0	18.0	516.0

1986

Changed from:	Farm land	Forest	Reclaimed	Total
Changed to:				
A. Urban-type use:				
1. Residential	70.0	15.0	1.0	86.0
2. Industrial	42.0	20.0	2.0	64.0
3. Public facilities	54.0	32.0	4.0	90.0
4. Leisure facilities	7.0	47.0	—	54.0
5. Other	33.0	—	6.0	39.0
Subtotal	206.0	114.0	13.0	333.0
B. Rural-type use:				
1. Agricultural	—	61.0	2.0	63.0
2. Forest	26.0	—	—	26.0
Subtotal	26.0	61.0	2.0	89.0
C. Other use:	25.0	30.0	—	55.0
Total	257.0	205.0	15.0	477.0

1987

Changed from:	Farm land	Forest	Reclaimed	Total
Changed to:				
A. Urban-type use:				
1. Residential	78.0	14.0	1.0	93.0
2. Industrial	45.0	16.0	3.0	64.0
3. Public facilities	56.0	30.0	4.0	90.0
4. Leisure facilities	8.0	59.0	—	67.0
5. Other	37.0	—	5.0	42.0
Subtotal	224.0	119.0	13.0	356.0
B. Rural-type use:				
1. Agricultural	—	54.0	2.0	56.0
2. Forest	24.0	—	—	24.0
Subtotal	24.0	54.0	2.0	80.0
C. Other use:	27.0	28.0	—	55.0
Total	275.0	201.0	15.0	491.0

1988

Changed from:	Farm land	Forest	Reclaimed	Total
Changed to:				
A. Urban-type use:				
1. Residential	77.0	7.0	—	84.0
2. Industrial	47.0	15.0	4.0	66.0
3. Public facilities	58.0	40.0	4.0	102.0
4. Leisure facilities	12.0	83.0	—	95.0
5. Other	41.0	—	6.0	47.0
Subtotal	235.0	145.0	14.0	394.0
B. Rural-type use:				
1. Agricultural	—	51.0	1.0	52.0
2. Forest	24.0	—	—	24.0
Subtotal	24.0	51.0	1.0	76.0
C. Other use:	32.0	28.0	—	60.0
Total	291.0	224.0	15.0	530.0

1989

Changed from:	Farm land	Forest	Reclaimed	Total
Changed to:				
A. Urban-type use:				
1. Residential	83.0	14.0	—	97.0
2. Industrial	56.0	17.0	3.0	76.0
3. Public facilities	64.0	36.0	5.0	105.0
4. Leisure facilities	16.0	81.0	—	97.0
5. Other	44.0	—	10.0	54.0
Subtotal	263.0	148.0	18.0	429.0
B. Rural-type use:				
1. Agricultural	—	57.0	5.0	62.0
2. Forest	34.0	—	—	34.0
Subtotal	34.0	57.0	5.0	96.0
C. Other use:	36.0	33.0	—	69.0
Total	333.0	238.0	23.0	594.0

1990				
Changed from:	Farm land	Forest	Reclaimed	Total
Changed to:				
A. Urban-type use:				
1. Residential	85.0	11.0	—	96.0
2. Industrial	62.0	20.0	5.0	87.0
3. Public facilities	64.0	37.0	3.0	104.0
4. Leisure facilities	16.0	65.0	—	81.0
5. Other	48.0	—	8.0	56.0
Subtotal	275.0	133.0	16.0	424.0
B. Rural-type use:				
1. Agricultural	—	50.0	—	50.0
2. Forest	29.0	—	—	29.0
Subtotal	29.0	50.0	—	79.0
C. Other use:	38.0	32.0	—	70.0
Total	342.0	215.0	16.0	573.0

Note: The data refer to farm land, forests, and reclaimed land and do not include transformation from grassland. Transformation of farm land to public facilities partly includes agricultural use such as farm roads, agricultural waterways, etc. Transformation of forests does not include those with less than one hectare. 1970-72 figure for "other" urban-type includes transformation of farm land to leisure facility use.

Source: Estimates by the Land Agency reported in *Land White Paper*.

played an important role, although the area affected is relatively small compared with the other sources of supply. Reclaimed land was typically turned to industrial use, and this is the usual case after the mid-1970s, whereas in earlier years reclaimed land was used for agricultural purposes. This pattern should be interpreted in the context of the concentration of economic activities along the coast in order to facilitate transportation by sea. This was also practically the only choice for an industrial location given the fact that nearly 80% of the territory is mostly mountains and hills. The area reclaimed amounted to 116.1 square km between 1955 and 1965.[2] The figures for the second half of the 1960s typically stood around 35 to 40 square km. The peak was reached in 1975 when 76 square km were added (51 square km according to the Geographical Survey Institute). Land reclamation has been tapering off in recent years, and the annual addition of land is down to around 15 square meters in recent years. As a result, the total land area of the country has increased from 377,118 square km in 1965 to 377,728 square km in 1990. However, the fact remains that reclamation runs directly counter to conservation of the natural environment.

9.3 Farm land

Farm land constituted about 14% of the total land area in 1990. Of the total farm land of 53,430 square km in 1990, paddy fields amounted to 28,460 square km, area under trees, vines, etc. about 4,750 square km, and the remaining was ordinary farm land. The farm land reached a peak of 60,860 square km in 1961 and has exhibited a continued decrease since then. Before this period, agricultural land was used extensively for multiple crops. For instance, in 1960, the rate of utilization of farm land (= total area planted/farm land) stood at 133.9, meaning that the same farm land was used 1.3 times during the year. Thus, the total area planted reached 81,300 square km, whereas the farm land area was around 60,700 square km in that year. The ratio dropped to 123.8 in 1965, 108.9 in 1970, and 100.3 in 1978, and the 1990 ratio was about 101.1.[Land Agency, 1993, p.6] The farm land area itself had expanded through clearing forests and, to a certain extent, by irrigation, as we have seen above. Throughout the last three decades, farm land was shifted to urban-type use such as industrial and residential. Farm land which was dilapidated artificially increased in the process of industrialization. Starting from 68 square km in 1955, it gradually increased to 238 square km in 1960, 689 square km in 1965, and 1,009 square km in 1970. It peaked off in 1971 at 1,110 squre km. More recent trends are recorded in Table 9-3.[3]

Agriculture is not immune from environmental concerns. Nitrogen, phosphate, and plant nutrients in the form of synthetic chemical fertilizers are applied in the fields in increasing quantities. In fact, the rapid gains in agricultural productivity after WWII can be attributed to the availability of low-cost chemical fertilizers. Their application, however, has hazardous consequences for the water quality. Another environmental concern attributable to farming is the consumption of pesticides (Table 9-6). The application of certain pesticides is found to be responsible for affecting the quality of the soil and water, causing undesirable effects on the flora and fauna. Residues of pesticides remain on crops after harvesting and potentially affect human health adversely. Similar concerns have been raised concerning post-harvest preservatives which

Table 9-3b Transformation of land use, total, time series

Year	1966	1968	1970-72
Changed to:			average
A. Urban-type use:			
1. Residential	143.3	179.8	295.0
2. Industrial	56.5	65.0	100.0
3. Public facilities	136.3	130.0	156.0
4. Leisure facilities	22.9	21.9	76.0
5. Other	n.a.	n.a.	72.0
Subtotal	359.0	397.2	699.0
B. Rural-type use:			
1. Agricultural			336.0
2. Forest			199.0
Subtotal			535.0
C. Other use:			40.0
Total			1274.0

	1982	1983
Changed to:		
A. Urban-type use:		
1. Residential	125.0	86.0
2. Industrial	38.0	50.0
3. Public facilities	123.0	114.0
4. Leisure facilities	11.0	18.0
5. Other	37.0	36.0
Subtotal	334.0	304.0
B. Rural-type use:		
1. Agricultural	126.0	86.0
2. Forest	37.0	32.0
Subtotal	163.0	118.0
C. Other use:	62.0	40.0
Total	559.0	462.0

Sources: Same as Table 9-6.

(unit : square km)

	1975	1976	1977	1978	1979	1980	1981
A.							
1.	147.0	133.0	127.0	120.0	123.0	106.0	97.0
2.	65.0	59.0	49.0	52.0	56.0	59.0	57.0
3.	145.0	132.0	98.0	131.0	118.0	117.0	122.0
4.	75.0	51.0	24.0	12.0	4.0	6.0	10.0
5.	40.0	42.0	40.0	40.0	43.0	42.0	38.0
	472.0	417.0	338.0	355.0	344.0	329.0	324.0
B.							
1.	326.0	306.0	189.0	127.0	115.0	102.0	99.0
2.	88.0	69.0	53.0	53.0	47.0	45.0	41.0
	414.0	375.0	242.0	185.0	162.0	146.0	140.0
C.	52.0	63.0	69.0	59.0	62.0	62.0	60.0
	938.0	855.0	649.0	599.0	568.0	536.0	525.0

	1984	1985	1986	1987	1988	1989	1990
A.							
1.	92.0	95.0	86.0	93.0	84.0	97.0	96.0
2.	65.0	66.0	64.0	64.0	66.0	76.0	87.0
3.	106.0	112.0	90.0	90.0	102.2	105.0	104.0
4.	27.0	33.0	54.0	67.0	95.0	97.0	81.0
5.	40.0	38.0	39.0	42.0	47.0	54.0	56.0
	329.0	344.0	333.0	356.0	394.0	429.0	424.0
B.							
1.	123.0	84.0	63.0	56.0	52.0	62.0	50.0
2.	29.0	29.0	26.0	24.0	24.0	34.0	29.0
	152.0	113.0	89.0	80.0	76.0	96.0	79.0
C.	60.0	59.0	55.0	55.0	60.0	69.0	70.0
	541.0	516.0	477.0	491.0	530.0	594.0	573.0

Table 9-4 Overview of land use

(unit: 1000 square km)

Year	Total	Farm land: Cultivated	Meadows & pastures	Forests	Grassland
1963	376.9	60.9	4.6	250.8	6.6
1964	376.9	60.9	4.4	251.2	6.5
1965	377.1	60.2	4.1	251.6	6.4
1966	377.2	60.1	3.8	251.6	6.3
1967	377.2	59.6	3.6	251.2	6.1
1968	377.2	59.1	3.4	252.2	6.0
1969	377.3	58.7	3.3	252.4	5.8
1970	377.3	58.1	3.1	252.3	5.7
1971	377.3	57.6	2.9	252.2	5.6
1972	377.4	57.3	2.6	252.3	5.6
1973	377.4	56.6	2.5	252.3	5.5
1974	—	—	—	—	—
1975	377.5	55.7	1.8	251.8	4.1
1976	377.6	55.4	1.8	252.8	3.8
1977	377.6	55.2	1.7	252.7	3.7
1978	377.6	54.9	1.7	252.8	3.7
1979	377.7	54.7	1.7	252.8	3.6
1980	377.7	54.6	1.5	252.6	3.3
1981	377.7	54.4	1.5	252.4	3.3
1982	377.7	54.3	1.4	252.6	3.1
1983	377.8	54.1	1.4	252.6	3.0
1984	377.8	54.0	1.1	253.0	3.1
1985	377.8	53.8	1.1	252.9	3.0
1986	377.8	53.6	1.1	252.9	2.9
1987	377.8	53.4	1.0	252.8	2.9
1988	377.7	53.2	1.0	252.8	2.7
1989	377.7	52.8	1.0	252.6	2.8
1990					

Source : National Land Agency, Survey on Land for Public Facilities, reported in *Summary Statistics on Land.*

are used in some countries exporting farm products.

9.4 Forests

Forests comprise 2/3 of the land area in Japan, as indicated in Table 9-4 together with land use for other purposes, and the area has not decreased over the last 30 years. Details of the forest area and forest resources are shown in

	Water area, rivers, and channels	Roads	Land for building: Residential	Industrial	Offices, stores etc.	Other
63	11.1	7.9	6.4	0.8	0.6	27.2
64	11.1	8.1	6.7	0.8	0.6	26.9
65	11.1	8.2	6.9	0.9	0.7	27.0
66	11.2	8.3	7.1	0.9	0.8	27.1
67	11.2	8.3	7.3	1.0	0.8	27.2
68	11.3	8.5	7.5	1.0	0.9	27.3
69	11.3	8.6	7.8	1.1	0.9	27.4
70	11.1	8.8	8.1	1.2	0.9	28.0
71	11.1	8.9	8.4	1.3	0.9	28.4
72	11.2	9.1	8.8	1.3	1.0	28.2
73	11.2	9.3	9.1	1.4	1.1	28.4
74	-	-	-	-	-	-
75	11.3	9.7	9.4	1.5	1.3	30.9
76	11.3	9.7	10.0	1.5	1.4	29.9
77	11.3	9.9	10.3	1.5	1.4	29.8
78	11.4	10.2	10.4	1.4	1.6	29.5
79	11.4	10.3	10.7	1.5	1.6	29.4
80	11.5	10.4	10.8	1.5	1.7	29.8
81	11.6	10.5	10.9	1.5	1.8	29.8
82	11.6	10.7	11.0	1.5	2.0	29.5
83	11.6	10.8	11.1	1.5	2.0	29.7
84	13.1	10.5	9.2	1.5	4.2	28.1
85	13.2	10.7	9.4	1.5	4.2	28.0
86	13.1	10.9	9.5	1.5	4.3	28.0
87	13.2	11.0	9.5	1.6	4.4	28.0
88	13.2	11.2	9.6	1.6	4.5	27.9
89	13.2	11.3	9.7	1.6	4.6	28.1
90						

Table 9-7. Of the total area of 245,880 square km, approximately 40% is man-made, and the share of the artificial forest increased consistently until about the mid-1980s. The stock of forest resources has risen steadily, from about 1900 million cubic meters in the early 1960s to more than 3100 million cubic meters in 1990. The stock in artificial forests has shown a faster growth, from about 520 million cubic meters in 1960 (1/3 of the total) to 1600 million cubic meters in 1990 (1/2 of the total). The significance of the forests is proba-

Table 9-5 Farm land

(unit: square km)

	Total	Fields proper	Attached dykes	Paddy fields	Under trees, vines, etc.
1955	51400	—	—	28470	—
1956	60120	57030	3090	33200	—
1957	60440	57300	3145	33350	—
1958	60640	57490	3152	33450	—
1959	60730	57570	3158	33640	—
1960	60710	57550	3157	33810	—
1961	60860	57690	3163	33880	4509
1962	60810	57630	3171	33930	4641
1963	60600	57440	3159	33990	4806
1964	60420	57260	3157	33920	5017
1965	60040	56900	3144	33910	5258
1966	59960	56810	3142	33960	5425
1967	59380	56260	3123	34150	5606
1968	58970	55860	3105	34350	5773
1969	58520	55430	3087	34410	2902
1970	57960	54910	3046	34150	6002
1971	57410	54420	2980	33640	6160
1972	56830	53910	2922	33120	6267
1973	56470	53590	2880	32740	6324
1974	56150	53310	2841	32090	6368
1975	55720	52930	2793	31710	6280
1976	55360	52620	2743	31440	6152
1977	55150	52440	2711	31330	6040
1978	54940	52250	2684	31080	5958
1979	54740	52090	2653	30810	5919
1980	54610	51990	2621	30550	5870
1981	54420	51830	2593	30310	5813
1982	54260	51690	2571	30100	5738
1983	54110	51560	2546	29890	5679
1984	53960	51450	2518	29710	5598
1985	53790	51300	2490	29520	5490
1986	53580	51130	2460	29310	5380
1987	53400	50970	2460	29100	5260
1988	53170	50770	2400	28890	5110
1989	52790	50420	2370	28680	4870
1990	52430	50100	2340	28460	4750

Note: "Fields proper" refers to land for actual cultivation and "attached dykes" refer to ridges circling cultivated land. The data exclude Okinawa through 1973.

Source : Ministry of Agriculture, Forestry, and Fisheries. 1955 figures are from *Provisional Basic Survey on Farming* (1/5 samples), and *Crop Survey* thereafter.

Table 9-6 Application of pesticides in selected countries

	(index, 1975=100)					(quantity, tons)	
	1975	1980	1985	1988		1975	Most recent
Japan	100	96	88	n.a.		94070	83096('85)
United States	100	100	n.a.	n.a.		292000	373333('84)
France	100	187	228	225		41200	92500('88)
Germany	100	132	120	120	('87)	24981	29857('87)
Netherlands	100	n.a.	120	104		17439	19146('89)
Sweden	100	100	78	84	('86)	4554	5715('86)
United Kingdom	100	133	n.a.	n.a.		30300	40300('82)

Note: Quantities are in terms of active ingredients, and do not include diluents and adjuvants. UK data refer to Great Britain only.

Source : OECD [1991b], pp.260-263.

bly shifting from logging to preservation of the natural environment due to the shifting preference of the people. This trend has accelerated with the rise in wages in other sectors of the economy, which has made the forestry industry economically unattractive. The demand for forest products is increasingly met by imports, which is a topic in chapter 10. Thus, in Japan, the conventional method of forest management, which includes 'economy' and 'self-financing accounting', has developed conflicts with the increasing number of people who want to preserve nature. Larger numbers of people see the forests as resources for recreation activities, such as camping, golfing, and skiing. With the expansion of the road networks, especially the ones going across the mountain ranges to the other side of the Japan archipelago in contrast to the ones running parallel to the coast lines, the conflict between development and conservation has been aggravated. This is one of the areas where the direct application of cost-benefit analysis is not warranted. For one thing, the subjects bearing the 'costs' and 'benefit' are widespread in society and not limited to the ones within geographical proximity of the develpment plans. Another difficulty is when 'intrinsic value' comes to play a larger role. This is particularly true when the forest in question represents a preserved natural environment or when it provides a habitat for endangered species. Last but not least, when the global climate is at stake, far more importance must be attached to forest resources.

Table 9-7 Forest area and resources

	Total				Artificial forests	
	Area		Growing stock		Area	Growing stock
	Total	Private	Total	Private		
1955('54)	229609	138946			53090	
1960	244030	141510			61110	
1965	250990	141930	1892	785	76630	547
1970	252840	145590	1906	869	87420	581
1975	252630	147890	2186	1184	93770	798
1980	252790	147330	2484	1449	98950	1054
1985	252550	146840	2862	1752	102190	1361
1989	252120	146510	3138	1967	103270	1598
1990	245880	137940			102530	

Note: As a general rule, figures are as of April 1. For 1954 and 1957, figures are as of August 1. Figures after 1975 are as of the end of the fiscal year. The 1990 figure is from the World Census of Agriculture and Forestry as of August 1.

Source : Forestry Agency, *Current State of Forestry.*

9.5 Industrial and commercial land

Table 9-4 lists the land area being occupied by factories and offices, among various other uses. It can be seen that industrial land has doubled in size in the past three decades. The pace of expansion, however, seems to have slowed down substantially after the 1973-1975 period, when the Japanese economy suffered a setback following the oil crisis and started to undergo a fundamental structural change from rapid industrialization based on basic metals, petrochemicals, and heavy machinery to a new growth path based on electronics and services.[4] The rapid expansion of the land area devoted to offices, stores, etc., as listed in the table, provides proof of our previous point. Although the statistics seem to suffer from data discontinuity in 1983 when part of the residential land was reclassified for commercial use (such as offices, stores, etc.), the table shows that the quantity of industrial land has remained more or less constant around 1600 square km, whereas the land for commercial use has reached 4600 square km and is still increasing. It should be noted that, within the manufacturing sector, the parts dealing with basic materials tend to be overshadowed by those in the machinery sectors (general machinery, electrical machinery, transportation equipment, and precision instruments). The man-

(unit: area in square km, quantity in million Cubic meters)					
	Natural forests		Other		Bamboo forest
	Area	Growing stock	Area	Growing stock	
1955('54)	165976				1639
1960	170152				1638
1965	157860	1339	14670	5.8	1830
1970	147620	1317	16130	8.1	1370
1975	144370	1386	13010	1.8	1480
1980	139940	1428	12460	1.2	1440
1985	136660	1500	12220	1.6	1470
1989	135230	1538	12100	1.6	1510
1990	135180				1450

ufacturing sector related to consumer goods is also active in land acquisition. Comparing the land acquisition data in physical terms and in value terms, it can be seen that the machinery and consumer goods sectors are purchasing more expansive land, implying that they are urban-located. For instance, in physical terms (i.e., land area), they shared 21%and 35%, respectively, while in value terms, their shares stood at 28% and 40% in 1991.[Land Agency, 1993, pp.11-21]

9.6 Residential land

The ratio of public land supply to private has not changed very much over time; in 1970 the share of the public supply stood at 26.8other words, the institutional divisions are not particularly important in discussing the residential land supply, and we continue our discussion without referring to this distinction. The public supply here refers to that of the Housing and Urban Facilities Public Corporation (Jutaku Toshi Seibi Kodan) and local public bodies, including land zoning projects by said organizations, while private supply refers to private land owners and private land developers and includes land supply by land zoning projects by the private sector. First, let us look at the relative

Table 9-8 Supply of residential land

	Residential land supply:		Housing starts:	
	Total (ha)	Private (ha)	number of units (1000)	floor space (1000 m^3)
1955			257.3	14976
1956			308.6	17470
1957			321.0	18788
1958			337.9	19423
1959			380.5	21964
1960			424.1	25046
1961			535.9	31023
1962			586.1	32285
1963			688.7	38523
1964			751.4	43102
1965			842.5	49667
1966	12300	8600	856.5	53856
1967	14700	10700	991.1	66173
1968	16000	11400	1201.6	79178
1969	19100	14100	1346.6	90116
1970	21300	15600	1484.5	101068
1971	22800	17100	1463.7	101544
1972	23400	17900	1807.5	128745
1973	22800	18300	1905.1	146542
1974	19700	14900	1316.1	107238
1975	16300	11700	1356.2	112422
1976	15000	11000	1523.8	125281
1977	13500	9800	1508.2	126818
1978	13500	10000	1549.3	136249
1979	13400	9700	1493.0	136514
1980	12400	9000	1268.6	119102
1981	11800	8400	1151.6	107853
1982	11400	8100	1146.1	107637
1983	11200	8100	1136.7	799442
1984	10800	7900	1187.2	100227
1985	10200	7500	1236.0	103131
1986	10400	7400	1364.6	111004
1987	10300	7300	1674.3	132526
1988	10700	7700	1684.6	134531
1989	10300	7300	1662.6	135029
1990	10800	8000	1707.1	137489

Source: Residential land data are from Land Agency, *Land White Paper*. Housing start data are from the Ministry of Construction, *Annual Report on Building Construction Statistics*.

size of land for housing which is listed in Table 9-4 together with land for other purposes. In recent years, the total residential land reached 9700 square km, which is about 2.6% of the total land area and is somewhat comparable to land used for the road networks and twice that used for commercial purposes. Second, the annual supply of residential land is available from two alternative sources. One is based on the estimates by the Land Agency, which provides concerning the shifts of land use from farm land, forests, and reclamation to residential land. The other refers to the data series compiled by the Ministry of Construction. This source provides the institutional breakdown into the public and private supply. Both sources are essentially identical because the fomer estimates are based on the latter. The latter has an advantage in that they provide a continuous picture starting in the latter half of the 1960s and cover the very turbulent years during the first half of the 1970s. The former is actually a product of these turbulent years in land supply and, especially, land price hikes which prompted the Japanese government to start collecting statistics concerning land in a systematic manner. With a little exaggeration, one can say that, prior to these years, land was outside the boundary of public administration in Japan. The supply of residential land reached a peak in 1972 of 234 square km. The annual supply in recent years hovers around 100 to 110 square km.

Towards the end of our observation period, the supply of land tended to increase again. It is noted that there were financial factors at work, and the aspect of land as a form of asset once again affected land transactions.[Land Agency 1993, pp.144-147] As of the end of March 1984, outstanding bank loans to the real estate sector increased by 15.2% compared with the same period in the previous year. Subsequent years have shown an annual increase of 14.9%, 25.9%, 32.7%, 11.3%, and 15.3%. The comparable figure for 1991 has been restrained at 0.3%. This is evidence of the formation of a 'bubble' and its collapse, which plagued the Japanese economic scene in the early 1990s.

Coming back to the real side of the economy, it is indicated in Table 9-8 that the housing starts also reflect the economic trends. Conversely, one can say that housing investment is an important factor behind the economic cycles. This is especially so now that capital formation in plant and equipment has somewhat lost its impetus in Japan. Housing starts in 1990 exceeded 1700 thousand units before dropping off to around 1400 thousand units more re-

Table 9-9 Land use for land fill

	Number of land fill sites:					Area	Capacity:	
	Hills	Sea	Inland water	Low-land	Total	(square km)	Total	Remaining (million m^3)
1980	1600	36	50	796	2482	52.1	356.1	191.9
1981	1619	40	48	779	2486	53.6	403.2	181.6
1982	1612	36	46	778	2472	53.9	377.6	176.0
1983	1638	34	41	766	2479	55.1	382.7	170.8
1984	1633	37	38	731	2439	56.2	403.1	174.2
1985	1630	38	42	721	2461	60.9	410.1	195.7
1986	1629	36	39	707	2411	55.7	429.9	197.0
1987	1637	31	35	692	2395	53.1	423.9	195.1
1988	1628	35	34	676	2373	49.4	414.3	170.9

Note: The data refer to the fiscal year.

Source: Ministry of Health and Welfare, *Welfare White Paper*.

cently. The same table lists the total floor space of the housing start, which exhibits a similar trend to the housing start in terms of the number of units. The average floor space per unit remained practically unchanged over the period; however, it should be noted that this is actually a very important economic variable as well as being an essential component of the qualitiy of life indicators. Whenever the real income tends to increase relative to construction cost and land price, the average floor space tends to expand, and vice versa.[Uno 1978] The sensitivity to the economic conditions is most marked in the case of the number of units for sale, which tends to exhibit radical ups and downs. In the case of the number of units owned, the trend has been consistently upward. It stood at 119.4 square meters in 1980 and rose to 136.4 square meters in 1990.

9.7 Waste disposal sites

Land is also being used for the disposal of wastes. We have seen in chapter 6 that the proportion of household wastes undergoing incineration has been increasing, thus reflecting an improved provision for proper waste treatment. Recycling is another possibility which is being sought rigorously. However,

Table 9-10 Road network length and area in selected countries

	Road network length		Estimated area	Motorvehicles in use
	All roads (1000 km)	Expressways (km)	(square km)	(1000)
Japan	1127.4	4407	9900	55098
United States	6627.4	83964	87000	188015
France	805.5	6950	8740	27868
Germany	496.7	8721	5800	31588
Netherlands	103.0	2074	1110	5894
Sweden	133.7	999	1740	3888
United Kingdom	354.3	2993	n.a.	23558

Source : Data on area refer to all roads. Road network length data and motor vehicles in use are from *OECD Environmental Data,* 1991, pp.205-209 and refer to 1989; data on area occupied by all roads are from OECD, *Tranport and Environment,* 1988, p.35, and are estimates for early 1980s.

there remains the residue after incineration which has to be disposed of. There are also portions going to land fills directly. Table 9-9 shows the time series data on available sites for land fills. In recent years, nearly 50 square km of land has been devoted to this purpose. In addition, there are plans to prepare reclamation sites for the final disposal of industrial waste in the Tokyo Bay and Osaka Bay areas. It is reported that the sites will have a land area of 5-6 square km in Tokyo and about 3 square km in Osaka.[Ministry of Health and Welfare, 1991, p.296]

9.8 Road network length and area

Of the land used for social facilities, let us examine the case of roads. We have addressed ourselves to the impact of transportation activities on the environment in chapter 5, where the focus was on the stock of motor vehicles in use, the traffic volume of passengers and freight, and the consumption of energy for transportation purposes. In this section, we focus our attention on the stock of the transportation infrastructure, namely the land area devoted to road networks. In Japan, where motorization proceeded at a rapid pace, and the economic activities are concentrated in the Pacific Belt extending south-

Table 9-11 Densely inhabited districts

	Population (1000)	(%)	Area (square km)	(%)
1960	40830	(43.7)	3865.2	(1.03)
1961				
1962				
1963				
1964				
1965	47261	(48.1)	4604.9	(1.23)
1966				
1967				
1968				
1969				
1970	55535	(53.5)	6399.2	(1.71)
1971				
1972				
1973				
1974				
1975	63823	(57.0)	8275.4	(2.19)
1976				
1977				
1978				
1979				
1980	69935	(59.7)	10014.7	(2.65)
1981				
1982				
1983				
1984				
1985	73344	(60.6)	10570.7	(2.80)
1986				
1987				
1988				
1989				
1990	78152	(63.2)	11732.2	(3.11)

Note: See text for the definition of Densely Inhabited Districts (DIDs).

Source : Bureau of Statistics, *Population Census.*

ward from Tokyo to Nagoya and Osaka and beyond, the shortage of roads has been a serious problem. The construction of new roads and improvement of existing ones continued at an accelerated pace after 1970. By this time, the demand for financial resources for directly productive purposes had lessened somewhat, and the government often used road construction as a measure to promote the economy when recession set in. Table 9-10 compares the road network length and the estimated area occupied by all roads in selected countries. It can be seen that, even around 1980, Japan was lagging in terms of a road network, especially when comparisons are made with the number of motor vehicles. Table 9-4 lists the time series of land area used for roads in Japan. During the 1970s, the road area increased by 18% and in the 1980s by nearly 10%. The expressways have been extended from a mere 698 km in 1970 to 2579 km in 1980 and nearly 4500 km in 1990.

9.9 Urbanization

Table 9-11 shows one representative measure of urbanization. A densely inhabited district (DID) is defined in the Japanese Population Census as an area (1) using census enumeration districts as the basic unit of area, (2) composed of a group of contiguous census-enumeration districts with a high population density, in principle 4000 inhabitants or more per square kilometer within the administrative boundary, and (3) constituting an agglomeration of 5000 inhabitants or more as of the date of the census-taking. The DIDs were established for the first time in the 1960 Population Census. Because of the fact that this measure is available only for the census year, the data appear at five-year intervals. Despite this drawback, they have an advantage as no redrawing of the administrative borders among various cities and towns is needed. In the urbanization process, cities often expand in size, by their own expansion and by merging with neighboring towns. It also happens that the extent of urbanization in terms of the concentration of economic activities and population varies within the city limit. DID is a statistical concept intended to overcome such difficulties. The urban population in this definition has nearly doubled during the 30-year period from 1960 to 1990, increasing from about 41 million to 78 million. Because the total population continued to increase at the same time, the ratio of urban population to the total showed an increase from 44% to 63% during the same period. One can observe that the speed of urbanization tended to decline over the period; the 1960s saw a 36% increase in the DID population, whereas the figure for the 1970s and the 1980s stood at 26% and 12%, respectively. In terms of area, DIDs have increased from about 1% of the total land to more than 3%.

Table 9-12 Exposure of population to road traffic noise

(unit: %)

	Sound level in Leq (dBA) outdoor				
	>55	>60	>65	>70	>75
Japan	80.0	58.0	31.0	10.0	1.0
United States	37.0	18.0	7.0	2.0	0.4
France	44.0	25.0	13.0	4.0	0.4
Germany	34.0	17.0	8.0	3.0	—
Netherlands	40.0	18.0	6.0	0.6	—
Sweden	38.0	24.0	11.0	4.0	1.0
United Kingdom	50.0	25.0	11.0	4.0	0.6

Note: The data refer to the early 1980s except for the United Kingdom (England only) where data are from a 1973 survey. Figures are daytime leq (6-22h) except for the United States and Sweden where they are leq over 24 hours. Japanese figures are OECD estimates.

Source : OECD [1988b], p.44.

Thus, a situation has developed in which more than 60% of the total population are crowded into 3% of the land area. This calls for the provision of housing, schools, city parks, roads and railways, water supply and sewerage, and waste disposal, among various others. As this was not always easy, especially when investment had to be diverted to production purposes, a huge gap between human needs and what was actually made available was created. The last two decades have witnessed the narrowing of this gap to a considerable extent, as observed in the previous sections. Because of the rapid urbanization, urban planning tended to lag behind the existing reality. Coupled with the lack of effective zoning, Japanese city dwellers are still left with urban nuisance such as traffic noise. Table 9-12 compares the extent to which the population is exposed to road traffic noise. Figures from selected countries are also listed for comparison. According to an OECD report, the population exposed to high noise levels due to road trafffic (defined here as leq $>$ 65 dBA) numbered 35 million in Japan (total population 123 million) in the early 1980s, compared with 19 million in North America (population of the USA and Canada totals 275 milllion) and 53 million in OECD Europe (total population 430 million). A decibel (dB) is a unit of sound pressure level related to a standard reference level of 0.000002 Newtons per square meter. The decibel scale is logarithmic, so that a very wide range of audible sound can be described in terms of a small range of numerical expressions. A sound level of 0 db at 1000 Hertz is just audible to a person with good hearing, while 60 db is described as corresponding to a "busy street through open windows" and 80 db as a "busy crossroads". A sound level of 120 db would cause pain in the ear. A decibel A-weighted [db(A)] is a unit of measurement of sound in which greater emphasis is given to the medium and high frequencies to which the human ear is most sensitive. It gives a good correlation with the subjective impression of loudness. The Leq (or "equivalent sound level") gives the average sound level over a given period, e.g. a full day.[OECD 1988b, pp.40-41, 60]

FOOTNOTES

[1] See OECD [1991b], p.277, for trends in pollution prevention investment by industry. As for technological development and legislation concerning motor vehicle exhaust emissions, see OECD [1988b], especially pp.89-91, pp.100-109, and p.129.

[2] The figures in this section are estimates by the Land Agency based on data supplied by the Ministry of Agriculture, Forestry, and Fisheries, Ministry of Transportation, and Ministry of Construction. Data on reclamation are also available from the Geographical Survey Institute. The two series differ slightly, presumably reflecting the timing of registration.

[3] The figures on farm land dilapidated artificially are from the *Crop Survey* by the Ministry of Agriculture, which is more comprehensive than the Land Agency data on transformation of land, which is the basis of Table 9-3. Although the former source provides figures which are about 40% larger than the latter, the trends of both series are basically very similar.

[4] More detailed data on the new acquisition and usage of industrial land are available in MITI's *Census of Manufacturers*.

Chapter 10

Resource Endowment and International Linkages

10.1 Introduction

This chapter is a reminder that an environmental analysis cannot be concluded within a national boundary. Resources have to be imported where the domestic supply is not sufficient to meet the demand, and Japan is a typical example. In other words, we have a situation where the final demand such as consumption and investment in Japan triggers a series of interindustry transactions within the economy, which then turns to foreign suppliers for raw materials. International linkages of environmental impact through international trade has to be sought in this context. Another route of international linkage would be through direct foreign investment which sometimes helps to relocate resource-intensive industrial activities abroad. A typical case in Japan is the aluminum industry, which has moved all of its refining processes to areas where energy is less expensive. The third source is transborder flow of pollutants. Acid rain is a typical example. In Japan, acid rain caused by foreign polluters has been detected, although the situation is presumably much worse on the European continent and North America due to the geographical proximity of countries. Fourth is global warming. The emission of GHGs will have global consequences, and the scope is much wider than the one discussed above. Another aspect of this issue involves the tropical rain forests and other vegetation which absorb CO_2 and help prevent the greenhouse effect. Seen in this way, the global community benefits from the existence of tropical rain forests, whereas their distribution is probably inversely related to the current levels of income. This chapter only touches on the first issue, which is the international linkages of Japanese society and economy through trade flows.

10.2 The case of wood products

Theoretically, the most consistent way to introduce linkages through international trade would be to expand the input-output framework regarding trade. In other words, import (and export) vectors included in the usual tables can be further disaggregated by trading partners. In the input-output tables, international transactions are already disaggregated at the individual industry level so that domestic supply, imports, and exports can be distinguished. The further provision of a trade matrix is a desirable way to connect the framework with the international trade matrices. International trade matrices, in turn, show the exporting countries and importing countries, identifying who is selling to whom (or who is buying from whom). The matrices are prepared for individual commodities and products. Such data are regularly prepared by the United Nations and the OECD. On the side of the exporting countries, mining, logging, harvesting, fishing and other activities bring various resources from the natural environment domain into the economic domain. We are concerned here with mineral and wood resources. The former represents the case of depletable resources, and the latter, undepletable ones. Coming back to the case of the Japanese economy, the import matrix for 1990 is provided in chapter 12 (see Table 12-3). The import matrix has a sectoral classification which is identical to the transaction table (Table 12-1) and the final demand table (Table 12-2), with detailed information on the trading partners for each industrial sector. In essence, this matrix corresponds to the import vector, which is included among the final demand matrix but lists only the commodity trade to the exclusion of service trade, which is not recorded in the customs statistics. The sectoral classification is basically the same as the standard 36-sector scheme which I have used in previous works, with some additional information on environment-related activities. Now, let us focus on one of the activities listed, say wood products, which fall under forestry. In Japan, the annual input-output tables are compiled, starting in 1973, by the Ministry of International Trade and Industry, of which the ones starting in 1977 provide an import matrix. Thus, one can use a time series of import matrices with consistent sectoral classification with the transaction tables and final demand tables. Table 10-1 is a list of such time series comparison for wood products. Tables 10-3 and 10-6 correspond to energy resources and metals, respectively. The data prior to 1977 have been supplemented by the final demand tables without detailed information as to trading partners. The tables also provide physical data regarding the trade volumes of corresponding imports (which are recorded in current prices). In an analysis pertinent to the share in total imports or in an aggregate economy on the side of the importers, or the economic importance on the part of the exporters, all have to be gauged in value terms. Thus, when it comes to resource depletion or the impact on the environment, physical data are probably more appropriate.

Table 10-2 in turn describes selected indicators of the global aspects of resources and production of wood products. Panel (a) concerns the production of wood products. Nearly one-half of the annual consumption of forest resources is attributable to firewood, and a large portion of this category is directed to the local market. Industrial use, which represents the other half of the demand, is subdivided into sawwood and plywood, pulp and wood chip, and others (not shown). According to the FAO, 84% of the demand for fuel is attributable to developing countries, whereas 77% of industrial use is attributable to developed industrial countries [FAO, Yearbook of Forest Products]. Panel (b) concerns the forest resources. Needless to say, panel (a) focuses on the annual flow of wood products, whereas panel (b) focuses on the stock of forest resources. The quantity measurement of vegetation is not an easy task because of the variety of types and intensity. In a survey conducted jointly by the FAO and the UNEP, woody vegetation is classified into:

(a) closed forest where a large portion of the earth is covered by trees;

(b) open forest where trees cover at least 10% of the earth surface, with dense grass covering the rest;

(c) forest fallow where wood has been cleared for cultivation but is left unattended thereafter, and is expected to return to forest again in due time;

Table 10-1 Import matrix, wood products

	USA	America, others	EC	Soviet & EE	Europe, others	East Asia
a. Round wood (I-O sector code: 0220)						
1955						
1960						
1965						
1970						
1975						
1976						
1977	302650	14314	3	142455	21	3828
1978	254389	8863	42	108752	0	3987
1979	432397	16446	39	156739	6	6637
1980	424580	27300	621	157025	3	5508
1981	263598	13518	323	98146	1	3443
1982	275610	15548	444	95825	1	3914
1983	222161	24670	349	94534	6	2886
1984	204968	40304	43	81263	2	2583
1985	213924	43126	112	81719	2	2452
1986	164925	31368	88	73322	—	2646
1987	201698	44128	153	56551	—	4729
1988	218913	33478	272	73566	1	9618
1989	288780	29556	625	79301	58	8894
1990	297844	22084	787	69651	—	7858
b. Sawlogs, veneer logs and wooden chips (I-O sector code: 2510)						
1955						
1960						
1965						
1970						
1975						
1976						
1977	108432	63206	864	6928	373	24104
1978	84636	59141	754	5923	424	23058
1979	135231	99695	866	8399	381	45448
1980	190652	144861	1290	11793	165	31374
1981	124739	89846	1023	8831	92	15320
1982	132580	113271	1062	7544	82	16203
1983	116117	95705	851	6371	232	13609
1984	113822	93365	849	6798	178	13266
1985	115733	101611	1029	6830	221	15592
1986	101475	83853	1008	6559	302	15795
1987	127738	114367	1027	6056	383	25004
1988	147892	136682	1470	7523	453	29425
1989	201950	189502	2115	10989	452	34499
1990	210844	207992	2826	11053	463	34100

Sources: *Input-Output Tables*, various issues.

(unit: current prices, million yen)

	Mid East	Asia, others	Oceania	Africa	Total	Physical quantity (1000 m^3)
55					22242	(2051)
60					58388	(6221)
65					163143	(15943)
70					501205	(39356)
75					673497	(35650)
76					913889	(41589)
77	0	407683	21379	3034	895367	(41880)
78	0	349451	16270	4058	745812	(42653)
79	0	727494	28273	7353	1375384	(44786)
80	0	643673	36307	9251	1304268	(37510)
81	0	397894	25895	3823	806641	(29220)
82	0	475765	28133	4802	900042	(30406)
83	0	353833	20237	4292	722968	(29795)
84	0	370870	23204	12300	735537	(28404)
85	0	296826	25403	4077	667641	(28900)
86	0	202164	17012	3547	495072	(28929)
87	0	293699	22608	7405	630971	(32292)
88	0	219655	20655	5533	581691	(29598)
89	0	252371	27622	7252	694459	(31298)
90	0	218830	32745	5723	655522	(28999)
						Sawlogs only (1000 m^3)
55						
60					3016	(166)
65					16117	(968)
70					108620	(3011)
75					234094	(2612)
76					273899	(3301)
77	0	18867	46429	3081	272284	(3585)
78	0	18129	40534	3265	235864	(3857)
79	0	33217	53289	3999	380525	(5116)
80	0	47387	78804	7072	513398	(5573)
81	0	31805	65209	6744	343609	(3898)
82	1	50921	69088	8546	399298	(4953)
83	0	46871	66538	9511	355805	(4667)
84	0	53861	69014	9231	360384	(4493)
85	0	80845	60790	9034	391685	(5176)
86	23	65491	50548	7110	332164	(5523)
87	9	137416	56868	6681	475549	(7397)
88	23	160218	56969	5679	546334	(8462)
89	44	261647	64217	6355	771769	(9624)
90	18	248422	70268	7281	793267	(9082)

Table 10-2 World forest resources and production of wood products

a. Production of wood products

(unit: million m^3)

	World total	Firewood	Industrial use		
			Total	Lumber & plywood	Pulp & chips
1965	2213.9	1082.6	1131.2	681.3	237.6
1970	2390.0	1113.3	1276.7	760.4	313.5
1975	2579.1	1283.0	1296.1	766.6	322.6
1980	2926.5	1475.9	1450.5	877.0	370.6
1981	2931.4	1518.2	1413.1	835.1	372.4
1982	2926.4	1551.6	1374.7	803.2	361.9
1983	3037.7	1580.2	1457.4	875.1	369.6
1984	3125.8	1613.8	1511.9	907.3	386.5
1985	3164.3	1645.9	1518.4	911.8	385.6
1986	3252.3	1678.4	1573.9	959.3	393.6

b. Closed forest resources

(unit: million m^3)

	Total	Coniferous trees	Broad-leaved trees
US	23396	15184	8212
Canada	22958	18310	4648
America, others	78584	1191	77393
Europe	10103	7040	3063
Soviet Union & EE	72064	57209	14855
East Asia	26364	211	26153
Mid East	—	—	—
Asia, others	3865	1021	2844
Oceania (PNG)	1817	6	1811
Africa	38782	66	38716
World total	276117	100233	175883

c. Annual depletion of tropical rain forests

(unit: 1000 ha)

Tropical Africa	1332
Tropical America	4076
Tropical Asia	1826
Total	7234

Note: Resource data refer to closed forests available for economic use as of 1980. Countries are limited to those with forest area of one million ha and with appropriate forest statistics. Tropical Asia includes PNG.

Sources: Production data are from FAO, *Yearbook of Forest Products*. Resource data are compiled from FAO/UNEP, *Tropical Forest Resources Assessment Project*, ECE/EAO, The Forest Resources of the ECE Region, and various country data. Modified here from Tominaga,ed., *Handbook on Resources* (J).

(d) shrubland where at least 10% of the earth's surface is covered by vegetation up to 7 meters in height.

In 1980 when the survey was conducted, the area corresponding to closed forest was estimated to be 2.95 million ha, open forest 13.7 million ha, and forest fallow and shrubland an additional 1.03 million ha. This corresponds to 22%, 10%, and 8% of the land surface, respectively. Panel (b) shows the growing stock of closed forest resources by regional breakdown. In the following, an attempt has been made to maintain a more or less consistent geographical classification which corresponds to the import matrices. It is shown that the world's total forest resources coming from closed forest are somewhere around 280 billion cubic meters. One has to quickly add that the precise measurement of a growing stock of trees, which should include the economically usable part of the trees and exclude branches and treetops, is a difficult task, and the availability of statistics is limited to a selected number of countries for this reason. It is estimated also that the total growing stock of the world would amount to 365 billion cubic meters.

About one-third of the growing stock is coniferous, while the remaining two-thirds is broad-leaved. Nearly three-quarters of the latter is concentrated in the tropics in Africa, Latin America, and Southeast Asia. Panel (c) of Table 10-2 focuses on this point, revealing that more than 7200 ha of closed forests are being lost annually in the tropics. According to the survey by the FAO and the UNEP, the total loss including open forest amounts to more than 11,000 ha. Of this, 6200 ha (4100 ha of closed forest and 2100 ha of open forest) are permanently turned into farmland or residential land, whereas the rest is lost to forest clearance in the tropics. Needless to say, this damage to the tropical rain forests is causing global concern. To the extent that this damage is triggered by the increased demand for wood products in the North, it ceases to be a policy matter for the South alone.

10.3 The case of energy resources

Table 10-3 summarizes the import of energy resources into Japan from various countries and regions. The table is similar to Table 10-1 in its construction and is based on the import matrix which comprises a part of the input-output

Table 10-3 Import matrix, energy resources

	USA	America, others	EC	Soviet & EE	Europe, others	East Asia
a. Coal (I-O sector code: 1101)						
1955						
1960						
1965						
1970						
1975						
1976						
1977	295654	174898	8539	57992	5	5887
1978	134226	134479	6524	36107	0	7802
1979	222074	133485	7026	34247	0	15923
1980	356280	153597	7	32813	1	27732
1981	430848	152585	5	20859	60	43679
1982	524453	189331	6	23566	6	53727
1983	290222	180509	11	26266	24	50957
1984	264056	263887	36	28285	15	49399
1985	223002	281012	27	47006	17	46950
1986	132801	196647	33	45905	19	33682
1987	85281	157008	41	43468	32	27305
1988	102869	176138	68	49910	24	27659
1989	105119	185578	132	61887	24	32144
1990	109673	199805	424	67301	40	41280
b. Crude oil (I-O sector code: 1301)						
1955						
1960						
1965						
1970						
1975						
1976						
1977	0	9616	0	1657	0	175674
1978	1	8221	0	971	0	158988
1979	0	13791	0	1746	0	221708
1980	1	196106	0	4534	0	440171
1981						
1982	53	410788	0	1869	0	580277
1983	1	427826	0	4948	0	494761
1984	0	468398	0	3922	0	555611
1985	0	376813	0	5456	4	526795
1986	5	196607	15386	1361	0	207466
1987	0	178382	0	896	0	226808
1988	199	128430	0	821	0	205323
1989	205	141416	0	355	0	210493
1990	253	176268	0	1346	0	321977

Source: Same as Table 10-1.

(unit: current prices, million yen)

	Middle East	Asia, others	Oceania	Africa	Total	Physical quantity
						(1000 t)
55					20236	(2861)
60					50935	(8292)
65					97302	(17080)
70					363800	(50172)
75					1024865	(62107)
76					1057879	(60930)
77	0	12693	374803	29469	959464	(61095)
78	0	5866	302241	23760	651005	(52284)
79	0	5209	330867	25144	773975	(58615)
80	0	7204	392835	39836	1010305	(68323)
81	0	5082	506056	58368	217542	(78348)
82	0	3389	542693	95569	1432740	(79091)
83	0	3118	532288	77129	1160524	(74666)
84	0	6322	557741	89614	1259355	(87818)
85	0	7657	548316	99011	1252998	(93077)
86	—	4460	361819	68876	844242	(91449)
87	2	3023	316769	45563	678492	(92762)
88	—	3437	296897	34320	691322	(104485)
89	—	4827	382793	35929	808433	(105472)
90	2	8076	438073	36310	900984	(107938)

	Middle East	Asia, others	Oceania	Africa	Total	Physical quantity
						(1000 kl)
55					53500	(8501)
60					174290	(31120)
65					423699	(84143)
70					921433	(197108)
75					5831691	(263373)
76					6289383	(267754)
77	4902770	1238557	4134	34685	6367093	(278017)
78	3860748	935025	4923	11387	4980264	(270650)
79	5559722	1545887	4811	29836	7377501	(281203)
80	8686827	2465975	0	220757	12011371	(254447)
81					11687743	(227444)
82	7954868	2336386	888	157764	11442893	(212259)
83	6664777	1785630	4437	123766	9506146	(207005)
84	6510366	1720239	9657	52026	9320219	(213201)
85	5833971	1370648	78908	110198	8302793	(195945)
86	2367581	594895	22368	25314	3430983	(188839)
87	2029715	556913	6714	15323	3014751	(184702)
88	1601120	450228	3953	22057	2412131	(192162)
89	2058767	535554	4158	10933	2961881	(204914)
90	3133133	780617	35137	21116	4469847	(225251)

Table 10-4 World proven reserves and production of crude oil

a. Time series (unit: billion bbl)

	Proven reserves	(growth,%)	Annual production	(growth,%)	Reserve production ratio(year)
1955	187.0	(19.4)	5.6	(8.2)	33.1
1960	265.7	(7.4)	7.6	(6.3)	34.7
1965	365.0	(6.6)	11.0	(7.5)	33.1
1965	343.5	(0.6)	11.6	(7.4)	29.6
1970	620.7	(29.2)	17.4	(9.4)	35.7
1974	720.4	(13.4)	21.2	(0.5)	34.0
1974	710.8	(—)	20.4	(—)	34.9
1975	657.8	(-7.1)	19.4	(-5.0)	34.0
1976	641.9	(-2.5)	20.9	(8.0)	30.7
1977	646.0	(0.6)	21.8	(4.3)	29.6
1978	641.6	(-0.7)	22.0	(1.0)	29.1
1979	641.6	(0.0)	22.8	(3.7)	28.1
1980	648.5	(1.1)	21.8	(-4.5)	29.7
1981	670.7	(3.4)	20.4	(-6.5)	32.9
1982	670.2	(-0.1)	19.3	(-5.2)	34.6
1983	669.3	(-0.1)	19.4	(0.5)	34.4
1984	698.6	(4.4)	19.7	(1.6)	35.4
1985	700.1	(0.2)	19.5	(-1.1)	35.9
1986	697.5	(-0.4)	20.4	(4.5)	34.2
1987	887.4	(27.2)	20.3	(-0.3)	43.6
1988	907.4	(2.3)	21.0	(3.6)	43.1
1989	1001.6	(10.4)	21.7	(2.9)	46.2
1990	999.1	(-0.3)	22.0	(1.6)	45.4

b. Regional distribution of reserves (unit: million bbl)

US	25860
Canada	6133
America, others	125027
Europe	18822
Soviet Union & EE	60100
East Asia	13088
Middle East	660247
Asia, others	31635
Oceania	1822
Africa	58837
World total	1001572

Note: Proven reserves are at the end of the calendar year. Annual growth figures prior to 1965 refer to the five-year average. Regional distribution refers to 1989.

Source: Data for the 1965-1974 period from *BP Statistical Review of World Energy and Oil* cited in Tominaga, ed. *Handbook on Resources* (J) and more recent data from Gas Journal cited in MITI, *International Trade White Paper*,

framework. In the import matrix for 1990 (Table 12-3), coal and crude oil are provided under sector 02 (mining). The table here is based on a time series of such matrices.

Along with the value of imports, physical quantity is also listed in the last column of the table. This is expected to serve as a linkage between the economic sphere to material flow and the natural environment sphere. The data also allow analysis of the impact of price fluctuations on the importers' side as well as on the exporters'. A typical case is crude oil. Between 1970 and 1975, there was a sharp increase in the value of imports compared with a rather moderate increase in quantity. In fact, the value of crude oil per 1000 kl in 1970 was calculated to be 4675 yen, which rose to 16,501 yen in 1975. In the 1980s, crude oil imports remained more or less stable in terms of quantity, but declined considerably in terms of value. The price has declined from 51,378 yen per 1000 kl in 1981 to 19,844 yen in 1990. On the side of the exporting countries, we have to look into the reserves of energy resources. Tables 10-4 and 10-5 provide some data on crude oil and coal, respectively.

Harris [1993] clarifies some conceptual ambiguities surrounding the usage of the following terms: resource base > mineral endowment > resources > potential supply > reserves. "An ore reserve is simply a known quantity of mineral resource that is economically producible under existing economic conditions. ... Resource exists only within an economic and technological reference. ... When the term resource is qualified only by a specified price, it is understood to refer to the sum of the element or compound in all (known plus undiscovered) deposits producible at that price (ignoring discovery costs), given current and near feasible technology. ...When the set of deposits that comprises resource is restricted to only those that also are economically discoverable, the sum of the element or compound in these deposits consitutes the stock known as potential supply. ... Resource base and endowment ... are not defined by economics or technology. Resource base is the total amount of an element or compound that is present in the earth's crust within the region of interest. ... Mineral endowment can be considered a component of resource base ... that meet specified minima of concentration and size and a maximum depth within the crust." Tables 10-4 and 10-5 provide reserve-production (R/P) ratios for crude oil and coal. As Harris rightly points out, reserve is an economic variable. "This reflects the planning function of man-

Table 10-5 World reserves and production of coal

a. Production

(unit: million ton)

	Hard coal
1955	1598.2
1960	1966.2
1965	2014.6
1970	2134.1
1975	2357.8
1980	2729.8
1981	2740.7
1982	2822.9
1983	2828.6
1984	2966.5
1985	3099.9
1986	3248.0
1987	3335.2
1988	3453.9
1989	3474.1
1990	(3478.0)

b. Regional distribution of reserves

(unit: million TOE)

	Proven amount in place	Proven recoverable reserves	Production	Reserve production production ratio (year)
North America	229281	126960	581.6	218
America, others	3907	2607	12.1	216
Europe	93660	36934	254.8	145
Soviet Union&EE	202386	138875	721.8	192
Asia	37339	14357	169.4	85
CPE Asia	214300	99450	667.4	149
Oceania	48584	27481	85.5	322
Africa	124963	59081	136.3	434
World total	954420	505745	2628.9	192

Note: Data on reserves refer to around 1980.

Source: Coal production figures are from UN, *Yearbook of World Energy Statistics*. Reserve figures are from the World Energy Congress (J), cited here from the *Handbook on Resources* (J).

agement, who typically desire to have enough reserves to support production capacity for the planning horizon, e.g. 5 - 10 years. Thus as proven reserves are produced, more may be added by additional drilling. For that reason, it is useful to consider a quoted reserve as the current working inventory. Change in product price or factor cost alters the magnitude of reserve." Thus, the R/P ratio is not an appropriate indicator of the sustainability of the resource base, despite frequent reference to it as such. In fact, table 10-4 reveals that as production (flow) increased over time, so did proven reserves (stock or working inventory), with the result that the R/P ratio remained more or less unchanged. For an obvious reason, this relation cannot hold in the very long run. Our society is consuming energy resources at an accelerating rate, and ultimately depletable resources will be literally drained, under current price and technology. At higher prices and with new technology, new sources of energy can be drawn into the economic sphere. The cases range from oil shale and oil sand, synthetic fuels, geothermal, wind and solar, and biomass, to nuclear fusion. The analysis of the technological feasibility of these new sources of energy is provided in chapters 3 and 8.

10.4 The case of metals

Lastly, table 10-6 describes the case of metals including iron ore and nonferrous metal ores. Table 10-7 provides a summary of information on the world reserves of iron ore and of the nonferrous metal bauxite. A regional breakdown of reserves is also provided.

There are two points to be made. First, in the case of iron ore, the physical quantity of imports has matched very closely the aggregate economic activity. In fact, the production of iron and steel in Japan took off with its industrialization drive in the 1950s and continued to expand at a faster pace than the economy as a whole. However, with the structural change of the economy which started in the mid-1970s and continued more distinctly in the 1980s, the production level has tapered off [for details, see Uno 1989a]. Service orientation and electronics were the factors contributing to the trend. There can be different types of economic development reflecting factor endowment (i.e., availability of labor, capital, and resources), but it is generally true that if an economy develops depending on a fairly large domestic population, iron and steel are required after achieving a certain per capita income level and before reaching some saturation point. This typically reflects active capital formation and the diffusion of consumer durables. Second, although the demand for

Table 10-6 Import matrix, metals

	USA	America, others	EC	Soviet & EE	Europe, others	East Asia
a. Iron ore					(I-O sector code:	1220)
1955						
1960						
1965						
1970						
1975						
1976						
1977	0	202719	0	5360	0	385
1978	0	153224	0	3937	0	290
1979	0	225951	0	2322	1	337
1980	0	275334	0	763	0	151
1981	0	280531	0	0	0	139
1982	0	312671	0	0	0	0
1983	0	249950	0	0	0	0
1984	2	256551	0	0	0	0
1985	1	241760	0	0	0	0
1986	0	153149	0	0	3383	15
1987	0	128908	0	0	3993	166
1988	0	116563	1	0	2602	530
1989	2	140003	2	0	2568	763
1990	1	166067	0	0	2356	845
b. Nonferrous metal ores					(I-O sector code:	1220)
1955						
1960						
1965						
1970						
1975						
1976						
1977	13496	165753	346	3359	1246	9165
1978	13958	123436	98	1002	477	4703
1979	49679	205763	7525	760	406	7931
1980	73898	250059	4109	630	3522	8753
1981	67723	201308	912	1342	1519	6557
1982	54986	184317	1229	2586	4188	6303
1983	20313	177113	4129	2604	6959	4187
1984	20869	187745	4403	3190	7973	4399
1985	38111	172502	6079	5754	6379	3345
1986	29947	103326	8340	2266	55	3641
1987	37588	102464	8321	1070	44	7927
1988	55548	129910	9529	3296	337	9901
1989	67305	144967	13048	3845	447	14476
1990	69849	180594	18330	5760	1140	10419

Source: Same as Table 10-1.

(unit: current prices, million yen)

	Middle East	Asia, others	Oceania	Africa etc.	Total	Physical quantity (1000 t)
55					29354	(5459)
60					76548	(15036)
65					188391	(39018)
70					443948	(102090)
75					670393	(131752)
76					708360	(133758)
77	0	121593	319581	55949	705587	(132614)
78	0	96132	234893	39347	527823	(114690)
79	0	115847	267315	51640	663413	(130276)
80	0	140436	325053	54881	796618	(133721)
81	0	134577	316384	49267	780898	(123379)
82	0	148925	381524	56139	899259	(121817)
83	0	127782	326230	42870	746832	(109181)
84	0	129006	331575	40497	757631	(125372)
85	0	139607	307010	44583	732961	(124513)
86	103	101665	187850	24718	470883	(115234)
87	0	85369	142526	20576	381538	(112034)
88	0	83043	146797	15983	365519	(123377)
89	0	92781	177586	19111	432816	(127709)
90	0	105504	194299	19395	488467	(125290)
						Bauxite only (1000 t)
55					15127	(343)
60					53427	(1094)
65					103689	(1675)
70					374104	(3660)
75					504280	(4600)
76					551392	(4276)
77	4747	122946	149869	34348	505275	(5318)
78	2680	105451	94108	18589	364002	(4742)
79	2457	167835	149775	29709	621840	(4597)
80	1186	266627	190152	42862	841818	(5707)
81	0	190481	143495	38438	651775	(4352)
82	1126	162474	130867	39713	587789	(3439)
83	767	149082	136590	38827	540578	(3579)
84	2766	117862	127330	43649	520186	(3861)
85	451	108820	122767	39086	503294	(3518)
86	1138	69263	77867	32410	328253	(2307)
87	657	66831	81995	30855	337752	(1872)
88	2340	95576	133001	38578	478016	(2148)
89	7395	124235	167756	43890	587364	(2269)
90	4408	104699	158896	52207	606302	(2302)

Table 10-7 World reserves and production of metals

a. Time series (unit: million ton)

	Iron ore		Bauxite	
	Reserves	Production	Reserves	Production
1955	n.a.	379.0	n.a.	16.8
1960	n.a.	489.8	n.a.	27.6
1965	n.a.	623.9	5852	37.2
1970	n.a.	773.8	8430	59.6
1975 ('74)	97000	888.8	15500	77.3
1980	105000	879.1	22700	93.1
1981	108000	851.4	22800	n.a.
1982	108000	775.6	22400	78.9
1983	98200	746.1	22400	78.6
1984	98200	822.5	22490	92.3
1985	98400	897.7	22300	89.6
1986	n.a.	909.4	23200	92.6

b. Regional distribution of reserves

	Iron ore	Bauxite
US	5900	40
Canada	9800	—
America, others	12000	6750
Europe	3300	650
Soviet Union and EE	25300	1200
Asia, others	8300	1200
Oceania	20200	4600
Africa	7400	5900
Other market economies	6200	2900
World total	98400	23240

Note: The unit for iron ore reserves is short ton (or American ton, 1 st = 2000 lb = 1102 kg). Data on iron ore reserves are for 1985 and on bauxite, 1986.

Source: Data on reserves are from the US Bureau of Mines, *Mineral Commodity Summaries*, and those on production from Metallgesellschaft and WBMS. Tominaga, ed., *Handbook on Resources* (J).

aluminum is growing in Japan, imports of bauxite have dropped considerably. Japan is now heavily dependent on imported aluminum. This is a result of the higher energy prices in this country after a series of oil price hikes which prompted the aluminum industry to shift to imports from countries where electricity is much cheaper (often hydro-power generation). Thus, industrial relocation is a factor behind the changing economy-environment interaction.

10.5 Resource endowment and sustainability

The aim of this chapter is a limited one, as was stated at the outset, which is to demonstrate, starting from the environmental accounting of a single economy, the empirical feasibility of establishing linkages across national boundaries in an attempt to arrive at a global picture of sustainabillity. Our focus in this chapter has been limited to the international repercussion through commodity trade. Even in this respect, one can envisage detailed resource endowment data for an individual country providing information on the value and physical quantity of various resources under given economic and/or technological conditions. This aspect can be properly included in the integrated economic-environmental statistical system. We have not dealt with it in this series of work because Japan is at the consuming end of world resource flows, herself endowed with practically no natural resources. Thus, the research represents a typical standpoint of the industrial North. There are at least three additional channels through which economic-environmental interrelations have to be analyzed beyond national boundaries. In the case of the impact of direct foreign investment (and part of official development assistance) which transfers production capacity overseas, often involving technology transfer, the environmental accounting in the host country can probably take place as a part of capital formation in relevant industrial sector(s). Transborder flow of pollutants, such as acid rain or contamination of international rivers, will require quite a different approach in that an analytical framework with a spatial dimension has to be prepared. As to the global warming issue, a group of country frameworks will work as a driving mechanism for the global climate model. In other words, the individual country framework, hopefully linked together on a global level through international trade, etc., will provide the resource requirement, CO_2 emission, and other relevant data for the global climate model. The latter, again, falls beyond the scope of this work.

Part V

Measuring Quality of Environment

Chapter 11

Composite Measures of Quality of Life

11.1 Introduction

This paper attempts to update the composite measures of quality of life developed in the 1970s. In so doing, the latest features developed in the System of National Accounts (SNA) are explicitly recognized and incorporated in the conceptualization as well as in the empirical estimation. Some of the new features of SNA which are particularly useful for our purpose include consumption expenditure according to purpose of government , non-profit institutions, and household; capital formation and stock by type of capital goods; stock of consumer durables ; and national assets and liabilities. Our final aim is to reveal the potential for the expanded use of National Account data in various social, economic, and environmental analyses. Methodology is discussed in order to utilize SNA and other supporting data sets in a modeling framework designed to focus on some of the recent policy issues.

The current flow of income and consumption is undoubtedly an important facet of our quality of life. This aspect is adequately dealt with in the framework of established SNA. However, quality of life is understandably more complex than can be measured by national accounts. This led to various attempts at social indicators, prevalent in the early 1970s, with their subsequent advantages as well as disadvantages. The social indicators approach certainly expanded the scope of analysis, but it did not lead to an aggregate measure. Another approach to the problem was to rearrange the items included in the gross national expenditure (GNE) and make them amenable to quality-of-life measurement.

The GNE modification approach was taken up by the Economic Council of Japan as net national welfare or NNW. "Net" in this case means net of investment in plant and equipment. As a personal viewpoint, it would be inappropriate to interpret the framework in a welfare context because the concept does not aim at measuring consumer surplus, which is essential in welfare measurement. In fact, the mere mention of "welfare" caused considerable confusion and criticism of the idea itself. One prominent economist dubbed NNW as "*N*o-one *(K)n*ows *W*hat," or "*N*aniga *N*andaka *W*akaranai" in Japanese. The framework actually deals with the *cost* of delivering the bundle of goods and services related to the quality of life. The framework itself is a useful tool, though, to come up with an aggregate measure of the quality of life.

Twenty years after its inception, this chapter reconsiders the framework of NNW and updates the empirical analysis which incorporates the following items:

(1) Government consumption (adjusted)
(2) Personal consumption (adjusted)
(3) Flow of services attributable to consumer durable goods
(4) Flow of services attributable to household-related social overhead capital
(5) Leisure time
(6) Nonmarket activities such as housewives' domestic services
(7) Loss due to urbanization
(8) Damages due to environmental pollution.

In section 2, we will review the historical context which gave rise to this framework. Section 3 introduces the updated account covering 1955 to 1990, taking empirical data from the Japanese economy. Section 4 is devoted to further discussion, especially the usefulness of the framework in addressing recent policy issues and the possibility of incorporating the framework in a multisectoral econometric model.

11.2 The background

In retrospect, the 1960s was an interesting era. After continued economic expansion and considerable prosperity, people were questioning the meaning of growth. For one thing, there was an emerging urban problem. Environmental disruption was becoming all too evident. There was a growing criticism about using GNP as a yardstick of economic and social success, although it never was intended as such. In late 1960s and early 1970s, there appeared many attempts to expand the National Accounting to incorporate new dimensions. The interest was shortlived, however. In 1973, the world community witnessed the first oil crisis and then became preoccupied with acute short-term stabilization problems like stemming inflation, securing employment, and restoring the fiscal balance.

In the 1990s, having lived through a decade of relative stability and prosperity, the world community is once again asking long-term questions. This time the central issues are sustainability and development. People are concerned about the depletion of our resource base, global warming, and acid rain. What if our grandchildren or great-grandchildren cannot inherit the economic and social system as the aftermath of growth itself? If economic growth is to be blamed for threatening sustainability, what is the implication for less developed countries which are striving for economic expansion? Thus, renewed interest in lifestyle , environmental accounting , pollution abatement cost, energy efficiency , and new technology reappeared.

Some people are more bold (and some more cautious) than others about introducing new ideas and modifying the framework of SNA. In a debate 20 years ago, Denison represented the cautious voice. "It would be enormously convenient to have a single, generally accepted index of the economic and social welfare of the people .... Some recent discussion seems almost to imply that such an index could be constructed." "Some suggestions to change the measurement of national product will derive from confusion between an output measure and a comprehensive welfare measure. Such proposals must be rejected. ... Efforts to do so can only impair their (i.e., GNP and NNP) usefulness for the very important purposes of both long-term and short-term analysis that they now serve well."[1]

Commenting on the environment, he wrote "To measure the state of affairs with respect to any aspect of the human and physical environment requires adequate and accurate data." "The idea of measuring the net gain from production by balancing the value of the deterioration of the physical environment caused by production against the value of greater output is attractive. ... But implementation of this suggestion would require an objective measurement of the value of the deterioration expressed as a dollar amount. Such a valuation does not exist. ..." But even if these data were available, he says, such an attempt would face two serious problems. "First, relations between environmental conditions and welfare are rarely linear, and nonlinear relationships are hard to establish. ... Second, weighting is required. ... The absence of any natural weighting scheme is an even greater obstacle to combining indexes of crime, water pollution, racial discrimination, and the like into a single index."

We should note three points in his argument. First is that, perhaps responding to a casual mention of "welfare" on the part of the proponents of a new measure, it is impossible to derive a welfare index. As I have touched upon earlier, Denison has a point here. Second, he is against "a single, generally acceptable index of welfare." We need not accept his criticism here because a coherent economic accounting system can include physical measures as well as monetary measures, and additivity is not essential. Third, his point on environment boils down to a lack of data, which is true but distinct from the desirability of environmental accounting if such data were forthcoming.

Among the proponents of modifying SNA to take account of the extra dimensions of economic and social reality, we cite here Sametz [1968], Juster [1973], and Nordhaus and Tobin [1972].

Sametz's point is as follows: "... Change in GNP over long periods of time is not a good measure of economic growth or welfare. One cannot ignore—as one can for cyclical problems—price and population changes, nor can one assume that composition of inputs (or costs) or of output is relatively unchanged. Worse, one cannot even assume unchanged tastes or constant social priorities. In short, since structural change is the essence of secular change, it must be accounted for." According to him, the minimum requirement is:

(1) An index of quality improvement parallel to the consumer price index—a "quality of goods deflator".

(2) A measure of the amount and value of change in leisure time—a "quantity of work deflator".

(3) Revisions and additions to GNP series

   (a) Development of imputed values for crucial nonmarket economic activities.

   (b) Reworking of the government account to remove "intermediary" or "necessary evil" social expenditures, assigning the remainder to consumption and investment.

   (c) Revision of private sector to remove "costs" from consumption just as we remove "maintenance" from investment.

   (d) The end product to be final consumption and investment regardless of originating sector and whether market priced or not.

In his suggestion, nonmarket-priced social costs (3-c above) include "water (air, noise, etc.) pollution, traffic congestion, and the like". Listing research projects which were underway at that time, Sametz saw the feasibility of actually implementing his idea in an empirical context.

Juster starts by saying: "The present system of U.S. national income and product account has, in recent years, been subjected to mounting criticism relating to conceptual framework, what is included and excluded from measured final output and income, the adequacy with which the accounts carry out their intended purposes, and whether they can and should be more specifically designed to measure changes in economic welfare ." Noting that "income (output) of the system is derived in one way or another from an implicit set of wealth accounts", he proposes a set of wealth accounts.[Juster 1973, p.42]

(1) Reproducible tangible wealth (structure and durable equipment). "To the extent that reproducible tangible assets are owned by enterprises, the income generated by these assets is counted. ... However, if such assets are owned by households or governments, they are not now counted as part of the wealth and are not viewed as producing income to the owners or to society." "The most questionable aspect of developing the necessary imputation is that of an appropriate rate of return on these assets in both household and government sectors."
(2) Reproducible intangible wealth (the stock of disembodied socially useful knowledge). This asset category is "meant to be coterminous with business and government outlays for research and development".

(3) Human wealth (the stock of skills and knowledge embodied in persons). "(Current treatment) mixes together capital and current account transactions which can and should be disentangled." "For the most part, these activities are unmeasured simply because they represent uncompensated uses of time on the part of individuals and families." He applies this principle in analyzing allocation of time, and time spent is divided into consumption (leisure) and investment, plus maintenance of the stock of human capital itself.

(4) Natural physical resource wealth (the stock of mineral, forest, water, climate, etc., assets). "The difference seems to be in difficulty of measurement, not in differences of concept. The point simply is that there are physical environmental assets that provide flows of direct consumption benefits to final users, and that deterioration in these environmental assets will result in a reduction in the flow of such benefits. ..."[Juster 1973, p.47] "My own preference is to subtract defensive outlays by consumers and not to add defensive outlays by enterprises."[Juster 1973, p.67]

(5) Sociopolitical wealth (the stock of personal and national security, freedom, equity, privacy, etc.).

Nordhaus and Tobin write as follows. "A major question raised by critics of economic growth is whether we have been growing at all in any meaningful sense. Gross national product statistics cannot give the answers, for GNP is not a measure of economic welfare . ... An obvious shortcoming of GNP is that it is an index of production, not consumption. The goal of economic activity, after all, is consumption." "Our adjustments to GNP fall into three general categories: reclassification of GNP expenditures as consumption, investment, and intermediate; imputation for the services of consumer capital, for leisure ,and for the product of household work; correction for some of the disamenities of urbanization ."[Nordhaus and Tobin 1972, p.4]

Thus, their measure of economic welfare (MEW) consists of the following items.[Nordhaus and Tobin 1972, p.10]

(1) Personal consumption (national income and product account)
(2) Private instrumental expenditures (deduction)
(3) Durable goods purchases (deduction)

(4) Other household investment (deduction)
(5) Services of consumer capital imputation
(6) Imputation for leisure
(7) Imputation for nonmarket activities
(8) Disamenity correction (deduction)
(9) Government consumption
(10) Services of government capital imputation
(11) Total consumption = actual MEW

In addition to theoretical exposition, which is in line with the two authors cited above, Nordhaus and Tobin provide an empirical account of their framework for the 1929-1965 period in the USA . Disamenity correction is based on their view that the higher earnings of urban residents to some extent represent compensation for the disamenities of urban life. They ran a cross-sectional regression among American cities to estimate the income differentials necessary to hold people in an urban environment.[Nordhaus and Tobin 1972, p.13] Otherwise, their estimation is based on statistical datasets drawn from various sources.

The framework employed by the NNW Measurement Committee of the Economic Council of Japan was strongly influenced by Nordhaus and Tobin's MEW and Sametz's Welfare GNP.[Economic Council 1973, p.39] Before turning to updating the NNW estimation in the next section, let us briefly observe what the committee had to say regarding the background of their work. "It is frequently argued, concerning national income, that it is merely intended as an indicator of economic activities and effective demand and not as an indicator of economic welfare at all. However, we find it to the contrary, when we look back to the historical background in the development of the national income concept or its estimation. For instance, the national income or 'national divident' in Pigou's *Economics of Welfare*, not to mention Adam Smith's 'annual produce', manifests an emphasis on economic welfare rather than effective demand. And anyone who has ever studied Simon Kuznets' *Commodity Flow and Capital Formation* based on the commodity flow method may easily understand that Kuznets intended for the national income series to reflect changes of economic welfare in the long-range point of view. In this sense, the committee wishes to make it clear that we attach importance to such a background and the history of national income estimates."[Economic Council 1973, p.4]

The thrust of the discussion was beginning to take a concrete shape by the mid-1970s. By that time, "welfare measurement" had ceased to be simply a manifestation of individual views and was starting to have an official nature. The United Nations Statistical Office in 1977 published a report whose purpose was "to make a critical review of the concepts, methodologies and empirical applications of monetary measures of welfare that may be used to supplement national accounts and balances, with special references to the use of such measures for international comparison and to their feasibility for government statistical work."[UN 1977, p.1] "There has been a tendency to treat GNP as the sole indicator of successful performance in the achievement of economic and social aims. ... For these reasons, critics have rightly emphasized the urgent need to give much greater weight, in the statistical inventory as in economic policy and analysis, to other measures of performance of a society - to employment, to the distribution of income and wealth, to indicators of social conditions, of education, of health, of housing. More recently, especially in developed countries, the costs of high rates of material progress have been publicized in terms of pollution, environmental damage and the disamenities of a modern urbanized - and particularly a motorized - society."

They have relied heavily on the US and Japanese estimates as well as the then ongoing exploration into a system of social and demographic statistics (SSDS) and social indicators .[2]

The listing below summarizes the topics discussed in this document as well as the coverage of major accounting frameworks such as SNA and SSDS. The distinction between market vs nonmarket activities is also made.[UN 1977, p.71]

As to the labeling of the alternative aggregate, the report rightly states: "It is recognized by all that it is not practicable to make a direct measure of the welfare of a community in monetary or in any other terms. The best that can be done is to measure a number of factors that are generally supposed to contribute to or detract from welfare." [UN 1977, pp.6-7]

A recent development which is indispensable in our context is the effort towards developing an environmental accounting . The publication of *Our Common Future* marked a renewed interest in the environmental aspect of the global community. The key word was "sustainable development ."[World Commission on Environment and Development 1987, p.43]

	SNA	SSDS	Market(M) nonmarket(NM)
(1) Household economy			
Productive activities:			
1.1 Own-account food, etc.	final (a)	(a)	NM
1.2 Housework	excluded	(b)	NM
1.3 Students	excluded	(b)	NM
1.4 Volunteers	excluded	(b)	NM
1.5 Leisure activities	excluded	(b)	NM
1.6 Instrumental activities (travel to work)	final (a)	(b)	M
1.7 Enterprise subsidies to households	inter-mediate(a)	(a)	M
Durable consumer goods:			
1.8 Stock	incl. in assets	Yes	NM
1.9 Flow of services	excluded	No	NM
(2) Government expenditures on goods and services			
2.1 Intermediate (e.g., law and order)	final (a)	Yes(a)	M
2.2 Final (education, health)	final (a)	Yes(a)	M
(3) Environmental variables			
3.1 Expenditures on pollution abatement and control	intermediate or final(d)	No	M&NM
3.2 Hypothetical costs of restoration	excluded	No	
3.3 Damage costs	part in-cluded (a)	No	M&NM
3.4 Input-output balances	excluded	No	NM
(4) Measurement of assets			
4.1 Commercial natural resources	included as assets	No	M
4.2 Public goods (air, water)	excluded	No	NM
4.3 Human capital	excluded	(f)	NM

Abbreviations:

(a) Included but not generally specified separately.

(b) Included in the form of time-budget.

(d) Intermediate or investment for enterprises; final (current or investment) for government.

(f) The stock of educational attainment of the population is measured in non-monetary terms only.

(c) and (e) are not relevant in the present context.

Thus, Pronk and Haq , summarizing one of many conferences leading to the Earth Summit in Rio, advocated the 'green GNP.' "Traditional economic indicators, such as GNP and GDP, are inadequate measures of sustainability. They measure production but provide little information about people or about the state of their living environment. If a deteriorating environment causes disease, resulting in increased health expenditures and thus increased GNP, the higher GNP would be interpreted as a higher level of development—even though people and their environment are worse off. Similarly, current income measures do not factor in inevitable future costs of current depletion of resources." Hence, the need to prepare environmentally sensitive national accounts ("green GNP") to reflect the impact of environmental damage on national and global output.[Pronk and Haq 1992, p.23] A similar feeling is expressed in a recent publication by the World Resources Institute .[Repeto et al., 1979, p.1] "... The SNA does classify as gross capital formation, expenses incurred in 'improving' land for pastures, developing or extending timber-producing areas, or creating infrastructure for the fishing industry. SNA records such actions as contributing to recorded income and investment, although they can destroy ... valuable natural assets through deforestation, soil erosion, and overfishing. ... The national accounts thereby create the illusion of income development, when in fact national wealth is being destroyed."

Peskin et al. observed that "frameworks, which are structurally very different, can rely on similar data sets. Therefore, with respect to the 'data framework' function of national accounting, some of the differences between these approaches may not be as great as they may first appear."[Peskin and Lutz 1990, p.21][3] Reflecting UNSO's efforts to provide a recommendation on environmental accounting, various countries including Japan are now initiating new forms of estimates.[4]

It may not be appropriate to provide a more comprehensive bibliographical survey of environmental accounting and related fields within the short span of this research. Interested readers are referred to Ahmad et al.[1989], Constanza [1991], UN[1989][1990][1993], Bartelmus, Stahmer, and van Tongeren [1991], among others.

11.3 Empirical measurement

Let us now have a closer look at the framework of composite measures of the quality of life. Tables 11-1a and b summarize the revised estimates of

NNW in current prices and 1970 constant prices, respectively. Readers are reminded of the restraint in using the word "welfare" touched upon earlier in this paper. In updating the estimates, I have followed the original method where it was deemed appropriate. However, I have shifted to new data sources wherever they became available. In the following, the method of estimation is summarized and the latest empirical figures are provided for the 1955-1990 period. The estimates supersede my earlier work.[5]

NNW government consumption: Whereas government budgetary data were employed in the past estimates, the new estimates are based on SNA data which classify government expenditures by purpose. Classifications of the purposes of government in SNA include the following[UN 1969, pp.87-89; Economic Planning Agency , *Annual Report on National Accounts*]:

(1) General administration
(2) Defense
(3) Education
(4) Health
(5) Social security and welfare services
(6) Housing and community amenities
(7) Other community and social services
(8) Economic services
(9) Other purposes

Of these, (3) education, (4) health, and (5) social security and welfare services are included in the NNW update for the 1970-1990 period (Table 11-2). Law courts, police, fire protection, etc. are included under the heading (1) general administration. These are, however, not included in NNW estimates on the grounds that they are defensive expenditures . In other words, they do not contribute to the improvement of well-being in a positive manner but simply nullify the potential damage caused by crimes, fires, etc. Housing and community amenities (item 6) are dealt with under the relevant headings below, such as housing and environment. Other community and social services (item 7) may be included under NNW government consumption; we did not do so in order to maintain consistency with earlier calculations.

SNA provides a classification of the purposes of private nonprofit bodies serving household.[6] Whereas the Blue Book distinguishes eight categories,

Table 11-1 Measures of Quality of Life "NNW"

a. Current prices	1955	1960	1965
1. Government consumption adjusted	394	617	1532
2. Private consumption adjusted	5371	8254	16750
3. Service of consumer durables	101	207	764
4. Service of social capital	102	148	255
5. Leisure time	(714)	0	2355
6. Extra-market activities	987	1378	3140
7. Loss due to urbanization (deduction)	121	247	505
8. Environmental pollution (deduction)	54	647	2569
Total NNW	6780	9710	21722
Reference:			
GNE	8399.1	19306.6	32772.8

b. Constant prices	1955	1960	1965
1. Government consumption adjusted	1199	1374	2254
2. Private consumption adjusted	10429	14706	22168
3. Service of consumer durables	91	195	755
4. Service of social capital	169	210	300
5. Leisure time	(3077)	0	4680
6. Extra-market activities	1876	2388	4068
7. Loss due to urbanization (deduction)	435	719	920
8. Environmental pollution (deduction)	119	1123	3376
Total NNW	13210	17031	29929
Reference:			
GNE	18298.6	27769.4	43050.7

Source: Estimates by the present author. Some of the figures for 1955-1970 have been adopted from the *Economic Council* [1973] and Kanamori, Takase, Uno[1977]. See text for details.

the Japanese framework provides three, which are listed in Table 11-3. Unlike earlier estimates of NNW, we have included these in the estimates. Therefore, the exact definition of item (1) in the summary table is the total of the final consumption of government and nonprofit institutions serving household. Together with the classification of household goods and services, which is discussed below, the SNA framework allows us to compile data on resource allocation by social purpose.

NNW private consumption: In the case of private consumption expenditures, purchases of consumer durables (see "services attributable to durable

(unit: billion yen)

	1970	1975	1980	1985	1990
1.	2761.7	7134.2	11499.3	14779.2	18454.8
2.	34408.6	77638.3	128470.5	170529.0	219552.0
3.	2286.8	5130.2	8477.9	11963.4	16651.8
4.	444.0	1324.5	2936.8	4176.7	6193.4
5.	8323.6	31675.8	46087.6	58808.8	93264.4
6.	7220.0	20138.0	29682.1	34475.5	40122.9
7.	1188.0	3068.8	4687.9	5394.5	7297.0
8.	6236.6	8901.2	8206.1	6619.6	6424.6
	48020.1	131071.0	214260.2	282718.5	380517.7
	73188.4	148169.9	240098.5	321555.9	428667.5

	1970	1975	1980	1985	1990
1.	2671.7	3396.2	4261.9	4838.0	5314.5
2.	34408.6	46243.5	55985.7	64802.9	79533.4
3.	2286.8	3709.2	5734.1	8553.8	13876.5
4.	444.0	802.4	1297.7	1745.9	2370.8
5.	8323.6	12464.8	12464.2	13218.6	17475.2
6.	7213.0	11703.8	12571.5	13079.3	13953.5
7.	1188.0	1226.5	1283.6	1230.0	1392.2
8.	6236.6	5461.2	3893.4	2826.6	2580.8
	47923.1	71632.2	87138.1	102181.9	128550.9
	73188.4	90907.3	113914.9	137305.5	183042.6

consumer goods" below) and "defensive expenses" are subtracted (Table 11-4). Commuting expenses, cost of obligatory payments including ceremonial services, etc. fall under the latter category. Starting with the private consumption figures from the SNA, adjustments were made based on the data from household survey concerning the "annual expenditures for individual items per household" and the number of households from the population census. Aggregate figures were obtained by simply multiplying the annual expenses for these items with the number of households.

The SNA classifies household expenditures by their type, some of which are common to the consumption of government and of nonprofit institutions.

Table 11-2 Government final consumption according to purpose

Purposes	1970
1. General public services	1556.0
2. Defense	560.2
3. Education	2077.9
4. Health	255.9
5. Social security and welfare services	219.3
6. Housing and community amenities	216.5
7. Other community and social services	73.5
8. Economic services	675.4
9. Other purposes	12.4
Total	5647.0
Deflator (PCG)	=100.0

Note: The data in this table refer to fiscal year.

Table 11-3 Consumption of private nonprofit institutions

Purposes	1970
1. Education	121.5
2. Medical and other health services	87.1
3. Miscellaneous purposes	229.2
Total	437.8
Deflator (PCG)	=100.0

Note: The data in this table refer to fiscal year.

Goods are subdivided into durables, semidurables, and nondurables, and there is an additional item covering various sorts of services. Expenditure items are also classified according to purpose.[7] Our table, therefore, shows the functional classification in the bottom half, enabling us to link it with consumption by purpose of both government and nonprofit institutions serving household.

Services attributable to consumer durable goods: The capital stock possessed by households consists of housing and consumer durables. The former is included in the personal consumption as imputed rents in the SNA (see Table 11-4 above). The latter is estimated as a part of SNA in Japan for five major categories, which are shown in Table 11-5. Unlike earlier estimates, we need not compile stock data based on current purchases. The SNA figures on the stock of consumer durables are on a net basis (i.e., net of depreciation). Net stock is assumed to generate a flow of services over its life span at a rate

(unit: billion yen, current prices)

1975	1980	1985	1990
4143.4	6460.5	8137.0	10553.0
1285.0	2066.4	2961.1	3944.4
5744.5	8990.8	11261.8	13760.6
628.3	908.1	1212.2	1678.1
724.3	1207.9	1726.2	2341.7
782.8	1397.4	1795.9	2433.6
241.5	472.2	717.6	1026.8
1671.6	2531.7	3107.7	3597.5
40.1	87.5	118.6	194.2
15261.5	24122.4	31038.0	39529.8
210.0	269.8	305.5	347.3

Source: Economic Planning Agency, *Annual Report on National Accounts.*

(unit: billion yen, current prices)

1975	1980	1985	1990
68.2	385.2	605.7	783.9
-31.1	7.3	-26.7	-109.5
542.6	1438.9	1969.4	2573.1
579.7	1831.5	2548.3	3247.6
210.0	269.8	305.5	347.3

Source: Economic Planning Agency, *Annual Report on National Accounts* .

equal to depreciation at a compounded interest rate.[8]

Table 11-6 has been prepared in order to provide a more vivid picture about the diffusion of consumer durables. Under the macro-aggregate, one can observe how the Japanese households started to own durable goods as the income level rose and new products became available. One advantage of this data set is that, rather than showing the percentage of households which possess a particular item, it shows the number of units possessed. The difference between the two approaches, needless to say, is that in the former case the maximum rate of diffusion is 100%, whereas in the case of the latter, it can take values exceeding 100% (here the figures are per 1000 households). This dataset is useful in tracing the changes in lifestyle.[9] In addition, given the data on energy efficiency of items such as refrigerators, television sets, air conditioners, and automobiles, the diffusion data can be useful in analyzing

Table 11-4 Adjusted private consumption

Items	1955	1960	1965	1970
SNA private consumption	5645.1	9065.2	18631.3	37521.3
imputed rent of owner-occupied dwellings				
By type of expenditures:				
1. Durable (deduction)	96.5	380.7	1031.3	2772.4
2. Semi-durable				
3. Nondurable				
4. Services				
By purpose:				
1. Food, beverages and tobacco				
2. Clothing and footwear				
3. Gross rent, fuel and power				
4. Furniture, furnishings, etc.				
5. Medical care and health expenses				
6. Transport and communication				
7. Recreation, education, etc.				
8. Miscellaneous				
Commuting expenses (deduction)	17.8	62.1	84.2	178.8
Ceremonial expenditures (deduction)	40.1	69.2	151.7	324.2
Adjusted private consumption	5371.1	8254.3	16750.9	32764.5
Deflator				

Note: For classification of household goods and services into durables etc., see United Nations, *A System of National Accounts*, Table 6.1. Commuting expenses and ceremonial expenditures are estimates based on household survey data. See text for details. Figures for 1955-1970 are from Kanamori (1978) based on old SNA data. Deflator shown is for total. Total household consumption includes adjustment for purchases of nonresidents.

energy demand in the household sector.[10] Some items such as television and audio equipment are deeply related to time budget.[11] Still others may be labor saving at home, affecting the female labor supply.

Services attributable to social capital: The stock of social capital such as housing, city parks, schools, water supply, among others is related to the well-being of households. Flows of services from these stocks are imputed and included (Table 11-7). The data source in this case is an estimate by the Japanese government which classifies social capital into 20 categories and 5 subcategories.[Economic Planning Agency 1976] Drainage and waste water disposal are related to the quality of life but are considered to be of a defensive nature. Roads, rail transportation, mail service, and telecommunications are

(unit: billion yen, current prices)

1970	1975	1980	1985	1990
37804.7	83904.7	139506.4	186234.6	241081.1
n.a.	n.a.	n.a.	22727.0	31699.7
2349.9	4836.2	6696.5	11442.9	16916.8
5373.7	11822.0	16308.7	22561.5	28295.3
14055.0	30234.3	45397.5	58361.3	67677.3
16028.5	36713.7	66347.3	92969.4	125073.7
11502.6	23781.8	34050.9	41537.2	49189.7
2923.9	6785.1	9315.8	12490.5	15308.6
6133.8	13020.3	24718.5	35082.3	46010.4
2893.2	5313.1	8043.2	11478.2	14669.7
2979.0	7504.1	13372.2	19548.7	25450.4
2934.5	7984.7	12698.3	18119.6	24450.4
3484.3	7327.5	11945.0	18259.3	23832.9
4955.7	11889.5	20606.1	28819.1	38983.6
347.5	781.3	1521.0	2209.0	2709.2
276.0	664.2	1330.5	2053.6	1903.1
34408.6	77638.3	128470.5	170529.0	219552.0
=100.0	167.9	229.5	263.2	276.1

Sources: Economic Planning Agency, *Annual Report on National Accounts*end, Bureau of Statistics, *Annual Report on the Family Income and Expenditure Survey*.

at least partially related to the well-being of the people, but the benefit received by invididuals is considered to be already included in the consumption expenditure above.

Leisure time: There are three items which are related to the time budget. One is leisure time which we discuss in this section. Second is the domestic service of housewives, and the third is time spent for commuting, which is part of the loss due to urbanization. Unlike earlier estimates of NNW which had to compile data from various sources, we are now able to rely on a time budget study conducted by the Japan Broadcasting Corporation. This survey has been conducted every five years since 1960 (with an extra survey in 1973), but suffers from changes in definition up to around 1970.[12] With the accu-

Table 11-5 Stock of consumer durables and imputed service

Items	1970
Net stock, end of the year:	
1. Furniture, fixtures, carpets, etc.	3042.8
2. Household appliances	2199.1
3. Transport equipment	2106.0
4. Radios, television sets, etc.	1810.5
5. Photographic equipment, musical instruments, etc.	606.2
Total	9764.7
Flow of services of consumer durables	2286.8
Deflator (PCD)	=100.0

Note: For derivation of flow of services, see text and footnote.

Table 11-6 Selected consumer durable goods owned by households

Items	1959	1964
Chest of drawers for 'kimono'	1555	1518
Chest of drawers	772	1086
Wardrobes	674	970
Dining table sets	—	208
Kitchen cabinets, sideboards, cupboards	—	1665
Vacuum cleaners	57	440
Washing machines	334	786
Refrigerators	69	637
Gas water heater	—	—
Microwave ovens	—	—
Sewing machines	731	919
Electric fans	347	922
Japanese electric warmers	—	1140
Kerosene stoves	444	485
Room air conditioners	—	23
Automobiles	('60) 12	68
Scooters, motorcycles	94	226
TV sets, monochrome	323	878
TV sets, color	—	—
Stereo phonograph sets	—	259
Video tape recorders	—	—
Cameras	439	799
Pianos	18	45

Note: The data are as of the end of November for ordinary households with two or more members. The survey has been conducted every five years since 1959. The sample size for 1989 was about 55,000. The data for monochromeTV sets and automobiles (1960) have been adopted from Economic Planning Agency, *Current Consumption Survey*.

(unit: billon yen, current prices)

1975	1980	1985	1990
6455.2	9526.8	12180.4	17946.3
5235.2	7820.9	9812.2	11719.5
5586.6	11620.6	17839.5	27309.8
3456.2	4796.8	7054.5	8381.3
1163.1	2434.6	4195.8	5744.2
21905.3	36199.7	51082.4	71101.1
5130.2	8477.9	11963.4	11651.8
138.3	147.9	139.9	120.0

Source: Economic Planning Agency, *Annual Report on National Accounts.*

(unit: per 1000 households)

1969	1974	1979	1984	1989
2416	1299	1278	1303	1288
—	1466	2027	2179	2125
1179	1413	1480	1601	1634
349	533	666	728	748
1442	1657	1880	2015	2105
787	1014	1146	1220	1277
975	1023	1054	1061	1082
944	1083	1125	1154	1200
426	695	786	743	553
—	129	301	541	761
979	1006	1030	1037	—
1282	1586	1747	1847	—
1262	1419	1527	1577	1043
1098	1461	1602	1629	1382
81	286	643	820	1131
278	512	670	837	1090
217	177	225	466	236
947	557	269	—	—
231	1066	1422	1685	1927
415	566	677	723	667
—	—	62	316	834
997	1127	1238	1331	1385
86	135	190	219	254

Source: Bureau of Statistics, *National Survey of Family Income and Expenditures.*

Table 11-7 Stock of social capital and imputed service

Items	1955	1960	1965
Transportation and communication:			
1. Roads	1764.8	2421.0	4484.5
2. Harbors	512.7	639.4	967.7
3. Airports	5.8	11.9	37.6
4. National Railways	2928.9	2922.6	3792.7
5. Japan Railway Construction Corp.	—	—	31.7
6. Subways	44.6	143.4	386.4
7. Telephone & Telegraph Corp.	762.8	1316.0	2607.3
Livelihood-related:			
8. Public housing	427.7	797.0	1420.3
9. Sewerage	473.4	610.1	902.7
10. Waste disposal	5.9	27.7	133.8
11. Water supply	428.0	702.2	1318.6
12. City parks	221.9	228.8	264.5
Education:			
13. Schools	2372.4	2549.7	3108.0
Land preservation:			
14. River improvement	1744.6	2078.6	2635.1
15. Forestry conservation	255.8	363.1	527.4
16. Coastal preservation	95.8	209.6	402.7
Agriculture, forestry, and fisheries:			
17. Agriculture, etc.	1958.8	2600.7	3749.5
Others:			
18. Postal administration	51.1	69.3	96.7
19. National forest	408.5	668.7	971.3
20. Industrial water supply	4.5	42.3	233.1
Total:	14468.0	18402.1	28071.6
Household-related stock	3450.0	4277.7	6111.4
Flow of services	169.7	210.4	300.6
Deflator (PIG)	60.3	70.8	85.2

Note: The capital stock data refer to gross stock. Totals include items not elsewhere classified. The 1990 figures are provisional pending official estimates.

mulation of data since then, we now consider this to be the most comprehensive time budget study available for Japan and can conduct time series comparisons with some caution. As for working hours, however, we can obtain more detailed and reliable data from the *Monthly Labor Survey*, which we use as a supplementary source.[13]

The definition of leisure can vary. If we observe the data over time, we realize that even sleeping hours are variable (people tend to sleep less now)

(unit: billion yen, 1970 prices)

	1970	1975	1980	1985	1990
1.	8334.0	15900.0	25301.5	36260.4	
2.	1531.2	2726.5	3900.2	5134.9	
3.	127.9	388.9	682.7	951.2	
4.	5118.0	6948.6	8857.3	9515.6	
5.	279.6	944.3	2031.3	2731.6	
6.	927.6	1638.4	2585.8	3504.9	
7.	5087.6	10228.2	15476.3	—	
8.	2591.5	3857.7	4883.0	5704.0	6662.9(p)
9.	1599.9	4065.2	8553.3	13830.4	
10.	277.2	724.1	1377.7	1937.4	
11.	2279.3	4932.9	7972.1	10106.9	12811.9(p)
12.	358.0	562.4	975.5	1693.0	2938.2(p)
13.	3800.5	6957.9	12546.0	17982.7	25775.0(p)
14.	3733.1	6300.5	9926.7	13967.6	
15.	763.1	1300.7	2107.7	2920.5	
16.	574.0	844.4	1238.1	1667.1	
17.	5808.4	10610.6	17690.2	24792.0	
18.	155.1	271.6	474.4	706.6	
19.	1255.1	1775.0	2659.6	3627.4	
20.	397.1	714.1	1027.6	1288.6	
	44968.5	94602.2	158409.0	205895.8	
	9029.3	16310.9	26376.6	35486.6	48188.0
	444.0	802.4	1297.7	1745.9	2370.8
	=100.0	165.1	226.3	239.2	261.2

Source: Economic Planning Agency, *Social Capital Stock in Japan.*

and that people are spending more time on "eating" and "taking care of oneself". A proper way of treating leisure, it seems to me, is to refer to the activities to which an increasing amount of time is spent as income goes up. Another point in measuring leisure is involuntary leisure caused by unemployment. In this study, we have ignored all the refinements.

Our procedure is as follows. We refer to the national average, although distinction by sex, age bracket, occupation, etc. are available. We use the

Table 11-8 Value of leisure

		1955	1960	1965
Leisure per person	(hours/year)	1305.2	1166.5	1339.0
Monthly wages	(1000yen)	18.343	24.375	39.360
Monthly working hours	(hours)	194.8	202.7	192.9
Hourly wages	(1000yen)	0.0941	0.1202	0.2040
Population aged 15 to 64	(million)	54.7	60.0	66.9
Aggregate value of leisure				
current prices	(billion yen)	—	0	2355.0
1970 prices	(billion yen)	—	0	4680.2

Note: Leisure here refers to free time excluding duplication of activities.

Source: Estimates by the present author.

Table 11-9 Time budget, adult male, weekdays

	1960	1965
Requisites:	9:48	10:01
Sleep	8:15	8:10
Meals	1:08	1:15
Taking care of oneself	:25	:36
Time spent for livelihood:	9:25	9:09
Work	8:10	8:07
Housekeeping	:38	:26
Commuting	:37	:36
Work-related activities	—	—
Free time:	—	—
Personal association	:35	:37
Leisure activities	:22	:29
Picnic and walking		
Sports		
Betting and games		
Hobby and taking lessons		
Moving n.e.c.		
TV-watching	:53	2:47
Radio listening	1:30	:31
Newspaper, journals, and books	:44	:47

Source: Japan Broadcasting Corporation, *National Time Budget Survey*.

time budget for weekdays, Saturdays, and Sundays. Time is allocated among requisites, livelihood, and free activities, and we define leisure as the last of the three (Table 11-9 for details). Whenever more than two activities are recorded

1970	1975	1980	1985	1990
1453.4	1572.1	1556.5	1561.7	1668.3
75.67	177.21	263.38	317.09	370.00
186.6	172.0	175.7	175.8	171.0
0.4055	1.0303	1.4990	1.8037	2.1637
71.5	75.8	78.8	82.5	85.9
8323.6	31675.8	46087.6	58808.8	93264.4
8323.6	12464.8	12464.2	13218.6	17475.2

(unit: hours & minutes)

1970	1975	1980	1985	1990
10:37	10:40	10:39	10:29	10:25
8:04	8:06	8:02	7:50	7:47
1:31	1:31	1:33	1:33	1:35
:58	:59	:59	:59	:55
9:23	8:55	9:05	9:17	9:10
7:54	7:15	7:19	7:40	7:26
:28	:27	:29	:26	:33
:42	:53	:55	:49	:51
:13	:13	:18	:17	:14
3:51	4:19	4:05	4:01	4:11
:27	:33	:29	:38	:40
:08	:06	:07	:12	:11
:05	:05	:04	:07	:07
:07	:10	:07	:07	:08
:07	:12	:11	:12	:14
:08	:06	:11	:13	:14
2:47	2:58	2:57	2:41	2:47
:35	:43	:45	:40	:35
:40	:44	:48	:48	:45

at the same time (watching TV while eating, for instance), the three major categories take precedence in the order listed above (eating is recorded in this example). An annual aggregate is obtained by considering the increasing

Table 11-10 Automobiles and traffic accidents

	1955	1960	1965
Traffic accidents (unit: cases):			
1. Number of cases	93981	449917	567286
2. Deaths	6379	12055	12484
3. Injuries	76501	289156	425666
Loss due to accidents (unit: billion yen):			
1970 prices	87.0	197.7	233.6
Current prices	24.3	67.8	128.3
Car fleet (unit: 1000):			
Trucks	693	1321	2870
Buses	35	58	105
for business	33	53	77
for private use	2	4	28
Passenger cars	158	440	1878
for business	47	76	151
for private use	111	364	1727
Light vehicles	531	1460	3058
Total motor vehicles	1502	3404	8123

Note: Total includes other types of cars such as those for special use, extra-large cargo, etc.

population (aged 15 to 64 years) but disregarding (for brevity only) paid holidays. Time budget data are then converted to money terms based on average hourly wages. A constant price series is then obtained, deflating by consumer prices. The result is reported in Table 11-8.

One additional consideration in our estimate is the fact that working hours tended to increase before full employment was achieved around 1960. We regard shorter working hours as involuntary and disregard leisure before 1960. In other words, we recorded the increment in free time after 1960 as 'leisure.'

Extramarket activities: The imputed value of domestic services of housewives corresponds to this item. According to a time budget survey, Japanese housewives spend 7 hours and 18 minutes on housekeeping during weekdays [Japan Broadcasting Corporation]. Compared with 1970, this figure is 39 minutes shorter. Judging from the increased automation at home which we observed above, it is not clear if this represents a real decline in services rendered. Incidentally, time spent on housekeeping by adult men is increasingly

1970	1975	1980	1985	1990
718080	472938	476677	552788	643097
16765	10792	8760	9261	11227
981096	662467	598719	681346	790295
408.2	260.7	233.0	257.5	304.0
408.2	614.6	804.0	1067.8	1490.1
				(1989)
5460	7381	8682	8306	8695
190	219	229	231	242
86	87	88	90	94
105	133	141	141	148
6777	14822	21544	25848	30882
217	243	250	253	257
6559	14579	21293	25595	30625
5968	5867	7297	12062	15975
18919	29143	38992	48241	57994

Source: Ministry of Transport, *Motor Vehicles Ownership*.

marked, particularly on Saturdays and Sundays. These factors, however, are not taken into account according to the convention followed in the earlier NNW estimates. The imputation is carried out based on the average wage of female workers. Covered are members of a female population who are 15 years or older and who are exclusively engaged in domestic work.

Loss due to urbanization: Economic development accompanies urbanization. In the pursuit of economies of scale and economies of scope, production activities tend to concentrate in urban areas. A positive effect of urbanization is that it improves efficiency, which is recorded as the growth of GNP. The negative side of urbanization is that it brings about various inconveniences, which go unrecorded in GNP because it is external to market activities. Examples are traffic congestion, car accidents, street crimes, and longer commuting hours, among others.

Following the original NNW estimates, we take up the problem of commuting and traffic accidents. A time budget study is the data source for our

Table 11-11 Loss due to environmental pollution

(unit: billion yen, 1970 prices)

	1955	1960	1965
Air pollution:			
SO_2	—	289.3	750.7
Soot and dust	34.6	89.3	177.4
Automobile exhaust gas	85.1	187.3	457.6
Water pollution:			
Industrial	—	299.3	630.7
Household, etc.	—	258.3	512.1
Waste:			
Industrial	—	—	948.4
Household, etc.	—	—	87.0
Total	119.6	1123.5	3563.9

Note: NOX emission from fixed origins and CO_2 are not included here due to unavailability of cost data thereof.

estimates on loss due to commuting.[14] According to the national average of hours spent for commuting, imputation is done based on average wage.

As for the loss caused by traffic accidents, deaths and injuries are included in the calculation. We have used the value of life calculated for the 1970 base year in the original NNW estimates, which stood at 10,884 thousand yen on the average for a death and 230 thousand yen for each injury. Such estimates depend on life expectations, labor force participation, and annual wage income.[15] Constant price series is derived based on the number of cases reported and converted to current price series based on the nominal wage index.

The year 1970 was the worst one as far as traffic accidents are concerned, with 16,800 deaths. By 1980, the situation had improved somewhat, with 8,800 deaths. Despite a considerably larger car fleet and longer distances travelled, the improved road conditions and provision of guard rails, etc. probably contributed to the decline.

Environmental pollution: The NNW Measurement Committee notes in their report that two methods are conceivable in estimating the social cost of environmental pollution.

(a) The direct estimation in money terms of damage caused by the above-stated discharge of polluting factors, such as that to health,

(unit: billion yen, 1970 prices)

1970	1975	1980	1885	1990
1727.9	1167.9 │	832.6	702.7	641.0
368.7	375.9 │	266.2	216.0	180.2
1151.1	1390.4	1121.1	668.6	738.8
1078.7	639.5 │	355.8	241.6	189.2
664.6	842.1 │	689.6	537.2	449.0
1138.9 │	920.6	452.7	339.1	283.2
106.7 │	124.8	175.4	121.4	99.4
6236.6	5461.2	3893.4	2826.6	2580.8

Source: Recent figures (marked by vertical lines) are estimates by the present author. Figures for 1955-1965 are from the Economic Council of Japan [1973]. 1970-1975 figures are from Kanamori, Takase, and Uno [1977] except for damage due to industrial and household waste for which the Economic Council figures are used.

human life, animals and plants, and properties.

(b) The method of estimating expenses necessary for the proper treatment of polluting factors overflowing into the present environment to an assumed "normal physical environmental level." It is a method to estimate expenses necessary to make the external diseconomies internalized.

While the former is more properly called environmental damage, the latter is judged a feasible method due to the fact that comprehensive, time-series surveys on the damage caused by individual polluting factors are still inadequate in Japan.

Environmental disruption was extremely strong in Japan from the mid-1960s to the mid-1970s. This necessitated the formulation of environmental policy in Japan discussed elsewhere.["Pollution Prevention Investment and Environmental Quality" in Uno 1987] A collection of statistical data was also initiated to support policy decisions. The cases in question included the emission accompanying various productive activities, such as industrial wastes, water pollutants, and air pollutants. The data were collected for each industrial sector for a particular year, typically around 1970, when the pollution problem was most threatening. This is one of the reasons for choosing 1970 as the base year.[17]The committee set a base year in which the environmental quality was deemed to have been acceptable. They chose 1955 for this purpose. Second, emission in excess of this initial period is estimated for subsequent years, and finally the removal cost is calculated. In this paper, we have followed the methodology established by the committee. The results are reported in Table 11-11.

Representative factors causing environmental pollution are classified as follows in this study:

- Air pollution
  - Stationary sources
    - Sulphur oxides
    - Soots and dusts
  - Mobile sources — Automobile exhaust (nitrogen oxides, carbon monoxides, hydrocarbons)
- Water pollution
  - Industrial water — BOD (biological oxygen demand)
  - Household water
- Wastes
  - Industrial wastes
  - Household wastes

Sectoral breakdown is available for most of the items for the base year.[16] However, time series data are available only for limited items. Thus, in our estimates of the cost of environmental damage, we need to take a detailed account in 1970 as the starting point and devise various ways to capture the time trend.

There are two points to be made in conjunction with the methodology. First, in contrast to a direct estimation of the cost of environmental degradation which we adopt here, there is an indirect method. This is the case in Nordhaus and Tobin [1972, pp.48-54]. See section 11-2 of this chapter. From an analytical point of view, their methodology can be more conprehensive than the direct method and can override the problem of lack of data concerning the cost of pollution. From a policy point of view, however, the direct method is preferable in that they explicitly recognize the sources of pollution, the quantity of pollutants discharged, and the policy leverage for improving the situation. The cases in question are the quantity of discharge of various pollutants, pollution prevention facilities, removal rate, and disaggregation by types of economic activities.

The second point is the concept of 'cost'. In our estimate this is the expenses required for the proper treatment of pollutants. This is the cost we have to incur in order to restore the quality of the environment to a level acceptable to society.[18] In other words, we refer to the measured cost of delivering a better environment. This treatment is consistent with the other parts of SNA. When we talk about consumption, it is the cost of delivering a bundle of goods and services measured at the margin, i.e., the area represented by marginal cost times the quantity. We do not measure consumer surplus which is represented by the area under the demand curve exceeding cost. In welfare economics, it is customary to use 'willingness to pay' as the measure of cost people are ready to incur in order to restore environmental quality. However, this is only good on the basis of individual decisions. It would be wrong to sum up a willingness to pay at the micro-level to arrive at the willingness of the society to pay as a whole. The point can be seen by noting that as the quality of the environment improves, the individual willingness to pay will almost certainly decline.

The convention followed in this paper is as follows: For each pollution category, we first examine the unit output coefficients of pollutants. These coefficients give the amount of pollutant generated in individual sectors in physical terms per unit of activity; for example, so many tons of industrial wastes per one billion yen worth of production. Second, we estimate the total quantity of pollutant output as the economy expands in its size and changes in its structure. This is achieved by multiplying the unit output coefficient vector with the amount of activity in each sector at different points in time. Third, based on the cost of removing a particular pollutant, the aggregate cost is estimated by imputation.

The items listed are not comprehensive enough to cover all of the problems related to the environment. Apart from additional aspects of the environment which are difficult to put into quantitative terms, there are aspects which can be treated in operational terms but are not recognized in the framework, which is already some 20 years old. One such case is the impact on global warming. Another would be radioactive wastes. These remain future tasks.

11.4 Social concerns, SNA, and modeling

Some 20 years have passed since the framework was first put into practice. In the meantime, there have been developments with regard to the data sources. The concept of satellite accounts, which is sometimes referred to as building blocks, has now been widely accepted in order to better reflect various social concerns in a framework providing detailed information in respective areas of interest but still being consistent with SNA. New technology has been developed, and the real resource requirement is changing rapidly in various segments of our society. The scope of our interest in the quality of life measurement has been widened to include not only regional or national concerns but, more importantly, also global issues such as sustainable development .

Should we be dealing with the social concerns, some of which we have discussed in section 11-2 of this paper? Is it worthwhile to try to combine such diverse aspects as floor space, refrigerators, automobiles, leisure, waste disposal, and pollution into a composite measure of the quality of life explored in section 11-3?

If we are to deal with such an extended scope of social concerns in an empirical context, we apparently need statistical data sets going beyond the production boundary of SNA. This is not to say that SNA has become irrelevant; on the contrary, SNA is the central core of satellite accounts and serves as the linkage among various accounts as well as providing a detailed description of the production sector. Here is an inventory of the statistical framework which can be useful for such an endeavor.

Capital stock: The most comprehensive account in SNA regarding the asset of a nation is the one dealing with closing stocks, capital transactions, and reconciliations.[19] The main items listed include the following. We have omitted the details of inventories (disaggregated to 5 categories) and financial assets (15 categories) as well as liabilities.

1. Tangible assets:
    (a) Inventories
    (b) Net fixed assets:
        i. Dwellings
        ii. Buildings other than dwellings

iii. Other structure

iv. Transportation equipment

v. Machinery and equipment

(c) Nonreproducible assets

i. Land

A. Residential

B. Agricultural

C. Others

ii. Forest

A. Forest land

B. Trees

iii. Underground resources

iv. Fishing ground

2. Financial assets

Within "machinery and equipment," we would like to see disaggregation into industrial sectors and by purpose (production, pollution prevention, R&D, for example). This can be supplemented by data from other sources.

There seems to be no objection, theoretical or otherwise, for including "consumer durable goods" among the net fixed assets in this account. Relevant data are already present in SNA. (See Table 11-5).

The entries regarding land can be supplemented by various data which describe land use by function, ownership, etc., in physical terms. Examples are agricultural land by types of crops, industrial land, residential land, and roads.

Forest and underground resources are important as segments of environmental accounting in addition to being economic assets.

Intangibles: One thing which is not included in the SNA account above is intangibles. No one would dispute the fact that intangibles such as industrial patents and computer software have become very important assets whose formation requires a substantial input of real resources. A theoretical justification for their recognition in SNA is that, like other assets, they generate a stream of pecuniary returns.[20]

Table 11-12 Assets of the nation

(unit: billion yen)

	1955	1960	1965
I. Tangible assets:	32504.7	59718.7	118719.9
(1) Inventories	3019.7	5965.0	11160.4
(2) Net fixed assets:	13738.3	19552.3	40159.3
1. Dwellings	2418.6	3330.8	7453.5
2. Buildings	3487.8	5166.8	10627.6
3. Other structure	4366.1	6668.6	12668.5
4. Transportation equipment	844.1	820.3	1822.6
5. Machinery and equipment	2621.8	3565.8	7587.1
(3) Nonreproducible assets	15746.6	34201.4	67400.2
1. Land	11813.9	29105.1	60451.8
a. Residential	4920.2	18333.3	44802.8
b. Agricultural	6179.0	8999.9	12100.1
c. Other	714.7	1771.9	3548.9
2. Forest	3419.8	4646.3	6562.0
a. Forest land	780.2	1441.5	1800.4
b. Trees	2639.6	3204.8	4761.6
3. Underground resources	452.8	378.1	254.1
4. Fishing ground	60.2	71.9	132.2
II. Financial assets	18917.3	48121.3	122850.8

Note: Figures are as of the end of calendar years. Details of inventories and financial assets are omitted here. Net fixed asset total includes 'subtraction item.'

Population and labor force: Another type of asset which is not explicitly recognized in SNA is human capital. The characteristics of human capital range from educational background, labor force participation, their skill in working force, life expectancy and health. The time budget is also closely related to this. Part of the resource allocation directed to the formation of human capital is already included in SNA. See "education" in the consumption accounts of government and private nonprofit institutions serving household in tables 11-2 and 3. One of the items among social capital stock is education, which can be seen in table 11-7. Although this social capital account is not formally a part of SNA, underlying data are already consistent with SNA.

Also, I have compiled data on the industry-occupation profile of the Japanese working force since 1955. Their educational attainment is also available from the census.[Uno 1989a] The data on demographic variables such as births

1970	1975	1980	1985	1990
313385.3	735190.9	1337695.7	1804355.7	3442238.0
22823.8	45656.8	64702.1	68093.9	77167.6
98117.4	285672.4	526588.1	687381.1	973645.9
20678.4	68143.0	133684.0	159173.8	218349.0
23631.3	64949.8	116321.4	156334.9	215279.1
28412.1	87982.1	185844.5	252748.4	360961.1
4773.8	12805.3	17487.2	19351.9	32763.3
20622.0	51792.3	73251.1	99772.0	149862.9
192444.1	403861.7	746405.5	1048880.7	2391424.5
181530.8	376405.7	705792.7	1004072.6	2338238.7
138558.5	289550.4	558161.6	813880.0	1987536.9
35616.7	64013.2	99315.5	121806.5	205068.1
7355.6	22842.1	48315.6	68386.1	145633.7
10194.4	26566.8	39009.7	43232.1	51565.2
3196.6	7198.9	9364.5	9626.1	12771.3
6997.8	19367.9	29645.2	33606.0	38793.9
379.6	316.5	567.2	880.8	883.1
339,3	572.7	1035.9	695.2	737.5
280050.8	655062.8	1297547.9	2131549.4	3664097.9

Source: Economic Planning Agency, *Report on National Accounts from 1955 to 1989,* and *Annual Report on National Accounts.*

and deaths as well as labor force participation can be arranged by age bracket. [Uno 1990b]

Resources: The stock of resources is recognized in the assets account for a nation. We would like to see details of the stock by types of resource, a task which remains to be done if we want linkages with environmental accounting . The depletion of natural resources is described in the assets account, albeit as an aggregate.

The flow of resources in the economic transaction can be seen in physical tables accompanying the input-output tables.[21] Statistics in physical terms is indispensable in order to trace material flow within an economic system. This has become all the more important because increasing efforts are directed at the conservation of resources in the production process and recycling of wastes after use.

If the resources in question are traded internationally, one is obliged to examine international repercussions. International linkages can be achieved by the use of trade matrices. [United Nations, *Yearbook of International Trade Statistics*] One would also want to have statistics on the stock of natural resources of various kinds in major supplying countries.

Energy: In a sense, energy is no different from other resources. However, judged from its importance as an input to productive activities and to maintaining the quality of life for the household sector, energy deserves special treatment. The point is strengthened by the fact that the sources of energy have much to do with regard to global warming. Emission of CO_2 through fossil fuel burning is the major contributor to global warming. An additional factor would be the fact that an increasing amount of energy is recycled via the recovery of heat in the production process and in incineration of wastes.

The supply and demand of energy becomes important in this context. In Japan, they are depicted in the *Comprehensive Energy Statistics*. Among the sources of supply, a major distinction can be drawn between fossil fuels and nuclear. Nonconventional sources of energy such as solar, geothermal, waste incineration, energy recovery, etc. are also listed.

The energy demand in most industrialized countries is equally divided among industry, household, and transportation. The industry demand for energy is rather straightforward, although one should note the fact that the energy intensity varies widely among the industrial sectors, calling for disaggregated analysis.["Energy Prices and Energy Requirements" in Uno 1987, pp.257-270] The household demand must be explained in terms of heating or cooling needs as well as diffusion of various consumer durables. We have discussed this point in section 3. Transportation activities range over market and nonmarket spheres. Increasingly important in most industrialized countries is the nonbusiness use of passenger vehicles. This is directly linked to the diffusion of automobiles among households. From the environment point of view, transportation is the major source of air pollution, a point made in section 3 above. The provision of social capital such as expressways and airports is also directly related to transportation activities. We have touched upon this point in discussing social capital.

Model building: The reason why we need to construct econometric models based on social, economic, and environmental accounting is threefold. First, based on statistical accounts, we often derive coefficients of various sorts.

These coefficients are usually not fixed but are variables which require explanation. Second, there are linkages among various accounts. Final demand induces intermediate demand. We also need to consider the interdependence of the social, economic, and environmental variables. Third, we want to improve the situation (whatever that might be) through policy means. Model building and simulation are indispensable for this purpose.

SNA has traditionally served as the major source of data in econometric model building, the purpose of which ranges from short-term (quarterly) analysis focusing on business cycles to long-term (annual) analysis focusing on productivity and growth. The focal point in either case has been the aggregate output, although the former tended to concentrate on effective demand or GNE and the latter on supply capability or GNP.[22] Although they are numerically identical on the aggregate, explanatory variables are altogether different.

If we are to respond to the wide social concerns, the scope of statistical accounting has to be expanded considerably. Whether the production boundary of SNA should be expanded or not is another issue. The fact remains, however, that statistical accounts covering social, economic, and environmental spheres need to be developed. The scope of the model building has to be expanded accordingly.

Environmental accounting is an approach which has emerged in recent years, the purpose of which in most general terms is to monitor the sustainability of our socioeconomic systems. The framework is also intended to respond to global issues such as climate change and resource depletion. The concept of sustainability is not operational in itself as described in the original text. We have shown that environmental burden can be broken down into four factors: (1) population density, (2) income per capita, (3) resource use per unit of income, and (4) environmental damage per unit of resource use.

One should note that the composite measure of the quality of life discussed in this chapter is more suitable to replacing the income term in the above formulation. Our measure takes account of the service flow of housing, consumer durables, social capital stock, transportation, pollution prevention, and other activities which fall outside the production boundary of SNA, making it a better proxy for what we consider the levels of living which we want to sustain in the longrun.

11.5 Final words

In lieu of a conclusion, I would only like to point to the fact that what appeared to be a 'misuse' of national account data has turned out to be a 'proper use' by the development of various data sets covering social, economic, and environmental spheres and, more importantly, by the development of SNA itself.

FOOTNOTES

[1] Denison [1971]. See also his comment on paper by Nordhaus and Tobin in Moss [1973].

[2] For SSDS, see United Nations [1975]. For social indicators, see, for example, United Nations [1978], OECD [1982a].

[3] Peskin and Lutz [1990], p.20. Country experiences covered in the table include Canada, France, Japan, The Netherlands, Norway, the USA, and Germany. The works by Peskin and Repetto are also included, as well as the framework by the United Nations Statistical Office.

[4] Economic Planning Agency of the Japanese government formed a research team in 1991, of which the present author is the chairman. Interim report (in Japanese) is being prepared, entitled *Report on Preliminary Research in order to Add Environmental Accounts to the SNA** (Kokumin Keizai Keisan Taikeini Kankyo Kanjowo Fukasurutameno Yobiteki Kenkyu Hokokusho), Japan Research Institute.

[5] Earlier estimates are available in Economic Council [1973], Kanamori, Takase, and Uno [1977](in Japanese), Kanamori [1980], and Uno [1989b].

[6] United Nations [1969], pp.89. For empirical data for Japan, Economic Planning Agency, *Annual Report on National Accounts*, various issues.

[7] United Nations, ibid., pp.105-108. The classification scheme is summarized below. The symbols D, SD, ND, and S stand for durables, semidurables, nondurables, and services, respectively.

1. Food, beverages and tobacco (ND)
2. Clothing and footwear (SD)

3. Gross rent, fuel and power
   - Gross rents (S)
   - Water charges (ND)
   - Fuel and power (ND)
4. Furniture, furnishing, and household equipment and operation
   - Furniture, fixtures, carpets, other floor coverings and repairs (D)
   - Household textiles, other furnishing, and repairs (S)
   - Heating and cooking appliances, refrigerators, washing machines and similar major household appliances, including fittings and repairs (D)
   - Glassware, tableware and household utensils including repair (SD)
   - Household operation except domestic services
     Non-durable household goods (ND)
     Household services excluding domestic services (S)
   - Domestic services (S)
5. Medical care and health expenses
   - Medical and pharmaceutical products (ND)
   - Therapeutic appliances and equipment (D)
   - Services of physicians, nurses and related practitioners (S)
   - Hospital care and the like (S)
   - Service charges on accident and health insurance (S)
6. Transport and communication
   - Personal transport equipment (D)
   - Operation of personal transport equipment
     Tires and tubes, parts and accessories, and repair charge (SD)
     Gasoline, oils and greases (ND)
     Other expenditure (S)
   - Purchased transport (S)
   - Communication (S)
7. Recreation, entertainment, education and cultural services

- Equipment and accessories, including repair
  Wireless and television sets and gramophones (D)
  Photographic equipment, musical instruments, boats and other major durables (D)
  Other recreational goods (SD)
  Parts and accessories for, and repairs to, recreational goods (SD)

- Entertainment, recreational and cultural services, excluding hotels, restaurants and cafes (S)
- Books, newspapers and magazines (ND)
- Education (S)

8. Miscellaneous goods and services (S and SD)

[8] The interest rate is assumed to be 5.5 , which is the rate applied to a 6-month saving deposit in 1970, the base year. Assuming a 5-year life span, this means that the flow amounts to 23.41 of the value of stock. A detailed description of durability and prices of consumer durables is found in Bureau of Statistics and Economic Planning Agency [1973], pp.436-441.

[9] Disaggregation by income class or by region is also possible.

[10] See Uno [1991a] where preliminary analysis is attempted along this line.

[11] Stock data in SNA are net of depreciation. However, the stock of social capital from Economic Planning Agency [1986] is in gross terms. Following the practice in original NNW estimates, we have assumed that net figures are 60 of gross. Imputation is then carried out assuming a life span of 25 years and an interest rate of 6.5% , which is the rate paid in the base year for local bonds. See Kanamori, Takase, and Uno[1977], pp.40-41.

[12] Japan Broadcasting Corporation, various issues. Notes for time series comparisons are found on pp.19-21 of the 1990 issue, volume on time quantity.

[13] The model COMPASS built by the present author has working hour equations at the disaggregated industry level which are based on working hours from this source.

[14] On weekdays, commuters are spending one hour and 32 minutes in the Tokyo metropolitan area, one hour and 20 minutes in Osaka, and one hour and 3 minutes in cities with more than 500,000 population. Japan

Broadcasting Corporation, ibid., 1990 issue, p.33. The use of this data set increased consistently in the fields related to time budget. However, in this case, alternative data provided in the *Housing Survey* are more detailed in, for example, giving the number of people classified by hours spent for commuting and by ownership of houses.

[15] Economic Council [1973], p.196 for value of life for each age bracket.

[16] See Kanamori, Takase, Uno [1977] for list of such data. See also Mine, Norikazu, "The Estimation of Some Basic Data about Industrial Pollution"(J), *The Journal of Japan Economic Research*, March 1976.

[17] Another reason is that various other statistics are available for this particular year. Most important are the details of the structure of private nonresidential capital stock. See Uno [1987], pp.87-106. *The National Wealth Survey* was carried out in that year, the third after WWII, and nothing is available thereafter. It was also the year of the population census. Detailed SNA series are also estimated starting in 1970, although major series have been made available retroactively starting in 1955.

[18] This approach is sometimes criticized on the grounds that the cost cannot be uniquely determined if it is measured against some environmental standards variable over time and across localities.

[19] In the Japanese account, this corresponds to a table entitled "closing stocks, capital transactions, and reconciliation of assets and liabilities for the nation." This account now dates back to 1955. For recent years, a distinction was made between private and public sectors. See Economic Planning Agency [1991] and *Annual Report on National Accounts*, various issues.

[20] As for R&D, see "Research and Development in an Input-Output Framework—A Methodological Exposition—" in Uno [1989] and "The Impact of Research and Development" in Uno [1991e]. For software, see Database Promotion Center, *Database White Paper.*

[21] See, for example, physical tables in *1985 Input-Output Tables* compiled by the Ministry of International Trade and Industry. For empirical analysis, see Uno [1991c] which examines the resources (some recycled) in physical terms.

[22] See "A Survey on Econometric Models and Statistical Databases" by the present author in Uno and Shishido, eds., *Statistical Data Bank Systems*, Amsterdam: Elsevier Science Publishers, 1988.

Chapter 12

Social, Economic, and Environmental Data Set

12.1 The basic approach

People are asking new questions about development. They are increasingly concerned with the quality of air and water and are alarmed by the depletion of natural resources. The concept of sustainable socioeconomic development has received increased attention in recent years. Along with many others, Nordhaus and Tobin in 1972 noted, “there are some socially productive assets (for example, the environment) that do not appear in any balance sheets. Their services to producers and consumers are not valued in calculating national income. By the same token no allowance is made for depletion of their capacity to yield services in the future.”[Nordhaus and Tobin 1972, p.13]

In chapter 1, we set out by saying that it is necessary to operationalize the concept of sustainability. In the following chapters, we examined various aspects of the sustainability issue involving the environment, economy, technology, and society. If one just looks at the economic impact on the environment, based on the experience of industrial development since the 18th century, the logical conclusion would be to halt further growth for the sake of preserving the environment and to safeguard our future generations. However, if one looks at the reality of the global community today, one is struck to find vast differentials in the well-being of its member societies.

It seems imperative to develop technology which is more resource-conserving and amenable to the environment, not to speak of accelerating the pace of diffusion of advanced technology across national boundaries. It is also imperative to start reexamining our lifestyle and find a way of preserving the quality of life while making our consumption pattern less resource-intensive. Thus, environment, economy, technology, and society are the four pillars on which we want to construct a statistical information system.

The report of the World Commission on Environment and Development, published in 1987, states referring to sustainable development : "Humanity has the ability to make development sustainable—to ensure that it meets the needs of the present without compromising the ability of future generations to meet their own needs. The concept of sustainable development does imply limits—not absolute limits but limitations imposed by the present state of technology and social organization on environmental resources and by the ability of the biosphere to absorb the effects of human activities. But technology and social organization can be both managed and improved to make way for a new era of economic growth."[WCED 1987, p.8] "But physical sustainability cannot be secured unless development policies pay attention to such considerations as changes in access to resources and in the distribution of costs and benefits."[WCED 1987, p.43]

The attempts by economists and statisticians to define the concept of sustainability in an operational term can be interpreted within this context. They noticed two major drawbacks of standard national accounting. These include (a) emerging scarcities of natural resources which threaten the continued expansion of the economy and (b) the environmental degradation from pollution and its effects on human health and welfare. National accountants and environmental specialists endorse the idea of developing a satellite system of environmental accounts. The endeavor is expected to develop into an internationally recommended manual of environmental accounting . The recent UN manual [United Nations 1992, 1993a] is the first of these attempts.

It is our conviction that the established statistical framework has its own purpose and structure, and we should not tamper with it unless we absolutely have to. In other words, we need not develop a new statistical framework every time we are confronted with a new problem. Needless to say, we have to be certain that the existing framework is capable of handling a new issue, and if it does not, we should not hesitate to modify the existing scheme. Even in this case, we believe we would benefit more by creating linkages with the established framework than by inventing something completely new. The cases in question include, among other things, the SNA (system of national accounts , which by the way has evolved over the past years in order to cope with new issues and to improve its coherence) and input-output framework. There are more than 60 countries in the world which have compiled input-output tables at least once in the past, and there is a significant number of countries which are doing so at regular intervals [Bureau of Statistics 1991, pp.140-141].

This is in line with what the UN manual says: "The input-output framework is the best suitable economic one for analyzing environmental economic relations because it could easily be extended to include flows of natural resources from the natural environment as inputs of economic activities and the flows of residuals of production and consumption activities as unwanted output delivered back to the natural environment. The starting point for the natural asset accounts of the SEEA are the non-financial asset accounts of the SNA comprising also non-produced natural assets." [United Nations 1993a, p.26].

There are conflicting points, however, particularly in the case where valuation of the economic use of the environment is involved. Although almost all experts agree that environmental damage has to be measured and deducted from the positive achievement, they tend to differ in methodology.

Bartelmus, Stahmer, and van Tongeren [1991] in their attempt to integrate environmental accounting in the SNA framework distinguished conventional GDP, environmentally adjusted DGP (GDP adj.), sustainable gross domestic product (SGDP), and sustainable net domestic product (SNDP). Environmentally adjusted GDP is derived from gross output by subtracting not only the intermediate inputs (this is what is traditionally done) but also expenditures on environmental protection services that are presently included in final consumption. Their intention is not only to reflect environmental protection as a production result, but also to include environmental degradation by households as a component of environmental cost.

We denote the relevant variables in their framework (modified here for our purpose) as follows:

C:	final consumption (households and the government)
CZ:	expenditures on environmental protection services
DK:	consumption of fixed capital
DZ:	depletion or degradation of environmental assets in economic activities
E:	exports
F:	final demand
I:	fixed capital formation (reproducible and renewable assets)
M:	imports
V:	GDP
$V*$:	environmentally adjusted $GDP(= GDPadj.)$
$V**$:	sustainable $GDP(= SGDP)$
$V***$:	sustainable $NDP(= SNDP)$

Then,

$$V = C + I + (E - M)$$
$$= F$$

Adjusting for expenditures on environmental protection services which are traditionally included in the final consumption, we obtain the environmentally adjusted GDP as follows:

$$V^* = FD - CZ$$

Sustainable gross domestic product (SGDP) is defined as the difference between gross output and cost including environmental costs. It reflects cost estimates which would have been necessary to avoid, restore, or replace decreases of environmental quantities and qualities during the reference period. In the case of reproducible assets, the consumption of fixed capital is estimated as the amount required to keep capital intact. In analogy, for the 'use' of the environment, the amount of money to keep the environment intact or, in the case of exhaustible resources, to provide for alternative investment which keeps the income stream intact is estimated. For the degradation of environmental quality, this concept can only be realized by setting specific quality standards and by taking into account the available knowledge about environmental protection techniques.

$$V^{**} = V^* - DZ$$

Finally, the sustainable net domestic product (SNDP) is derived by subtracting consumption of fixed capital from SGDP.

$V^{***} = V^{**} - DK$

Their framework is a typical one in this field today, as can be witnessed by the fact that Stahmer acted as a consultant in drafting the UN manual and Bartelmus and Van Tongeren were on the receiving end at the UN Statistical Office. The framework now has to be validated in an empirical context, and the crux of the problem lies in estimating the term DZ in the above formulation, which represents "depletion or degradation of environmental assets in economic activities."

The UN manual on environmental accounting distinguishes three different valuation methods:

(1) market valuation according to the concepts of the non-financial asset accounts in the SNA;

(2) maintenance valuation, which estimates the costs necessary to sustain at least the present level of natural assets; and

(3) contingent valuation, which could be applied especially in estimating the value of the consumptive services of the natural environment. [United Nations 1993a, p.28]

Related to this point, the SNA manual, which underwent a revision in 1993, states from a slightly different perspective: "There are now three main approaches to environmental accounting and they complement and overlap each other. The first one, generally referred to as natural resource accounting, focuses on accounts in physical terms. The second approach, ... identifies the actual expenditures on environmental protection and deals with the treatment of environmental cost to natural and other assets *caused* by production activities in the calculation of *net product*. ... The third approach is a welfare-oriented one. It deals with the environmental effects *borne* by individuals and by producers other than the ones causing these effects. The latter effects may often be much larger than the cost caused and do not affect net product but rather *net income* through transfers of environmental services. Of the three approaches, physical resource accounting is the most advanced in terms of its practical implementation. Experience with monetary satellite accounting is much more recent, and many controversies still surround this approach, particularly with regard to valuation. The least consensus exists with regard to the welfare approach to environmental accounting." [United Nations 1993b, pp.508-509] Willingness to pay or contingent valuation is useful in finding out social concerns and setting priorities among various policy issues. This methodology, however, is alien to convention in established economic accounting such as SNA which records the market transactions in terms of marginal cost of delivering a bundle of goods and services, not including the consumer surplus which is defined geometrically in microeconomics as the area under the demand curve exceeding the cost.

Our approach regarding this point has consistently been to exclude the consumer surplus portion and to distinguish the actual transaction in the market and the imputed value of benefits and costs of nonmarket activities. The latter is derived from conventional SNA rather than by modifying the SNA boundary itself. As for environmental degradation, urban congestion, and some other items for which market valuation is lacking, imputation was applied based on the market-determined price parameters such as treatment cost and wages, but this is done outside the SNA core. Refer to chapter 11 for details.

Here we compare the concept of NNW (net national welfare) with that of GDP. We denote relevant variables as follows:

C: personal consumption expenditures
CZ: personal 'defensive' expenditures
CG: government current expenditures
CGZ: government 'defensive' expenditures
IGn: government net investment
IGg: government gross investment
IPn: private net investment
IPg: private gross investment
DG: government capital consumption allowances
DP: private capital consumption allowances
TL: imputed leisure hours
TW: imputed housewive's services
$VKCD$: imputed services obtainable from stock of consumer durables
VKG: imputed service from social overhead capital
VKH: imputed service from housing capital

C' and CG' are items which are modified from GNP in accordance with the concept of NNW.

The interrelationship among GDP and NNW is as follows:

$$GDP = C + CG + IGg + IPg$$

$$NNW = C' + CG'$$
$$= (C - CZ) + (CG - CGZ) + VKCD + VKH + VKG + TL + TW$$

One word about the term NNW. This term was first used by the Economic Council of Japan [1973] in an attempt to come up with a measure of economic well-being. However, the term caused unnecessary criticism in referring to "welfare". There is a strong tradition in mainstream economics for opposing the measurement of welfare, not to speak of its interpersonal, intertemporal comparisons or aggregation. This tradition is theoretically understandable. After all, what they did in estimating NNW was not to measure welfare, for, if it were, they would have to measure the consumer surplus portion of consumption in the conventional sense before attempting to evaluate nonmarketed phenomena. Actually, they measured consumption at marginal cost , as usual, and they measured pollution prevention at marginal cost, just as usual.

With this in mind, we have used the word MQL (measures of quality of life) in this chapter.

12.2 The framework of SEEDS: a social, economic, and environmental data set

The conceptual framework is one thing, and empirical feasibility is another. In this section, we examine the availability of statistical data in Japan. Lacking an indigenous supply of raw materials while being the world's second largest industrial power, she represents a typical case of the North. Also, due to the fact that the island country consists mainly of mountains and steep hills, her 124 million population is concentrated in about 20% of the already very small geographic span, aggravating the environmental impact of human habitat and industrial activities. Thus, Japan experienced keen environmental disruption already in the 1960s and the 1970s. Coupled with a series of oil crises, Japan was obliged to undertake fundamental changes in industrial structure and technology. The case of Japan, therefore, is probably illustrative of the problematics of the Northern response to the environmental issue.

The prototype is compiled based on a social, economic, and environmental data set (SEEDS). There are three points to be noted about SEEDS. First, it is not a rigid statistical system as is indicated by its name "data set". It is rather a group of modules, or building blocks, each designed to be compatible with related ones, which can be assembled for a particular analytical need in order to respond to a particular policy issue. It presupposes a basic statistical framework such as the SNA and input-output tables. It originates from an attempt to provide statistical accounts based on which social indicators can be derived.[Uno 1974, 1978] In a sense, it is similar to the proposed system of social and demographic statistics (SSDS), which is now being developed into the social accounting matrix approach.[1] The social indicator approach appeared in the 1960s with the proper intention of reflecting the limits of typical economic accounts, but ended up simply being a wish list with respect to various social concerns. In reality, human aspiration is limitless, whereas what is attainable is constrained by the productive capacity of a society. SEEDS therefore was built around an input-output framework so that various competing policy objectives would somehow be reduced (either via a market mechanism or by a political process) to a set of final demands which can be met by the productive capacity in hand. In this sense, the input-output framework was the core account of SEEDS. Its main focus has been to seek the relations between the quality of life and the economy.

Second, the framework includes, together with monetary measures, physical measures which relate to the quality of life. It is often the case that physical measures are more directly related to how people live. As an example, the number of housing units and the floor space of each unit is probably more relevant from the quality-of-life point of view, although the monetary value of housing investment is useful in obtaining an aggregate demand in the macroeconomic analysis.

Third, from the start, it distinguishes among statistical accounts, indicators (summary of the accounts, often scalar), and policy models (explicit causal relations among indicators). Therefore, SEEDS has always been accompanied by an econometric model, with a name of COmprehensive Model for Policy ASSessment (COMPASS), designed to respond to various policy analyses. Precisely for this reason, SEEDS emphasizes time series data, together with accounting consistency at a given point in time. It is important to be able to simulate economic and social phenomena, distinguishing appropriate policy tools in the model system. The accounting system in itself does not lead to policy decisions.

Fourth, as it started about 20 years ago, it has a fairly long history. The model has evolved from a two-sector development model in 1973 which generated social indicators into a 25-sector multipurpose model. The supporting data set SEEDS was also extended in its scope and sectoral disaggregation. COMPASS was used to describe the Japanese industrial development from 1955 to 1980 and was subsequently used for various simulation experiments including the analysis of the exchange rate appreciation after 1985.[Uno 1987, 1988] A more recent attempt has involved integrating the impact of microelectronics such as computer, communication, and control technology (industrial robots, numerically controlled machines, etc.) and the investment pattern and nonprice competitiveness in foreign trade.[Uno 1989, 1991]

The basic skeleton of the statistical system is presented in Figure 12-1, taking account of the availability of empirical data and our basic approach to the problem, spelled out in section 1 above.

Financing is important in discussing the environmental maintenance effort. The allocation of purchasing power occurs before the allocation of real resources for a particular purpose. In the case of the environment, various measures can be financed by private industry out of their retained profits, or borrowed from financial institutions, or subsidized or borrowed from public institutions such as local government. Investment by local government can be financed by tax, bond flotation, borrowing, or subsidy from the central government. Part of the effort has to be borne by the household, in the form of an increased cost to finance an anti-pollution catalyst in a car or to pay for the extra cost of garbage collection. We have dealt with the financing part of the problem in an earlier article, and it is omitted in the present work.

The structure of the diagram is obvious. Vertically, it shows:
(1) opening stock of produced and nonproduced assets (the latter including natural assets and environmental quality such as air, water, etc.),
(2) current activities, which now describes not only the production of goods and services but also the use of the natural environment for socioeconomic activities,
(3) closing stock, with corresponding scope for current activities,
(4) measures of quality of life , composed of private consumption (adjusted), government consumption (adjusted), imputed services of stock of consumer durables, housing, and social capital, imputed value of leisure and housework,

Figure 12-1 Conceptual scheme of Social, Economic, and Environmental Data Set

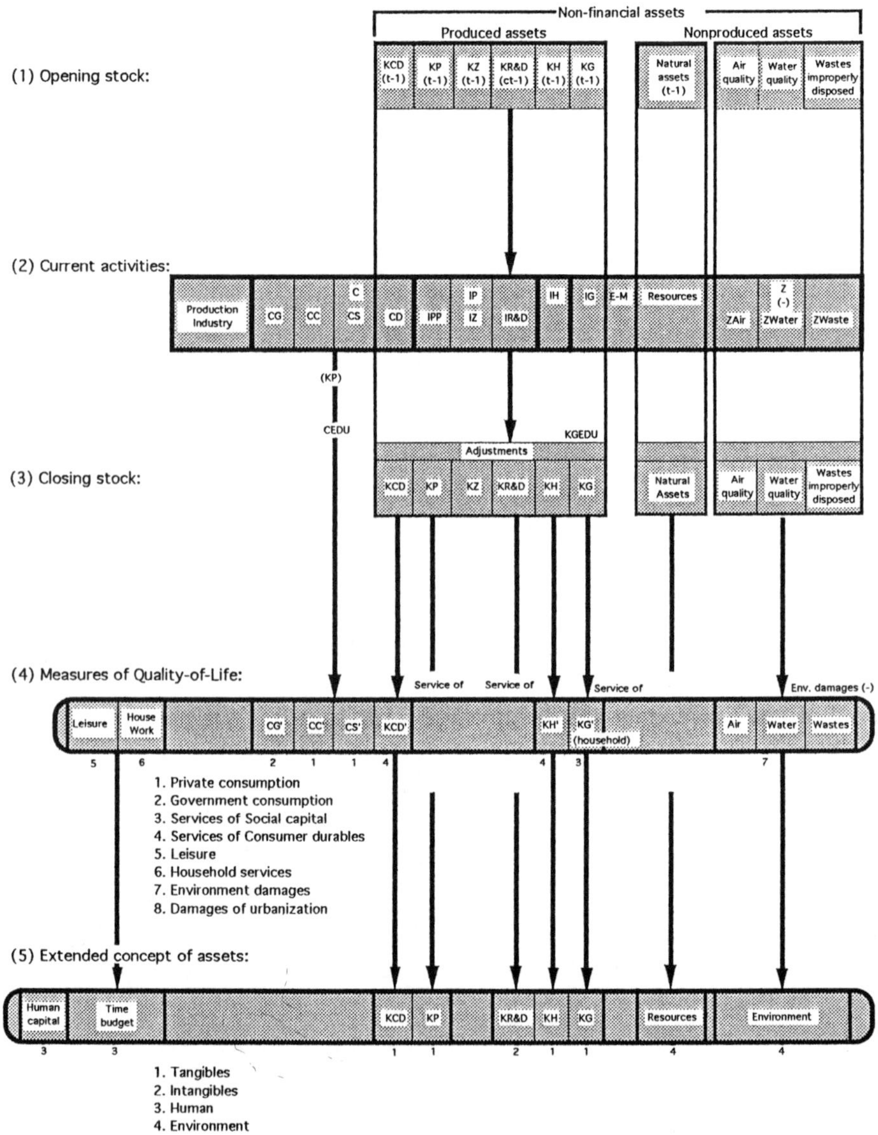

and imputed damage due to urban congestion and environmental degradation, and
(5) extended concept of assets including, in addition to the tangibles which have usually been the case, intangibles, human, environment, and institutional.

There are several points to be noted. First, the framework presents the stock (carry-over which exceeds the current period) as well as flow (activities in the current period). In conventional economic accounts, only the industrial plant and equipment are explicitly recognized as stock. The proposed framework includes, in addition, the stock of consumer durables and housing stock. Diffusion of automobiles and electric appliances has to do with the environment via an increased consumption of energy, harmful exhaust gas, and disposal problem after use. In the case of energy consumption, about 1/3 is attributable to the household sector (which incidentally continues to increase, reflecting the higher income), 1/3 to transportation (again, half of it is attributable to passenger cars owned by a household, which continues to increase), and 1/3 to industry (which tends to decrease, reflecting improved technology and structural change). Thus, household stock is too important to be neglected. The diagram also shows the stock of the natural environment, spanning mineral resources, forest resources, land, air, and water.

Ideally, the framework should also describe the accumulation of intangibles (technology, software, knowhow) and human capital (health, education, and skill). Environmental stocks are exploited in the current socioeconomic activities, and either consumed or accumulated. Accumulation can be in the form of plant and equipment, or in intangibles and/or humans. In this manner, one form of asset has been turned into another form of asset, leaving the society's net worth intact. If the development is to be sustainable, probably technology and quality of the workforce will be more important than tangibles in the sense that the former may reinforce and perpetuate the development process, whereas the latter is limited in its durability, likely a maximum of 20 years in the case of plant and equipment and 50 years in the case of social overhead capital. Some of the variables pertaining to this aspect are included in the proposed framework, but full implementation has not yet been achieved. The cases in question include a functional classification of activities of the household, government, and private nonprofit institutions serving household. From our perspective, they are important in that they relate to education and medical services. Another case in question involves research and development and further disaggregation of private investment in plant and equipment into pollution prevention, energy conservation, etc.

Second, physical measures are provided as well as monetary. This is now a standard procedure in the UN manual. It has been the case for SEEDS from the outset in order to deal with the variables which have to do with the quality of life. For instance, in the case of housing capital, the average floor space and the number of units are more relevant than the total value of dwellings. In the case of pollution prevention, the capacity of the pollution prevention facility has to be measured in physical terms rather than the amount of money which went into it. The proposed framework consists of physical tables which are prepared in parallel with monetary tables in order to explicitly describe the flow from natural environment to the economy, the flow within the economy, and the flow from the economy to the environment in the form of wastes.

Third, a portion representing current activities is very similar to the input-output framework. That portion describing the intermediate demand shows the input structure which is the production technology per se. Technological change such as energy conservation is described as the change in the input coefficients over time. The input-output table contains more than 500 sectors, but they are somewhat compressed in our case. Some of the environment-related activities are explicitly recognized, as described below. The portion pertaining to final demand is also similar to the usual input-output tables, and its only difference is that it is further disaggregated to explicitly describe the items related to the quality of life and environment. The cases in question include various consumer durables, pollution prevention investment , energy conservation investment , research and development, and provision of social capital (water supply, sewerage, waste disposal, forest preservation, flood control, in addition to road construction, etc.). This framework is also directly linked to emission factors pertaining to air pollutants, water pollutants, and waste output. These are the tables indicating the output of pollutants per unit of production (for example, one million yen). Given the final demand, the input-output framework generates a sectoral output, taking due account of interindustry repercussion. The unit emission tables then lead to the calculation of the output of pollutants corresponding to activity levels in individual sectors.

The SEEA proposed by the UN distinguishes several versions. One version is described as referring to "data according to the conventional concept of the SNA. These data are further disaggregated to reveal environment-related activities and the monetary flows and stocks connected. In this case, domestic production activities comprise only production activities of industries, and, therefore, produced assets contain only assets of industries. The use of products is limited to products of industries; and the use of assets, to the use of produced fixed assets of industries." Another version which aims to "describe different approaches of valuing imputed environmental costs imply the recording of additional costs associated with different economic activities (production, final consumption, use of produced assets), and with reverse sign, with volume changes of natural assets deteriorated by economic activities." "A third stage of development of the SEEA implies extensions of the production boundary of the SNA. The extended concept of household production activities is reflected in the SEEA matrix as an additional record of production activities ... and products. The corresponding extension of the concept of produced assets entails the introduction of asset accounts of consumer durables and the record of corresponding user costs. If environmental services are treated as production activities, a further extension of the concept of domestic production is necessary" [United Nations 1992, pp.48-50, also United Nations 1993a, pp.32-33].

On this point, the revised SNA explains, "In addition to the disposal services provided by the environment 'free of charge' in the case of degradation, actual expenses are incurred to avoid environmental degradation or eliminate the effects after degradation takes place. Increasingly, enterprises explicitly produce such services on a commercial basis. In many instances, however, the services are produced as the ancillary activities" [United Nations 1993b, p.516]

There is also an added factor which is the fact that, unlike earlier days when enterprises were retrofitting the existing facilities with pollution prevention equipment, today we have a situation where the production processes themselves have been altered to avoid or lessen environmental impact. Thus, the whole concept of investment for pollution prevention is becoming somewhat dubious.

In the current version, sewerage and waste disposal activities are made into independent sectors. In the prevention of air pollution typically due to SO_X

and NO_X, private enterprises are mostly responsible for the installation of purification equipment, and these are precisely where activities are ancillary. A similar situation prevails regarding research and development activities undertaken by enterprises. Unlike the activities of research institutions, which are recorded, business R&D are not fully recorded. The capital formation pertaining to R&D is included in private capital formation but is not earmarked by purpose, and current expenditure on R&D is also buried in the intermediate demand without demarcation. It is desirable to 'peel off' these expenditures from conventional activities. This remains a future task.

The Economic Planning Agency of the Japanese Government has recently initiated a project which aims at implementing the UN manual in the Japanese context. A preliminary study was carried out in 1991 which mainly consisted of a literature survey. A three-year project was then started, and an attempt is now being made to implement the parts of SEEA. Although I am serving as the chairman of the working group, the prototype presented is solely my responsibility and is not the outcome of a joint effort at the EPA. Rather, it is intended to demonstrate the validity of the approach represented by SEEDS. It is also intended to examine the validity of earlier attempts initiated by Japanese scholars, part of which is inherent in the current work.[2]

The System of Environmental Economic Accounts (SEEA) has been presented in a UN manual, but its empirical application has yet to be attempted by individual countries based on their historical experience and policy needs. The UN proposal is expected to evolve further as a result of continuing discussion based on country experience. The prototype presented here is one of these attempts.

12.3 Input-output framework as the core account

An attempt is made in this section to rearrange the input-output tables in order to focus on the interaction between the economy and the environment. This corresponds to the "current activities" indicated in Figure 12-1.

For illustrative purposes, let us start from the 36-sector classification of industry which has been followed in a series of the my work related to input-output framework and multisectoral econometric model building.[3] Some additional details are inserted in the table in order to illustrate environment-related activities, which include the following:

Sector number	Sector name
01	Agriculture, forestry, and fisheries
	Forestry
02	Mining
	Coal
	Petroleum
	Natural gas
07	Pulp and paper
	Pulp
12	Petroleum and coal products
	Petroleum products
	Coal products
25	Electricity, gas, water, and heat supply
	Electricity
	Gas and heat supply
	Water
	Sewerage
	Waste processing
32	Public service
	Research

Additional details are provided for forest products, fossil fuels, and sewerage and waste treatment which are deemed relevant to reflect environment-related activities in the interindustry transaction table. In addition, such sectors as chemicals (as inputs to agriculture), motor vehicles (as an addition to the existing stock of automobiles), and transportation (as an activity consuming 1/3 of the energy) are the ones particularly of interest for our purpose. Research activities are also related to pollution abatement and energy conservation.

These additions appear in tables 12-1, 12-2, and 12-3 with an asterisk. In the above, "sewerage" is a sub-sector of the water supply and is listed with double asterisks. It should be noted that these additional sectors are shown as subdivisions of the standard 36-sector industrial classification and not as

Table 12-1 I-O table with environmental aspects, 1990

		01 Agriculture, etc	* Forestry	02 Mining	* Coal	* Petroleum
01	Agriculture,etc	2091622	629930	2217	1692	1
*	Forestry	602673	595971	2217	1692	1
02	Mining	22	22	4435	0	179
*	Coal	0	0	126	0	0
*	Petroleum	0	0	0	0	0
*	Natural gas	0	0	1595	0	179
03	Food Processing	1381282	4170	70	14	14
04	Textiles	99054	12006	3871	374	10
05	Lumber and Wood	10437	4557	2918	495	0
06	Furniture, Fixtures	6421	630	1303	267	5
07	Pulp and Paper	200407	2	0	0	0
*	Pulp	2556	2	0	0	0
08	Printing, Publish	7774	658	4822	1060	40
09	Leather & Prod.	1972	493	4030	41	1
10	Rubber Products	16081	1445	8262	586	13
11	Chemicals	686821	24278	10834	792	1
12	Petroleum & Coal	432386	20383	123504	1589	184
*	Petroleum products	280709	18797	122621	1397	167
*	Coal Products	20	0	138	-31	0
13	Nonmetal Minerals	26294	4788	142	75	3
14	Iron & Steel Mat	0	0	0	0	0
15	Iron & Steel Prod.	485	12	5700	3216	0
16	Nonferrous Metals	0	0	0	0	0
17	Fabricated Metals	28245	8012	33182	5522	180
18	General Machinery	127859	7481	38439	2896	550
19	Electrical Mach.	4293	90	1135	2	0
20	Motor Vehicles	0	0	0	0	0
21	Other Transport	107409	8155	199028	776	81
22	Precision Inst.	724	319	19	3	0
23	Manufacturing nec	12272	587	4616	427	0
24	Construction	43309	7592	8208	801	66
25	El.,Gas,Water	61093	7708	42889	8894	692
*	Electricity	46932	5325	37223	8499	406
*	Gas & Heating	25	4	202	3	0
*	Water	13430	2379	2775	136	68
**	Sewerage	0	0	0	0	0
*	Waste Processing	706	0	2689	256	218
26	Wholesale Retail	593286	39888	114219	2764	171
27	Fin.,Ins.,Estate	710854	39277	212454	8709	883
28	Real Estate Rent	0	0	0	0	0
29	Transportation	463789	75988	87749	2953	293
30	Communication	10255	2025	6560	1429	63
31	Public Admin	0	0	0	0	0
32	Public Services	16225	675	7065	497	407
*	Research	10268	0	594	48	279
33	Private Services	45280	5157	68411	2029	468
34	Office Supplies	9293	743	3288	314	19
36	Activities nec	173970	3898	39861	3277	276
	Subtotal	7369144	910955	1039161	51480	4586
	Total Output	17237693	1622197	2087770	120934	10512

Note: An asterisk signifies environment related activities. Compiled from the Ministry of International Trade and Industry *Input-Output Tables 1990* [1993].

(unit: million yen, current prices)

	* Natural gas	03 Food Processing	04 Textiles	05 Lumber and wood	06 Furniture etc.	07 Pulp and Paper
01	8	10028453	418250	1258826	2899	39406
*	8	22772	887	1258786	2899	37816
02	1416	22	137	0	2	27906
*	0	22	110	0	2	13030
*	0	0	0	0	0	0
*	1416	0	4	0	0	114
03	14	6625603	41873	1520	15	11159
04	72	31987	4210467	6092	80872	21052
05	0	39121	2362	511769	557682	335301
06	63	17961	11951	1352	133791	4450
07	0	601552	113717	31294	91602	3105004
*	0	35754	7547	30709	13419	3023865
08	170	91537	28599	4107	12969	568436
09	0	1026	15128	1125	2033	234
10	93	3640	13012	850	6309	2771
11	8	387511	859089	119058	121795	291873
12	1108	839532	179302	21880	102807	311258
*	1017	164143	85719	10213	10122	180805
*	0	85	11	10	5	0
13	27	313545	2273	1853	108464	1236
14	0	0	0	0	0	0
15	76	12	0	15803	179827	0
16	0	39751	256	2701	25433	2259
17	332	802808	45602	50106	267268	6865
18	2051	190409	56784	20126	54125	43834
19	0	162	0	13	3755	0
20	0	0	0	0	0	0
21	612	34368	6127	7759	3403	7570
22	2	126	212	59	130	169
23	3	17864	206006	11572	35490	100
24	305	74109	22370	7669	9050	32720
25	5574	454895	245317	79900	46433	364340
*	5434	281831	193350	71376	37008	310246
*	0	46706	5827	55	1323	9574
*	96	119908	22978	3362	4617	18904
**	0	0	0	0	0	0
*	44	6450	23162	5107	3485	25616
26	1009	2070870	1316947	286290	221353	339474
27	3992	683282	587722	216890	201766	332697
28	0	0	0	0	0	0
29	1076	1136553	271975	186688	104461	425808
30	489	58681	31956	7887	14539	16455
31	0	0	0	0	0	0
32	384	150157	77456	8386	32968	50822
*	0	76653	25610	3472	19671	24331
33	2532	1143251	154984	34801	115862	142945
34	73	44947	36732	9885	9761	11190
36	1540	259581	67840	32393	86970	74556
	23015	26191450	9024418	2938640	2633820	6571862
	78585	41778324	13456348	4529669	4506089	9718296

"Repair, machinery" and "repair, auto" have been omitted.

	* Pulp	08 Printing, Publ.	09 Leather & Products	10 Rubber Products	11 Chemicals	12 Petroleum & Coal
01	39383	0	5794	59051	53368	125
*	37816	0	241	0	25392	0
02	27906	1	48	472	112784	5096665
*	13030	1	7	94	22039	579162
*	0	0	0	0	37	4500283
*	114	0	0	0	26206	1183
03	10993	14	116880	15	175164	1057
04	8469	29179	27192	147788	14857	47963
05	333515	695	10924	53	2205	2468
06	2391	7993	401	2410	13007	8303
07	1908339	1840472	29254	22943	440148	40179
*	1876564	1795625	1638	13462	97641	18075
08	153146	1495118	2095	9117	66234	17085
09	69	386	226970	473	1606	948
10	806	5958	33212	65576	20275	6464
11	211260	319642	14174	667976	7795872	2148717
12	186476	31545	53640	135450	1470170	2625322
*	165553	19508	3931	20847	1088441	513768
*	0	5	0	0	37202	28014
13	1236	164	148	8842	129966	91461
14	0	0	0	0	224	0
15	0	0	373	11079	478	8727
16	2081	15519	1961	15003	81703	7781
17	5015	3692	5862	78947	259672	83935
18	35711	38597	1325	37371	238560	252434
19	0	14	77	0	0	7309
20	0	0	0	0	0	0
21	4454	11855	1396	807	10654	6115
22	126	105	6	30	521	353
23	74	1495	17359	855	591	134
24	22742	16081	1976	5798	97427	38873
25	303089	117031	9372	66019	1094651	313681
*	258803	85754	7330	60941	888011	283356
*	5470	11097	84	2049	20125	3313
*	15988	3227	1407	1735	88401	21074
**	0	0	0	0	0	0
*	22828	16953	551	1294	98114	5938
26	167637	386095	71585	183970	589470	634115
27	190535	428563	29500	263117	1090122	575549
28	0	0	0	0	0	0
29	308392	284403	24640	77642	656528	434389
30	7109	54380	1886	7404	125017	57484
31	0	0	0	0	0	0
32	37576	30665	3234	39610	783783	72618
*	21832	10165	846	17417	670801	33003
33	40037	691293	16524	66037	1107794	273640
34	4222	27687	2671	6098	17080	16457
36	42023	135176	14217	47617	190154	45105
	4054798	5973804	724682	2027556	16639973	12915414
	5720011	11645571	1217260	3571325	25531785	21026076

	* Petroleum products	* Coal products	13 Nonmetal Minerals	14 Iron & Steel Mat.	15 Iron & Steel Prod.	16 Nonferrous Metals
01	0	125	503	564	22	276
*	0	0	0	564	22	0
02	4484155	612322	1041959	546707	15148	574091
*	5227	573932	73668	5087	294	2075
*	4500283	0	0	0	0	0
*	710	288	6557	3495	11194	2692
03	539	14	2881	14	61	28
04	623	2843	22136	612	8473	11545
05	0	0	41727	550	27025	6918
06	561	188	19680	925	4548	2958
07	0	0	115701	0	2960	1367
*	0	0	27925	0	1393	1289
08	2962	1177	33644	2904	8648	7059
09	31	404	1350	514	735	202
10	0	756	8237	4556	16230	1300
11	28701	5628	176668	77184	58115	113246
12	391978	137867	290509	564352	245523	65154
*	377689	109874	257081	13258	101250	49191
*	0	27993	16402	551094	144004	10144
13	525	16613	1205684	76281	55978	21701
14	0	0	0	2297991	4765953	0
15	0	0	99921	0	5472319	16868
16	406	1	24624	52085	180183	2941821
17	27825	5090	37477	145	38157	14688
18	89284	16096	188655	98010	159886	85791
19	0	0	314	0	0	1925
20	0	0	0	0	0	0
21	3685	139	23896	5476	16238	3526
22	33	4	233	39	43	53
23	3	92	5502	4	7100	4254
24	11048	7571	36633	37689	102321	33328
25	56312	15939	366192	446538	487418	318279
*	49300	12668	312363	430707	433806	300985
*	3	120	8990	1130	10168	4474
*	7009	2616	21607	10733	31401	10145
**	0	0	0	0	0	0
*	0	535	23232	3968	12043	2675
26	238718	56050	416442	89973	592586	205793
27	268715	17130	564573	217119	444115	249175
28	0	0	0	0	0	0
29	94764	112128	726558	331392	572317	283858
30	25807	1012	23140	1233	17312	16240
31	0	0	0	0	0	0
32	27297	5564	86758	27900	123412	61155
*	13997	2069	47365	6474	75299	41523
33	39288	6743	99584	45133	103321	58275
34	572	290	16019	1353	11180	5740
36	3520	11478	164571	126258	182715	77330
	5797338	1033250	5841715	5053487	13720000	5183916
	9531052	1354577	10205842	6556814	19449236	7231016

	17 Fabricated Metals	18 General Machinery	19 Electrical Mach.	20 Motor Vehicles	21 Other Transport	22 Precision Inst.
01	0	68	0	0	452	0
*	0	19	0	0	452	0
02	2469	14315	95	0	628	0
*	452	25	38	0	112	0
*	0	0	0	0	0	0
*	220	231	57	0	516	0
03	28	70	112	28	70	14
04	21102	34760	62378	24307	56092	5809
05	17009	15000	11523	1561	24901	2372
06	14426	10395	120413	9278	16281	1476
07	37679	35711	235434	1122	20850	27259
*	2664	9247	79357	4	7201	760
08	42784	91839	163260	13504	21738	12739
09	7148	5003	6695	6438	2992	50652
10	14884	362232	180915	340588	934799	16058
11	132201	189889	790746	99350	439978	44477
12	161168	280943	1500609	299821	933433	162786
*	92808	81078	143051	21191	100514	7016
*	1590	1767	489	456	2331	304
13	63882	236327	499876	170872	133785	59690
14	-38003	-27185	-12625	0	-33721	-1672
15	3590560	2819640	968427	0	1902799	71867
16	890153	532257	1758019	90683	769964	155857
17	856539	1115305	1131183	51765	406690	76204
18	269502	8123576	1069918	129199	795118	65147
19	61552	2393699	14065826	996984	1487504	212871
20	0	0	0	0	0	0
21	30225	47063	14914	9735210	10361918	2213
22	187	153206	35557	3767	12064	652760
23	3712	2077	15604	8058	11049	29102
24	26806	38964	70044	29309	74648	5672
25	307242	655809	803119	139056	412679	42899
*	269355	501983	648396	86761	323540	28150
*	11154	11142	36654	16984	16856	2235
*	17628	36186	38019	14123	41590	8020
**	0	0	0	0	0	0
*	9105	106498	80050	21188	30693	4494
26	530278	1623176	1912213	759907	1282445	154681
27	626400	1280794	1808235	271939	681354	206029
28	0	0	0	0	0	0
29	396083	779232	884844	199805	551873	83024
30	49930	124357	129357	7778	45513	23916
31	0	0	0	0	0	0
32	194206	909576	1736538	357777	733847	103819
*	79357	564483	1534078	339892	667883	84105
33	370778	735853	2173953	322178	740802	111521
34	32192	49882	65988	5571	42115	8483
36	215495	666475	734444	7437	248256	38360
	8928589	23300238	32937502	14083264	23112846	2426071
	16180863	38514137	51135206	17105562	33064922	4377563

	23 Manufac-turing nec	24 Const-ruction	25 El.,Gas, Water	* Electricity	* Gas & Heating	* Water
01	91808	180202	0	0	0	0
*	4383	105310	0	0	0	0
02	11957	1020936	1702671	1428293	274245	5
*	3	30	300086	227497	72456	5
*	0	0	413717	413717	0	0
*	0	0	988931	787142	201789	0
03	17734	42	56	14	14	14
04	73428	370868	15884	2053	502	3420
05	215859	3070433	233	230	3	0
06	16172	1012814	27179	8717	1104	7117
07	168910	270466	7984	0	0	2168
*	33178	226066	276	0	0	0
08	15045	148431	41862	8431	14406	10206
09	13056	8478	2341	362	1574	119
10	35303	376478	26079	0	0	2713
11	336205	433610	88494	4120	6131	35074
12	509250	2378095	1010163	709363	20284	196187
*	23396	836906	747836	542548	85351	41024
*	396	394389	101897	166814	-65067	4
13	30414	5868862	17471	878	34	15701
14	-521	-9929	0	0	0	0
15	49222	2765344	1922	0	0	1922
16	205280	724997	3105	2167	0	938
17	92636	8440776	8393	3560	1318	3291
18	16641	2315849	608341	463168	30099	82810
19	65744	1204292	283	44	0	239
20	0	0	0	0	0	0
21	12190	673805	60245	14343	1762	3449
22	78	284	123	0	0	102
23	357059	221817	2064	689	51	15
24	5660	174894	481371	293356	9310	164543
25	54142	730754	561468	123739	64120	333273
*	36134	431554	382543	15156	22134	322704
*	1318	91776	38688	381	30174	3736
*	7540	111170	34160	9307	5294	6169
**	0	0	0	0	0	0
*	9150	96254	106077	98895	6518	664
26	274289	4713854	366458	229399	36917	33835
27	163680	2566994	1165784	892855	79851	121211
28	0	0	0	0	0	0
29	184504	3576968	362449	232848	52142	15145
30	20035	306120	48782	11114	4010	11823
31	0	0	0	0	0	0
32	66294	424725	312033	159734	24894	110339
*	46015	106103	139132	125949	13124	59
33	174924	4758865	948495	505382	97087	305320
34	10142	60687	49804	11897	11430	7356
36	66888	898199	129320	61857	33606	19345
	3354014	49688968	8050801	5168599	764880	1487666
	5373990	88461108	21632668	12919486	1660140	3376029

	** Sewerage	* Waste Processing	26 Wholesale, Retail	27 Fin.,Ins., Estate	28 Real Estate Rent	29 Trans-portation
01	0	10431	0	0	1672	0
*	0	0	0	0	0	0
02	128	0	0	0	96	0
*	128	0	0	0	96	0
*	0	0	0	0	0	0
*	0	0	0	0	0	0
03	14	14330	28	14	6009	28
04	9909	216478	36677	158	113559	16669
05	0	95937	0	0	196197	145
06	10241	105708	80556	17221	26619	16559
07	5816	563398	47270	184	266049	13697
*	276	0	12961	65	74913	1796
08	8819	331807	787102	2328	137109	142208
09	286	13463	1452	4	3839	3702
10	23366	3143	0	0	26436	3233
11	43169	363	101	0	15102	43087
12	84329	1508210	102070	36762	1123954	31106
*	78913	1162578	47184	19945	1073077	26132
*	146	0	30	0	298	0
13	858	47790	0	0	1473	2
14	0	0	0	0	0	0
15	0	0	0	0	30244	0
16	0	1243	0	0	673	0
17	224	250632	1195	13	43828	352
18	32264	32260	65339	0	53743	95457
19	0	17431	0	0	5666	1907
20	0	0	0	0	0	0
21	40691	827333	35191	12142	1760194	26381
22	21	147206	1203	134	681	185
23	1309	63680	118	0	1226	13127
24	14162	486947	507822	2984759	461472	28410
25	40336	971222	307618	7664	957499	145895
*	22549	778853	160082	6259	565734	99030
*	4397	4485	47251	446	13911	11110
*	13390	98256	42915	959	91350	19437
**	0	0	0	0	0	0
*	0	89628	57370	0	286504	16318
26	66307	2599623	161776	23692	482608	45549
27	71867	10546950	5267052	1303737	2635794	318338
28	0	0	0	0	0	0
29	62314	2153101	297876	12198	3421104	251237
30	21835	1219717	825515	534	232316	284520
31	0	0	0	0	0	0
32	17066	213419	191522	97974	357439	144986
*	0	11391	26	0	30836	83012
33	40706	6957937	4325444	613899	1357033	1215528
34	19121	383822	183542	5555	101182	33200
36	14512	249163	561010	94056	171462	105822
	629656	30032730	13787451	5213014	13992194	2981302
	3677013	76567932	53328867	36679399	34850364	10261469

	30 Commu-nication	31 Public Admin	32 Public Services	* Research	33 Private Services	34 Office Supplies
01	0	209087	16850	976673	0	107418
*	0	2548	0	34214	0	30724
02	479	12303	3000	2289	0	41773
*	44	12303	3000	2161	0	7316
*	0	0	0	0	0	4517
*	7	0	0	0	0	2937
03	14	541222	5936	4556728	14	315447
04	62394	182398	14480	241118	54808	135130
05	1368	138	0	56025	0	12695
06	45784	252821	42635	317838	0	16080
07	6485	135464	36508	359771	904489	64238
*	1892	73033	28674	167741	190813	40211
08	360616	1579754	694155	4060167	0	164537
09	4250	4874	777	24313	0	17263
10	23521	44211	0	52463	23636	36781
11	14771	5060509	67410	449773	83522	226155
12	157725	504011	106412	958172	36920	253650
*	149104	395756	50724	638852	0	175601
*	0	500	0	8972	0	33071
13	4057	89668	15190	173496	10412	31113
14	2	0	0	6	0	47933
15	389	307	0	2827	35	46495
16	3582	42886	0	26781	810	43700
17	44238	18760	482	94702	407	47840
18	112558	429008	140829	883058	72210	65836
19	182564	8143	0	9345	0	235351
20	6883	0	0	0	0	0
21	491919	112113	10039	270855	0	209372
22	9819	162701	101	21261	0	15422
23	340347	246422	121374	390794	216599	27771
24	223005	472809	71649	534642	0	107957
25	858479	1672607	377097	1675887	0	155888
*	236723	922515	302030	854926	0	132146
*	10509	120410	11922	172079	0	18360
*	64745	406782	49017	377213	0	1488
**	0	0	0	0	0	0
*	546502	222900	14128	271669	0	3894
26	240152	2062265	233037	2534163	367480	425838
27	159681	1952626	843634	5680866	0	333353
28	0	0	0	0	0	0
29	422445	816748	300169	1202881	70206	204528
30	225164	525671	192675	2435762	0	203389
31	0	0	0	0	0	370404
32	2302	277727	34246	750417	0	60931
*	94	33	0	7958	0	12174
33	2347031	3256804	866000	11382563	0	595137
34	63532	343242	109844	203993	0	916
36	64772	590652	134925	249233	31695	0
	6480314	21607895	4439440	41050529	1873229	4620327
	20215739	58153354	5300556	90755975	1873229	6602821

	36 Activities nec	Intermed Demand
01	107418	15539187
*	30724	2131919
02	41773	10230410
*	7316	1018383
*	4517	4918554
*	2937	1045939
03	315447	13808238
04	135130	6487157
05	12695	5273551
06	16080	2345074
07	64238	9803060
*	40211	5993076
08	164537	10476978
09	17263	434695
10	36781	2713493
11	226155	22296908
12	253650	19441182
*	175601	8663632
*	33071	1333624
13	31113	9483222
14	47933	6988453
15	46495	18061670
16	43700	8641070
17	47840	14438109
18	44097	10275770
19	235351	20968159
20	0	6883
21	204686	21679965
22	15422	1219460
23	27771	2275840
24	107957	7282752
25	155888	14973975
*	132146	10245903
*	18360	750510
*	1488	1787285
**	383	341889
*	3894	2190277
26	425838	28672915
27	333353	43777104
28	0	0
29	204528	21939856
30	203389	7158805
31	370404	370404
32	60931	8507746
*	12174	4770064
33	595137	46555880
34	916	1873229
36	0	6841053
	4620327	431370320
	6602821	864884282

independent sectors. This is a convention in order to maintain consistency with other parts of the SEEDS.

Table 12-1 lists the intermediate transactions based on the sectoral classification scheme explained above. There is nothing much to add except to say that environment-related activities are highlighted, which enables us to gauge the magnitude of each activity, examine their input structure or "production technology", and look at the demand for them among the interindustry relations.

We have added considerable details for the classification of final demand. Input-output tables in many cases distinguish private consumption, government consumption, government investment, private investment, increase in stock, and exports among final demand.[4] It would be more useful if final demand is further disaggregated by its purpose or function. If this is actually done, the interindustry table can be made the central account of a satellite accounting system which provides detailed information on various social concerns such as pollution prevention, housing, medical care, research and development, etc.

In table 12-2, which relates to final demand, the industrial classification is of course identical to the previous table. What is new about the table is the detailed breakdown of final demand by purpose, including:

Private consumption

1. Food, beverages, and tobacco
2. Clothing
3. Fuel and light
4. Water
5. Housing
6. Furniture and household utensils
7. Medical care
8. Transportation and communication
9. Reading and recreation
10. Miscellaneous

Government consumption

1. Defense
2. Health and sanitation
3. Education
4. Others

Government investment

1. Roads
2. Harbors
3. Agriculture, etc.
4. Housing
5. Environment and sanitation
6. Welfare
7. Others

Private investment

1. Housing investment
2. Plant and equipment

of which:

3. Pollution prevention
4. Energy conservation
5. Research and development

The 1968 version of the SNA manual identifies eight purposes for government and household goods and services. The former include general public services, defense, education, health, social security and welfare services, housing and community amenities, other community and social services, economic services, and other purposes. The household goods and services include food, beverages and tobacco; clothing and footwear; gross rent, fuel and power; furniture, furnishing, and household equipment and operation; medical care and health expenses; transport and communication; recreation, entertainment, education and cultural services; and miscellaneous goods and services. Although such data are compiled in the Japanese SNA, the practice is not carried through in the case of an input-output framework.

The breakdown provided in table 12-2 is for illustration only. This kind of information was compiled once in the 1970s, and our table here is constrained by the unavailability of coverters by social purpose for recent years. In some cases, the figures do not add up to the total due to the omission of details.

Table 12-2 Final demand disaggregated to reflect environmental aspects, 1990

	C Outside household	Private consumption	1 Food, beverages, and tobacco
01 Agriculture,etc	125872	4389834	8628280
* *Forestry	1884	186703	144672
02 Mining	0	-942	28934
* *Coal	0	624	0
* *Crude oil	0	0	0
* *Natural gas	0	11367	0
03 Food Processing	714527	26418546	34333729
04 Textiles	125932	7872012	0
05 Lumber and Wood	12255	67592	0
06 Furniture, Fixtures	61955	722474	0
07 Pulp and Paper	23377	90458	0
* *Pulp	6010	-36415	0
08 Printing, Publish	51700	1095303	0
09 Leather & Prod.	3175	1219534	0
10 Rubber Products	9074	634890	0
11 Chemicals	215872	2557149	0
12 Petroleum & Coal	28742	2823246	0
* *Petroleum products	27511	2287019	0
* *Coal Products	166	4815	0
13 Nonmetal Minerals	94819	431899	0
14 Iron & Steel Mat	0	-32850	0
15 Iron & Steel Prod.	0	0	0
16 Nonferrous Metals	32090	361069	0
17 Fabricated Metals	69005	489359	0
18 General Machinery	0	46264	0
19 Electrical Mach.	145083	3930672	0
20 Motor Vehicles	0	5756135	0
21 Other Transport	0	418459	0
22 Precision Inst.	0	881339	0
23 Manufacturing nec	263483	2655980	0
24 Construction	0	0	0
25 El.,Gas,Water	3075	5226780	0
* *Electricity	804	2663639	0
* *Gas & heater	708	909328	0
* *Water	1563	1213945	0
** **sewerage	253	254856	0
* *Waste Process	0	439868	0
26 Wholesale Retail	1101369	39882205	11319193
27 Fin.,Ins.,Estate	0	9948371	0
28 Real Estate Rent	0	36681159	0
29 Transportation	183206	11010077	1093725
30 Communication	302	3155737	0
31 Public Admin	0	366968	0
32 Public Services	355700	32647608	0
* *Research	0	0	0
* *Repair, machine	14038482	32799794	0
* *Repair,auto	0	1653241	0
33 Private Services	5026	118724	2465223
34 Office Supplies	0	0	0
36 Activities nec	0	0	0
Subtotal	17948813	239921587	57869086

2 Clothing	3 Fuel & light	4 Water	5 Housing	6 Furniture & household utensils	7 Medicare
0	411993	0	0	0	0
0	0	0	0	0	0
0	212798	0	0	-15179	0
0	0	0	0	0	0
0	0	0	0	0	0
0	0	0	0	0	0
0	39838	0	0	0	0
9848109	0	0	0	0	0
0	35952	0	207061	0	0
0	0	0	0	1841820	0
0	0	0	0	0	44459
0	0	0	0	0	0
0	0	0	0	0	0
286754	0	0	0	0	0
400765	0	0	0	183844	0
709975	74819	0	0	0	942423
0	2230982	0	0	0	0
0	0	0	0	0	0
0	0	0	0	0	0
0	0	0	0	52286	0
0	0	0	0	-52286	0
0	0	0	0	0	0
0	0	0	0	-26986	0
0	0	0	342723	0	0
0	0	0	0	2803210	0
0	0	0	0	3697133	0
0	0	0	0	1907599	0
0	0	0	0	1224506	0
324757	0	0	0	62406	48843
364488	0	0	0	350822	167820
0	0	0	0	0	0
0	5495835	2399215	0	0	57609
0	0	0	0	0	0
0	0	0	0	0	0
0	0	0	0	0	0
0	0	0	0	0	0
0	0	0	0	0	0
5662533	975569	0	254662	4559011	505965
0	0	0	0	0	0
0	0	0	11069482	0	0
385218	237090	0	26180	274923	16281
0	0	0	0	0	0
0	0	0	0	0	0
0	0	0	0	0	3156024
0	0	0	0	0	0
0	0	0	0	0	0
0	0	0	0	0	0
0	0	0	0	0	1322524
0	0	0	0	0	0
0	0	0	0	0	0
17274354	9716824	2399215	11900110	16866487	6261953

8 Transportation & Communication	9 Reading & recreation	10 Education	Miscellaneous	Government Cons.	1 Defense
0	0	0		0	0
0	0	0		0	0
0	0	0		0	0
0	0	0		0	0
0	0	0		0	0
0	0	0		0	0
0	0	0		14	0
0	0	0		0	0
0	0	0		0	0
0	0	0		0	0
0	-662740	774848		0	0
0	0	0		0	0
0	4631475	0		5614	5614
0	0	0		0	0
0	0	0		0	0
0	0	0		85341	85341
0	0	0		69995	69995
0	0	0		0	0
0	0	0		0	0
0	0	0		0	0
0	0	0		0	0
0	0	0		0	0
0	0	0		0	0
0	0	0		238806	238806
0	0	0		0	0
0	0	0		225331	225331
0	0	0		52777	52777
0	0	0		1229967	1229967
0	1441074	0		0	0
0	1635658	1316147		1871	1871
0	0	0		45665	45665
0	0	0		1420668	18715
0	0	0		0	0
0	0	0		0	0
0	0	0		373800	0
0	0	0		373800	0
0	0	0		1046868	0
0	2595090	956092		89084	89084
0	0	0		0	0
0	0	0		0	0
18429994	111741	35275		-40575	26201
2155277	0	0		19089	19089
0	0	0		16094101	1182056
0	0	9080443		16684701	32938
0	0	0		533857	0
0	0	0		0	0
0	0	0		0	0
0	9511477	0		32938	32938
0	0	0		0	0
0	0	0		387780	387780
20585272	19265703	12164024		37543161	3743053

2 Health & sanitation	3 Education	4 Others	Government investment	1 Roads	2 Harbors
0	0	0	0	0	0
0	0	0	0	0	0
0	0	0	0	0	0
0	0	0	0	0	0
0	0	0	0	0	0
0	0	0	0	0	0
0	0	14	14	0	0
0	0	0	3584	0	0
0	0	0	5186	0	0
0	0	0	108757	0	0
0	0	0	0	0	0
0	0	0	0	0	0
0	0	0	0	0	0
0	0	0	0	0	0
0	0	0	0	0	0
0	0	0	0	0	0
0	0	0	0	0	0
0	0	0	0	0	0
0	0	0	0	0	0
0	0	0	0	0	0
0	0	0	-21991	0	0
0	0	0	0	0	0
0	0	0	0	0	0
0	0	0	17904	0	0
0	0	0	477348	69577	14683
0	0	0	2498058	0	0
0	0	0	73681	0	0
0	0	0	46296	0	0
0	0	0	190718	0	0
0	0	0	162888	0	0
0	0	0	24476929	7027314	581361
0	0	1401952	0	0	0
0	0	0	0	0	0
0	0	0	0	0	0
0	0	373800	0	0	0
0	0	373800	0	0	0
0	0	1046868	0	0	0
0	0	0	473738	2129	477
0	0	0	0	0	0
0	0	0	0	0	0
0	0	-66776	45765	709	358
0	0	0	0	0	0
0	0	18296310	0	0	0
3517794	13068774	65194	0	0	0
0	0	533857	0	0	0
0	0	0	0	0	0
0	0	0	0	0	0
0	0	0	0	0	0
0	0	0	0	0	0
0	0	0	0	0	0
3517794	13068774	17213540	28558861	7099732	596880

3 Agriculture etc.	4 Housing	5 Environment & sanitation	6 Welfare	7 Others	Private Investment
0	0	0	0	0	321261
0	0	0	0	0	0
0	0	0	0	0	0
0	0	0	0	0	0
0	0	0	0	0	0
0	0	0	0	0	0
0	0	0	0	14	14
0	0	0	0	3584	346646
0	0	0	0	5186	31804
0	0	0	0	108757	1395775
0	0	0	0	0	0
0	0	0	0	0	0
0	0	0	0	0	0
0	0	0	0	0	0
0	0	0	0	0	0
0	0	0	0	0	0
0	0	0	0	0	0
0	0	0	0	0	0
0	0	0	0	0	0
0	0	0	0	0	0
0	0	0	0	-21991	-130716
0	0	0	0	0	0
0	0	0	0	0	1912
0	0	0	0	17904	649677
16741	0	0	0	376345	15878715
0	0	0	0	2498058	14390468
0	0	0	0	73681	5887900
0	0	0	0	46296	2039028
0	0	0	0	190718	1248975
0	0	0	0	162888	768224
2726941	1562169	3164321	945298	8469523	56701427
0	0	0	0	0	0
0	0	0	0	0	0
0	0	0	0	0	0
0	0	0	0	0	0
0	0	0	0	0	0
0	0	0	0	0	0
548	0	0	0	470581	4199882
0	0	0	0	0	0
0	0	0	0	0	0
274	0	0	0	44422	588245
0	0	0	0	0	0
0	0	0	0	0	0
0	0	0	0	0	0
0	0	0	0	0	0
0	0	0	0	0	0
0	0	0	0	0	0
0	0	0	0	0	0
0	0	0	0	0	0
0	0	0	0	0	0
2744506	1562169	3164321	945298	12445955	104319223

1 Housing investment	2 Plant & equipment	3 Pollution prevention	4 Energy conservation	5 R&D	6 Increase in stock
0	104233	5107	6462	1876	90639
0	0	0	0	0	1618
0	0	0	0	0	66790
0	0	0	0	0	2825
0	0	0	0	0	66494
0	0	0	0	0	1217
0	0	0	0	0	34726
0	64143	3143	3976	1154	29110
0	16035	785	994	288	8533
0	833869	40859	51699	15009	22645
0	0	0	0	0	4505
0	0	0	0	0	-1405
0	0	0	0	0	47056
0	0	0	0	0	9358
0	0	0	0	0	10536
0	0	0	0	0	154763
0	0	0	0	0	154963
0	0	0	0	0	129765
0	0	0	0	0	6985
0	0	0	0	0	56413
0	906031	44395	56173	16308	-5211
0	0	0	0	0	133540
0	617384	30251	38277	11112	84705
0	697563	34180	43248	12556	187540
0	22057450	1080814	1367561	397033	629514
0	12067053	591285	748157	217206	399817
0	11104895	544139	688503	199887	56745
0	1579541	77397	97931	28431	249535
0	785761	38502	48717	14143	82688
0	48107	2357	2982	865	84829
24139468	25865988	1267433	1603691	465586	0
0	0	0	0	0	0
0	0	0	0	0	0
0	0	0	0	0	0
0	0	0	0	0	0
0	0	0	0	0	0
0	0	0	0	0	0
0	5676726	278159	351956	102180	126385
0	0	0	0	0	0
0	0	0	0	0	0
0	801797	39288	49711	14432	46210
0	0	0	0	0	0
0	0	0	0	0	0
0	0	0	0	0	0
0	0	0	0	0	0
0	0	0	0	0	0
0	0	0	0	0	0
0	0	0	0	0	0
0	0	0	0	0	0
0	0	0	0	0	0
24139468	80179754	3928807	4971144	1443232	2735670

(unit: million yen, current prices)

7 Exports	8 Final demand	9 Total demand	10 Imports	11 Import Tax	12 Domestic output
53089	4980695	20519882	-3215059	-67130	17237693
7784	197989	2329908	-706097	-1614	1622197
13982	79830	10310240	-7697879	-524591	2087770
151	3535	1021918	-900984	0	120934
1	66495	4985049	-4469847	-504690	10512
2	12586	1058525	-960041	-19899	78585
214526	27382269	41190493	-2978312	-436810	37775315
769381	9146665	15633822	-1987452	-190022	13456348
10923	136293	5409844	-855808	-24367	4529669
43834	2355440	4700514	-193626	-799	4506089
268125	386465	10189525	-468246	-2983	9718296
206053	174243	6167319	-444871	-2437	5720011
59438	1253497	11730475	-84900	-4	11645571
47770	1279837	1714532	-452172	-45100	1217260
482155	1136655	3850148	-262378	-16445	3571325
2555688	5483472	27780380	-2187652	-60943	25531785
698112	3705063	23146245	-2023883	-96286	21026076
373430	2817725	11481357	-1858700	-91605	9531052
24487	36453	1370077	-15400	-100	1354577
482234	1065365	10548587	-338652	-4093	10205842
15109	-175659	6812794	-252574	-3406	6556814
1704589	1838129	19899799	-445545	-5018	19449236
508316	988092	9629162	-2379744	-18402	7231016
620471	2033956	16472065	-288361	-2841	16180863
5683745	22715586	32991356	-1157007	-3157	31831192
11301049	32665147	53633306	-2497139	-961	51135206
6246287	18020748	18027631	-922069	0	17105562
4393746	7147064	28827029	-863382	-511	27963136
1380125	3783845	5003305	-624059	-1683	4377563
572067	4507471	6783311	-1392141	-17180	5373990
0	81178356	88461108	0	0	88461108
10549	6661072	21635047	-2379	0	21632668
10112	2674555	12920458	-972	0	12919486
289	910325	1660835	-695	0	1660140
148	1589456	3376741	-712	0	3376029
0	628909	970798	0	0	970798
0	1486736	3677013	0	0	3677013
3000352	48783931	77456846	-888914	0	76567932
385409	10333780	54110884	-782017	0	53328867
461	36681620	36681620	-2221	0	36679399
3686756	15519684	37459540	-2609176	0	34850364
32831	3188870	10347675	-86206	0	10261469
0	19845335	20215739	0	0	20215739
69433	49757442	58265188	-111834	0	58153354
14594	548451	5318515	-17959	0	5300556
780871	47619147	94175027	-3419052	0	90755975
104	1653345	5102290	-504	0	5101786
0	123750	6682945	0	0	6682945
0	0	1873229	0	0	1873229
1188089	1188089	8029142	-1426144	-177	6602821
47291807	478319122	909689442	-43123066	-1682094	864884282

Table 12-3 Import matrix, 1990

(unit: million yen,current prices)

		US	Other America	EC	Russia EE
01	Agriculture,etc	1155918	388143	43167	90382
*	Forestry	302113	23109	4658	69653
02	Mining	247305	756779	36791	77256
*	Coal	109673	199805	424	67301
*	Crude oil	253	176268	0	1346
*	Natural gas	23658	0	0	0
03	Food Processing	834974	288487	450254	67594
04	Textiles	84187	14782	432334	7306
05	Lumber and wood	218231	209862	7659	11079
06	Furniture, fixtures	34150	956	47460	1250
07	Pulp and paper	215238	150319	19249	855
*	Pulp	207346	149519	15309	804
08	Printing & publishin	34650	532	37887	257
09	Leather & products	19114	5283	148831	1229
10	Rubber products	42325	2012	53514	151
11	Chemicals	658586	118915	705148	23025
12	Petroleum & coal	157876	20628	53639	22373
*	Petroleum products	104098	19408	25263	22322
*	Coal products	275	115	1299	0
13	Nonmetal minerals	66490	5861	125287	5417
14	Iron & steel materia	21253	51399	5542	42504
15	Iron & steel product	33688	57213	20343	21169
16	Nonferrous metals	464861	428428	259447	191277
17	Fabricated metals	79764	2071	74991	449
18	General machinery	357998	10064	455520	3909
19	Electrical machinery	1347279	47108	288720	859
20	Motor vehicles	97216	3940	783455	175
21	Other transport	625384	25856	137860	443
22	Precision inst.	189451	11204	115744	643
23	Manufacturing nec	190886	9056	262650	11044
24	Construction	0	0	0	0
25	Electricity, gas & water	0	0	0	0
*	Electricity	0	0	0	0
*	Gas & heat supply	0	0	0	0
*	Water	0	0	0	0
**	Sewerage	0	0	0	0
*	Waste processing	0	0	0	0
26	Wholesale & retail	0	0	0	0
27	Fin., ins., estate	0	0	0	0
28	Real estate rent	0	0	0	0
29	Transportation	0	0	0	0
30	Communication	0	0	0	0
31	Public admin.	0	0	0	0
32	Public services	0	0	0	0
*	Research	0	0	0	0
*	Repair, machinery	0	0	0	0
*	Repair, auto	0	0	0	0
33	Private services	1982	3	777	11
34	Office supplies	0	0	0	0
36	Activities nec	223517	9726	496699	6098
Subtotal		7585147	2623872	5070585	587040

Other Europe	East Asia	Mid East	Other Asia	Oceania	Africa etc.	Total
15360	410209	7169	687831	276083	136235	3210497
31	40901	96	225663	32749	7124	706097
7926	420842	3200465	1881248	918494	150773	7697879
40	41280	2	8076	438073	36310	900984
0	321977	3133133	780617	35137	21116	4469847
0	0	58047	803341	73873	1122	960041
51377	547065	12147	244719	319585	45784	2861860
27243	1032359	13815	272763	57104	6354	1948247
1058	66707	20	262779	70404	7984	855783
5744	54764	88	48151	390	90	193043
41820	11061	0	5144	16676	7466	467828
41102	6602	0	3574	13152	7463	444871
3011	3576	61	3327	491	19	83811
2561	124610	223	23127	3062	1767	329807
19952	102000	137	40818	1407	7	262323
183547	200344	89645	123314	26536	14046	2143106
5954	178974	949537	464819	27981	30555	1912336
1153	120820	949367	449722	24763	30237	1747153
2	11185	0	0	2224	300	15400
9852	102618	1353	17905	2639	1230	338652
14610	42035	1011	22063	16505	35652	252574
9988	258451	10552	21484	8907	3749	445544
204597	123238	51044	173038	354606	129177	2379713
15433	75982	273	36448	1894	308	287613
138600	117601	6633	59729	6857	96	1157007
50668	466028	6377	283398	4999	44	2495480
31086	1297	6	138	592	2614	920519
13463	38437	67	7458	14321	23	863312
128344	47809	1449	45179	1842	17	541682
44776	204718	96161	210587	3101	737	1033716
0	0	0	0	0	0	0
0	0	0	0	0	0	0
0	0	0	0	0	0	0
0	0	0	0	0	0	0
0	0	0	0	0	0	0
0	0	0	0	0	0	0
0	0	0	0	0	0	0
0	0	0	0	0	0	0
0	0	0	0	0	0	0
0	0	0	0	0	0	0
0	0	0	0	0	0	0
0	0	0	0	0	0	0
0	0	0	0	0	0	0
0	0	0	0	0	0	0
0	0	0	0	0	0	0
0	0	0	0	0	0	0
0	0	0	0	0	0	0
61	26	0	31	9	0	2900
0	0	0	0	0	0	0
16710	53382	8932	79102	15041	1741	910948
1046422	4685581	4462097	5017517	2149512	583998	33811771

In providing a detailed breakdown of the final demand items, we have used converters estimated for 1965, provided in Uno [1989], in our standard 36-sector classification scheme. Information on the environment-related activities added in the interindustry table is lacking, and no details are provided (zeros are inserted). Nevertheless, the table clearly indicates a different environmental impact of household consumption activites by purpose. Based on the survey on household expenditures, we can obtain the changing structure of household consumption over time, such as increased spending on energy, consumer durables, transportation, etc. It is strongly hoped that expansion of the input-output tables in this direction is pursued by the appropriate statistical offices.

Table 12-3 is the import matrix for 1990. The export matrix is also available but is omitted here because, in the Japanese case at least, import of raw materials is more directly related to the environment than her export, which largely consists of manufactured goods. The table shows the country of origin of Japanese imports. The table relates to tradable goods; nontradable activities (services) have null values. This table corresponds to "imports" in table 12-2, the difference being the noninclusion of service imports in table 12-3.

The accounting framework includes the portion which relates to the current input-output relation, the only modification being the explicit inclusion of environment-related activities. We can apply a standard I-O analysis regarding this part of the framework. It can also be used in conjunction with the matrix of fixed discharge coefficients dij computed as Zij/Xj for the base year, where Z is a vector of residuals with elements Zi (i: type of residual) and Xj, gross production of industry j. Thus, the framework possesses a theoretical structure which is widely used in the recent literature [Kneese and Sweeney, 1993].

12.4 Reconciliation tables, 1970 and 1990

In earlier chapters, data have been collected for the 1970-1990 period in numerous tables, of which 1970 and 1990 data were fitted into the reconciliation tables presented here.

Environmental accounting has emerged as an important tool in describing the interaction between human activities and the environment. Rather than focusing on a particular aspect of the environment, this approach attempts to fit various aspects into a statistical accounting framework in a consistent manner, taking the national accounts as a starting point. Economic development has traditionally been measured based on national accounts. Value added, national income and expenditures are well-established concepts in this respect. The usefulness of the national accounts lies in the fact that the concepts are geared toward market transactions.[5] The main thrust of the accounts is placed on the flow of goods and services in the current period.[6] In contrast, the statistical system to describe the interaction between the environment, economy, technology, and society has to be more complex. Thus, there is a need to construct a concise table which contains much of the information into a single, consistent framework. This is the purpose of presenting a reconciliation table.

The basic format of the reconciliation table is similar to the scheme shown in figure 12-1. However, for the purpose of presentation, the table here is somewhat like a transpose of the diagram, where current activities are placed vertically in the center of the diagram.

Table 12-4, which shows the empirical application of SEEDS, consists of four parts, namely, (a) supply of goods and services, (b) demand for goods and services and accumulation, (c) technology, and (d) measures of the quality of life.

The sources of data are listed in parentheses immediately after the item named. T refers to the table numbers in preceding chapters. Other abbreviations are listed at the end of the table. The 1990 data also show deflators, which were employed in obtaining 1970 constant price figures from 1990 current price data. Although for some purposes the current price figures are desirable, constant price figures are more useful in environmental analysis simply because one would have to deal with physical quantities at the same time. At any rate, the 1990 price figures can be calculated, if required, by the use of deflators. Deflators are also useful in estimating the cost of the environmental preservation efforts.

The first part is directly derived from the input-output framework. Actually, the 1990 data are taken from tables 12-1 and 12-2, showing only intermediate demand, final demand, imports, and domestic output for each sector. In Japan, the input-output tables have been compiled since the 1950s at five-year intervals with varying sectoral disaggregation.[7] The tables with comparable

Table 12-4 Empirical application of SEEDS (social, economic, and environmental data set), reconciliation table

1970 (unit: bil. yen, 1970 prices)

a. Supply of goods and services

Modified JSIC code	Industrial sectors	Sources	Inter-mediate demand	Final demand	Imports	Domestic output
	Total	(I-O)	161657.5	84307.0	-7725.3	161657.5
(01) 01&	Agriculture, etc.	(")	6811.9	1986.6	-1638.0	7113.5
#	Forestry	(")	1557.0	29.1	-509.0	1076.9
(02) 10&	Mining	(")	3156.4	68.4	-2146.0	959.2
#	Coal	(")	495.0	20.4	-363.8	151.7
#	Petroleum	(")	919.0	8.8	-804.8	6.4
#	Natural gas	(")	24.6	0.9	-8.3	17.2
(03) 18	Food processing	(")	2990.1	7166.7	-401.3	9620.5
(04) 20&	Textiles	(")	2965.5	3342.8	-152.9	6134.5
(05) 22	Lumber & wood	(")	2316.1	99.5	-108.5	2304.3
(06) 23	Furniture, fixtures	(")	567.6	577.2	-2.2	1142.1
(07) 24	Pulp and paper	(")	2566.9	132.6	-74.6	2621.1
#	Pulp	(")	555.9	-521.0	-61.5	439.6
(08) 25	Printing & publishing	(")	1447.3	431.0	-25.9	1852.4
(09) 29	Leather & products	(")	103.5	98.5	-11.0	189.3
(10) 28	Rubber products	(")	493.8	252.0	-5.5	739.7
(11) 26	Chemicals	(")	5099.1	1454.1	-406.3	6110.9
(12) 27	Petroleum & coal prod.	(")	2834.1	465.5	-265.0	3018.7
#	Petroleum products	(")	2222.3	429.5	-263.1	2372.7
#	Coal products	(")	611.8	36.0	-1.8	646.0
(13) 30	Nonmetal minerals	(")	2420.3	269.9	-18.3	2669.8
(14) 31 A	Iron & steel material	(")	4256.2	-35.7	-217.4	4001.7
(15) 31 B	Iron & steel products	(")	6270.6	1020.3	-6.9	7283.5
(16) 32	Nonferrous metals	(")	2271.4	-17.6	-390.5	1856.7
(17) 33	Fabricated metals	(")	2718.6	1086.8	-25.8	3776.5
(18) 34	General machinery	(")	3603.5	5096.6	-343.9	8323.6
(19) 35	Electrical machinery	(")	3512.5	4394.9	-245.9	7632.3
(20) 36M	Motor vehicles	(")	1478.4	2990.5	-27.7	4430.7
(21) 36 S	Other trans. equip.	(")	1381.4	1975.5	-145.0	3211.6
(22) 37	Precision instrument	(")	477.8	746.9	-107.5	1103.4
(23) 38&	Manufacturing nec	(")	1321.5	1001.9	-89.8	2219.6
(24) 15	Construction	(")	1343.3	14915.4	0.0	16258.7
(25) 70	Elec., gas, & water	(")	1773.0	854.9	-0.0	2627.9
#	Electricity	(")	1440.0	416.9	0.0	1856.9
##	Nuclear		n.a.	n.a.	n.a.	n.a.
#	Gas and heat supply	(")	120.0	122.5	0.0	242.6
#	Water	(")	178.7	132.8	0.0	311.5
##	Sewerage	(")	0.0	26.5	0.0	26.5
#	Waste processing	(")	127.5	212.0	0.0	339.5
(26) 40	Wholesale & retail	(")	5947.5	8461.2	-119.1	14289.6
(27) 50&	Fin., ins., estate	(")	3177.7	6574.4	-20.8	9731.3
(28) 50R	Real estate rent	(")	1062.1	30.0	-9.0	1083.2
(29) 60T	Transportation	(")	2992.2	3588.8	-396.3	6184.7
(30) 60C	Communication	(")	974.1	290.7	-5.7	1259.2
(31) 97	Public administration	(")	0.0	2313.1	0.0	2313.1
(32) 75P	Public services	(")	149.8	5341.5	0.0	5491.4
#	Research	(")	34.8	85.7	0.0	120.5
(33) 75M	Private services	(")	2574.8	6772.1	-93.7	9252.8
(34)	Office supplies	(")	814.2	24.3	0.0	838.6
(35)	Packing materials	(")	411.3	0.7	0.0	412.0
(36)	Activities nec	(")	3287.4	533.5	-223.2	3597.6

1990 (unit: bil. yen, 1970 prices)

a. Supply of goods and services

Modified JSIC code	Industrial sectors	Inter-mediate demand	Final demand	Imports	Domestic output	Deflators ('70 =1.00)
	Total	360714.1	199490.8	17985.1	360714.0	2.3977
(01) 01&	Agriculture, etc.	6886.9	2207.4	-1424.8	7639.7	2.2563
#	Forestry	659.5	61.2	-218.4	501.7	3.2326
(02) 10&	Mining	4373.0	34.1	-3290.5	892.4	2.3394
#	Coal	360.9	1.2	-319.3	42.8	2.8208
#	Petroleum	2414.9	32.6	-2194.3	5.1	2.0367
#	Natural gas	406.4	4.8	-373.0	30.5	2.5733
(03) 18	Food processing	6036.2	11969.8	-1301.9	16513.0	2.2876
(04) 20&	Textiles	3362.9	4741.6	-1030.2	6975.7	1.9290
(05) 22	Lumber & wood	2436.0	62.9	-395.3	2092.3	2.1648
(06) 23	Furniture, fixtures	956.8	961.0	-78.9	1838.5	2.4508
(07) 24	Pulp and paper	5149.1	202.9	-245.9	5104.6	1.9038
#	Pulp	2926.6	85.0	-217.2	2793.3	2.0477
(08) 25	Printing & publishing	3584.9	428.8	-29.0	3984.7	2.9225
(09) 29	Leather & products	188.8	555.9	-196.4	528.7	2.3019
(10) 28	Rubber products	1226.9	513.9	-118.6	1614.8	2.2115
(11) 26	Chemicals	14191.9	3490.1	-1392.4	16250.8	1.5711
(12) 27	Petroleum & coal proc	6787.8	1293.6	-706.6	7341.2	2.8641
#	Petroluem products	2924.2	951.0	-627.3	3216.9	2.9627
#	Coal products	497.6	13.5	-5.7	505.4	2.6796
(13) 30	Nonmetal minerals	3885.7	424.2	-138.7	4181.8	2.4405
(14) 31A	Iron & steel material	4081.0	102.5	-147.4	3829.0	1.7124
(15) 31B	Iron & steel prod.	10595.1	1078.2	-261.3	11409.1	1.7047
(16) 32	Nonferrous metals	6265.6	716.4	-1725.5	5243.2	1.3791
(17) 33	Fabricated metals	6997.9	985.7	-139.7	7842.5	2.0632
(18) 34	General machinery	6586.0	11693.8	-555.4	17722.9	2.0839
(19) 35	Electrical machinery	25944.1	40417.0	-3089.7	63270.4	0.8082
(20) 36M	Motor vehicles	3.7	9814.1	-502.1	9315.7	1.8362
(21) 36S	Other trans. equip.	10610.5	2731.9	-324.2	13017.9	2.6614
(22) 37	Precision instrument	977.2	3721,2	-500.0	3508.1	1.2478
(23) 38&	Manufacturing nec	963.8	1908.9	-589.5	2275.9	2.3612
(24) 15	Construction	2596.8	28949.8	0.0	31547.0	2.8041
(25) 70	Elec., gas, & water	4291.1	1908.8	-0.6	6199.3	3.4895
##	Electricity	5200.1	1357.4	-0.4	6557.0	1.9703
##	Nuclear	—	—	0.0	1668.8	1.9703
#	Gas and heat supply	350.6	425.2	-0.2	775.6	2.1404
#	Water	743.2	660.9	-0.2	1403.9	2.4047
##	Sewerage	117.8	216.8	0.0	334.7	2.9001
#	Waste processing	471.3	319.9	0.0	791.3	4.6462
(26) 40	Wholesale & retail	14107.2	24001.9	-437.3	37671.7	2.0325
(27) 50&	Fin., ins., & estate	15355.5	3624.7	-274.2	18705.9	2.8509
(28) 50R	Real estate rent	0.0	11846.1	-0.7	11845.4	3.0965
(29) 60T	Transportation	6957.9	4921.8	-827.4	11052.3	3.1532
(30) 60C	Communication	3738.6	1665.3	-45.0	5358.9	1.9148
(31) 97	Public administration	89.9	4818.8	0.0	4908.7	4.1183
(32) 75P	Public services	2142.0	12527.6	-28.1	14641.5	3.9718
#	Research	1536.1	176.6	-5.7	1707.0	3.1051
(33) 75M	Private services	11863.4	12134.4	-871.2	23126.6	3.9243
(34)	Office supplies	1088.4	0.0	0.0	1088.4	1.7210
(35)	Packing materials					
(36)	Activities nec	2853.1	495.4	-594.7	2753.8	2.3977

1970

b. Demand for goods and services and accumulation of non-financial assets

Item	Sources	Opening stock	Current expenditures	Adjustment	Closing stock
GDP	(SNA)		73344.9		
1.Consumption: government	(SNA)		5647.0		
General public services	(")		1556.0		
Defense	(")		560.2		
Education	(")		2077.9		
Health	(")		255.9		
Social security, welfare	(")		219.3		
Housing & community	(")		216.5		
Other community services	(")		73.5		
Economic services	(")		675.4		
Other purposes	(")		12.4		
2.Consumption: Non-profit	(SNA)		437.8		
Education	(")		121.5		
Health	(")		87.1		
Other purposes	(")		229.2		
3.Consumption: private	(SNA)		37807.2		
Food & beverages	(")		11502.6		
Clothing & footwear	(")		2923.9		
Gross rent, fuel & power	(")		6133.8		
# imputed rents (dup.)	(")		n.a.		
Furniture, equipment	(")		2893.2		
Medical & health	(")		2979.0		
Trans. & communication	(")		2934.5		
Recreation, education	(")		3484.3		
Miscellaneous	(")		4955.7		
Durables (dup.)	(SNA)		2349.9g		
Durables (dup.)	(SNA)	8731.8n	2175.9n	-1143.0	9764.7n
# Furniture, carpets	(")	2682.3n	331.1n	29.4	3042.8n
# Household appliances	(")	1996.6n	618.1n	-415.6	2199.1n
# Trans. equipment	(")	1976.1n	360.2n	-230.3	2106.0n
# Radio and TV sets	(")	1553.3n	781.5n	-524.3	1810.5n
# Cameras, musical inst.	(")	523.5n	84.9n	-2.2	606.2n
4.Fixed capital formation	(SNA)		26043.2g		
Reproducible tangibles	(SNA)	78361.9n	15286.4n	4469.1	98117.4n
Dwellings	(")	16169.5n	3903.8n	605.1	20378.4n
Buildings	(")	19484.1n	2632.5n	1514.7	23631.3n
Other structures	(")	23232.6n	3595.9n	1583.6	28412.1n
Transportation equipment	(")	3437.2n	1561.4n	-224.8	4773.8n
Machinery & equipment	(")	16038.3n	3592.7n	991.0	20622.0n
Subtraction item	(")				
Nonreproducible tangibles	(SNA)	162223.4n	909.2n	29311.5	192444.1n
Land	(")	151873.7n	757.5n	28899.6	181530.8n
# Residential	(")	114239.3n	448.0n	23871.2	138558.5n
# Agricultural	(")	31223.1n	309.5n	4084.1	35616.7n
# Others	(")	6411.3n	0.0n	944.3	7355.6n
Forests	(")	9681.6n	98.8n	414.0	10194.4n
# Forest land	(")	2917.6n	98.8n	180.2	3196.6n
# Trees	(")	6764.0n	0.0n	233.8	6997.8n
Subsoil resources	(")	381.0n	52.9n	-54.3	379.6n
Fishing ground	(")	287.1n	0.0n	52.2	339.3n

1990

b. Demand for goods and services and accumulation of non-financial assets

Item	Opening stock	Current expenditures	Adjustment	Closing stock	Deflators ('70=1.00)
GDP		170373.7			2.4918
1. *Consumption: government*		11390.7			3.4695
General public services		3059.2			"
Defense		1136.8			"
Education		3946.7			"
Health		487.4			"
Social security, welfare		673.9			"
Housing & community		700.6			"
Other community services		294.7			"
Economic services		1035.2			"
Other purposes		56.0			"
2. *Consumption: Non-profit*		837.2			3.8019
Education		189.4			"
Health		-28.9			"
Other purposes		676.7			"
3. *Consumption: private*		85276.2			2.7836
Food & beverages		18392.1			2.6575
Clothing & footwear		5178.2			2.9522
Gross rent, fuel & power		16165.6			2.8426
# imputed rents (dup.)		11136.0			2.8426
Furniture, equipment		7012.8			2.0903
Medical & health		9401.8			2.7092
Trans. & communication		8900.5			2.7380
Recreation, education		9917.9			2.4099
Miscellaneous		12715.2			3.0575
Durables (dup.)		13942.7g			1.2042
Durables (dup.)	54551.4n	12325.8n	-7834.0	59043.2n	"
# Furniture, carpets	13557.1n	1829.0n	-493.3	14892.8n	"
# Household appliances	9774.4n	2606.0n	-2597.9	9782.5n	"
# Trans. equipment	20225.0n	4776.4n	-2395.0	22606.4n	"
# Radio and TV sets	6626.3n	2322.5n	-1975.3	6973.5n	"
# Cameras, musical inst.	4368.5n	791.8n	-372.5	4787.8n	"
4. *Fixed capital formation*		62312.6g			2.1943
Reproducible tangibles	404980.8n	31640.7n	6461.5	443083.0n	2.1943
Dwellings	72333.2n	5328.9n	624.5	78286.6n	2.7822
Buildings	66711.1n	4241.7n	2321.6	73274.4n	2.9328
Other structures	106620.8n	5892.7n	3842.8	116356.3n	3.0918
Trans.equipment	21436.1n	4603.6n	-1173.1	24866.6n	1.3022
Machinery & equipment	119870.8n	17779.5n	-4307.3	133343.0n	1.1309
Subtraction item	625.6n	983.4n	-68.2	1540.8n	2.1943
Nonreproducible tangibles	733011.7n	1490.7n	77695.7	812198.1n	3.0074
Land	715887.2n	1216.8n	77168.9	794272.9n	"
# Residential	608067.6n	423.4n	66855.2	675346.2n	"
# Agricultural	62878.7n	793.4n	2807.9	66480.0n	"
# Others	44940.7n	0.0n	7505.9	52446.6n	"
Forests	16589.6n	243.7n	610.9	17414.3n	"
# Forest land	4055.4n	243.7n	-167.9	4131.2n	"
# Trees	12534.2n	0.0n	748.9	13283.1n	"
Subsoil resources	287.5n	30.0n	-56.5	261.0n	"
Fishing ground	247.3n	0.0n	2.4	249.7n	"

5.Inventory	(SNA)	20043.2n	2573.2n	207.4	22823.8n
6.Exports	(SNA, I-O)i		8272.7		
7.Imports	(SNA, I-O)i		7488.5		
8.Reproducible tangibles					
Private	(SNA, PBC)	49954.5n	20152.7g	—	64230.4n
# Dwellings	(SNA)		4746.4g		
# Plant and equipment	(SNA, PBC)i		15406.3g		
9.Reproducible tangibles					
Government	(SNA)	28407.4n	5890.5g	—	33887.0n
Social capital stock	(SCS, MHA)	40983.3g	6171.7g	—	44968.5g
# Trans. & communication					
(1) Roads	(")	7425.2g	1262.7g	—	8334.0g
(2) Harbors	(")	1366.8g	180.6g	—	1531.2g
(3) Airports	(")	98.3g	32.1g	—	127.9g
(4) National railways	(")	4883.4g	615.7g	—	5118.0g
(5) Railway Const. Corp.	(")	213.3g	66.4g	—	279.6g
(6) Subways	(")	807.1g	114.1g	—	927.6g
(7) Telephone & Telegraph	(")	4474.5g	697.8g	—	5087.6g
# Livelihood-related					
(8) Public housing	(")	2320.2g	600.0g	—	2591.5g
(9) Sewerage	(")	1412.4g	228.7g	—	1599.9g
(10) Waste disposal	(")	249.1g	31.1g	—	277.2g
(11) Water supply	(")	2058.3g	230.6g	—	2279.3g
(12) City parks	(")	329.5g	46.4g	—	358.0g
# Education					
(13) Schools & cultural	(")	3620.3g	554.7g	—	3800.5g
# Land preservation					
(14) River improvement	(")	3432.9g	376.7g	—	3733.1g
(15) Forestry conservation	(")	709.0g	53.9g	—	763.1g
(16) Coastal preservation	(")	538.1g	46.1g	—	574.0g
# Agriculture, forestry, etc.					
(17) Agriculture	(")	4140.9g	369.5g	—	4498.8g
(18) Forestry	(")	729.3g	55.6g	—	771.3g
(19) Fisheries	(")	503.0g	39.8g	—	536.6g
# Others					
(20) National forest	(")	1173.7g	59.1g	—	1225.1g
(21) Industrial water supply	(")	355.1g	43.0g	—	397.1g

5.*Inventory*	45896.2n	1466.1n	1201.3	48563.6	1.5837
6.*Exports*		43253.9			1.4898
7.*Imports*		25924.4			2.2550
8.Reproducible tangibles					
# Private	278478.6n	52194.6g	-24061.8	306611.4n	2.0812
## Dwellings		8899.2g			2.8337
## Plant and equipment		44411.9g			1.8787
9.Reproducible tangibles					
# Government	117369.0n	10672.4g	—	126884.2n	2.6334
Social capital stock	176590.7g	13479.9g	—	190070.6g	
# Trans. & communication					
(1) Roads	41324.1g	3605.8g	—	44929.9g	
(2) Harbors	5497.5g	282.9g	—	5790.4g	
(3) Airports	1212.2g	118.3g	—	1330.5g	
(4) National railways		(privatized)			
(5) Railway Const. Corp.	2592.7g	92.8g	—	2685.5g	
(6) Subways	3720.1g	118.3g	—	3838.4g	
(7) Telephone & Telegraph		(privatized)			
# Livelihood-related					
(8) Public housing	8551.2g	737.5g	—	9288.7g	
(9) Sewerage	14559.6g	1096.2g	—	15655.0g	
(10) Waste disposal	1933.9g	278.9g	—	2212.8g	
(11) Water supply	9333.1g	520.4g	—	9853.5g	
(12) City parks				2938.2g	
# Education					
(13) Schools & cultural	16566.0g	1153.6g	—	17719.6g	
# Land preservation					
(14) River improvement	16998.7g	1154.8g	—	18153.5g	
(15) Forestry conservation }		}	}		
(16) Coastal preservation	1756.5g	60.1g	—	1817.6g	
# Agriculture, forestry, etc.					
(17) Agriculture	16834.1g	984.8g	—	17819.0g	
(18) Forestry	2488.7g	145.6g	—	2634.3g	
(19) Fisheries	2814.1g	164.6g	—	2978.8g	
# Others					
(20) National forest					
(21) Industrial water supply	1255.8g	36.6g	—	1292.4g	

1970

c. Technology					
Item	Sources	Opening stock	Current expenditures	Closing stock	
1. Pollution prevention	(T6-1)		193.0g		
# Heavy oil desulphur.	(T6-1,-4)	24.6	9.9g	—	34.5
# Flue gas desulphur.	(T6-1,-5)	1.3	3.6g	—	4.9
# Flue gas NOX removal	(T6-1,-6)	—	—	—	—
# Ind'l water processing	(T6-1,-7)	10.0	30.0g	—	39.0
# Sewerage treatment	(T6-1,-7)	36.3	17.1g	—	53.4
# Municipal waste	(T6-1,-9)	12.8	16.5g	—	29.3
# Industrial waste	(T6-1,10)	—	—	—	—
2. R&D expenditures	(T8-1)		1195.3	('76 in '70p	1414.0)
of which fixed capital form.	(STA)		273.4		
Environmental protection	(")		16.9	('76 in '70p	51.8)
Energy	(T8-1,-2)		n.a.	('76 in '70p	77.0)
# Fossil energy	(T8-2)		n.a.	('76 in '70p	6.0)
# Solar, etc.	(")		n.a.	('76 in '70p	3.9)
# Nuclear energy	(")		n.a.	('76 in '70p	62.5)
# Energy conservation	(")		n.a.	('76 in '70p	3.9)
# Other energy	(")		n.a.	('76 in '70p	0.8)
3. Energy conservation invest.	(T8-3)		n.a.		
4. Alternative energy invest.	(T8-4)		n.a.		

1970

d. Quality of life		
Measures of quality of life	(T11-1)	47923.1
1.Gov. consumption adjusted	(T11-1)	2761.7
2.Private consumption adjusted	(T11-1)	34408.6
3.Service of consumer durables	(T11-1)	2286.8
4.Service of social capital	(T11-1)	444.0
5.Leisure time	(T11-1)	8323.6
6.Extra-market activities	(T11-1)	7213.0
7.Loss due to urbanization	(T11-1)	-1188.0
8.Environmental pollution	(T11-1)	-6236.6
# Air pollution:		
## SO2	(T11-11)i	-1127.9
## Soot and dust	(T11-11)	-368.7
## Automobile exhaust	(T11-11)	-1151.1
# Water pollution:		
## Industrial	(T11-11)i	-1078.7
## Municipal	(T11-11)	-664.6
# Wastes:		
## Industrial	(T11-11)i	-1138.9
## Municipal	(T11-11)	-106.7

Note: (g) signifies gross accumulation and (n) net of depreciation.
(i) denotes that industrial breakdown is available.
(#) and (##) indicates sub-items.

1990

c. Technology

Item	Opening stock	Current expenditures		Closing stock	Deflators ('70=1.00)
1. Pollution prevention		755.9g			
# Heavy oil desulphur.	269.8g	0.6g	—	270.2g	
# Flue gas desulphur.	538.4g	11.4g	—	545.7g	
# Flue gas NOX removal	135.5g	23.1g	—	150.2g	
# Ind'l water processing	766.5g	56.3g	—	802.5g	
# Sewerage treatment	1536.1g	131.6g	—	1620.3g	
# Municipal waste	1064.4g	209.1g	—	1174.2g	
# Industrial waste	—	8.8g	—	—	
2. R&D expenditures		3893.4			3.1051
of which fixed capital form.		1047.8			1.8781
Environmental protection		75.6			3.1051
Energy		294.3			"
# Fossil energy		28.9			"
# Solar, etc.		9.0			"
# Nuclear energy		129.4			"
# Energy conservation		119.1			"
# Other energy		7.6			"
3. Energy conservation invest.		137.4			1.8781
4. Alternative energy invest.		32.3			1.8781

1990

d. Quality of life

Measures of quality of life	128550.9	
1.Government cons. adjusted	5314.5	3.4730
2.Private consumption adjusted	79533.4	2.7610
3.Service of consumer durables	13876.5	1.2000
4.Service of social capital	2370.8	2.6120
5.Leisure time	17475.2	5.3369
6.Extra-market activities	13953.5	2.8754
7.Loss due to urbanization	-1392.2	5.2413
8.Environmental pollution	-2580.8	2.4893
# Air pollution:		
## SO2	-641.0	
## Soot and dust	-180.2	
## Automobile exhaust	-738.8	
# Water pollution:		
## Industrial	-189.2	
## Municipal	-449.0	
# Wastes:		
## Industrial	-283.2	
## Municipal	-99.4	

PBC: Economic Planning Agency, *Gross Fixed Nonresidential Private Business Capital in Japan.*

SCS: Economic Planning Agency, *Social Capital Stock in Japan.*

SNA: Economic Planning Agency, *National Accounts.*

STA: Science and Technology Agency, *Indicators of Science and Technology.*

sectoral classification to the current version have been available since the 1960s. They also contain physical tables and data on the generation and reuse of scraps, providing a basis for extension in this regard. Annual tables have been available since 1973.

The second part of the reconciliation table concerns the current expenditures as well as the opening and closing stock where applicable. The main part has been obtained from the SNA. Functional classification is provided for consumption by government, nonprofit institutions , and household. Some details are provided for stocks of consumer durables owned by household. The classification scheme follows the SNA.

The tables reported here do not fully reflect the extra dimensions to be added to the conventional economic accounts and, in that sense, interim in their nature. First of all, there is the problem concerning production boundary. "The case of environmental protection activities and products is even more complex. Some establishments specialize in the production of environmental protection services for delivery to other units (e.g., waste disposal, sewage treatment) or goods which are used for the environmental protection (e.g., filters). ... Moreover, an important part of environmental protection activities is internal to establishments. They are ancillary activities in the central framework and have to be externalized if one wants to measure the environmental activities more broadly." "The above-mentioned re-analyses take place within the limits of the central framework's boundary of production. For certain objectives, however, the production boundary itself may be changed. ... Generally speaking, the scope of non-market activities may be extended considerably."[United Nations 1993b, pp.490-491]

It should be noted that the assets are valued net of depreciation in the SNA. This is appropriate from the point of view of measuring the value of assets for it should represent the present value of the future flow of profits to be derived from the existing assets, which should also equal its resale value. However, from the point of view of environmental analysis, old machines will consume just the same amount of energy as the new ones, will emit the same amount of pollutants as the new ones, etc., etc. Thus, gross stock which accounts for physical replacement but not depreciation in the economic or business accounting sense is a better indicator. The table provides both gross and net values for the current expenditure for comparison.

Fixed capital formation figures are also from the SNA, which is subdivided into reproducible tangibles and nonproducible tangibles. Both are subdivided into types of assets. Nonreproducible tangible assets have a direct bearing on the environment, consisting of land (residential, agricultural, and others), forests (forest land, trees), and subsoil resources. Gross capital formation figures are available for individual items but are not presented in the table.

One problem here from the point of the environment would be the boundary of assets. The SNA manual under revision states, "The most important amendment introduced into environmental accounting as compared with the SNA is the extension of the asset boundary. In the SNA, natural assets are included only if they provide economic benefits to the owner, a characteristic that manifests itself through being controlled by an institutional unit. This often means explicit ownership, subject to government legislation in the case of natural forests, and/or availability of a market price. These assets are referred to in the SNA as economic assets. In SEEA, the asset boundary is defined to be much wider. It includes in principle all natural assets; some may directly participate in production activities while others may be affected by the environmental impacts of economic activities." [United Nations 1993b, p.513]

When one tries to respond to today's environmental concerns, however, one would have to deal with a natural environment for which the establishment of ownership is not feasible. Global climate change is a typical example. CO_2 emission may lead to global warming, but no ownership is established to the atmosphere, and human beings have used the atmosphere as a free sink since the start of the Industrial Revolution. The same thing can be said of the oceans. Although water rights are often established, river and lake waters are not effectively owned or controlled by any institutional unit.

Given the present level of scientific knowledge, there may be differing views as to the global climate change or the degree of damage done by acid rain. There are difficulties in establishing causal relations, and there are wide variation as to the valuation of the damage done. For the time being, rather than including damage estimates in the core account, we have opted to list them separately in the measures of quality of life described below. Even then, we would have to resort to physical measures.

Reproducible tangibles are then divided into private and government. Of the private expenditures, gross capital formation figures are provided for dwellings and plant and equipment. Industrial breakdown is available for the flow of stock of plant and equipment in gross terms, which are used in production analysis in SEEDS and the multisector econometric model COMPASS. Functional classifications, such as pollution prevention, R&D, labor-saving measures, and capacity expansion, are available from the survey data, but their appropriateness on the national economy level needs careful examination.[8] The current expenditure on pollution prevention equipment is provided in detail below, together with estimated accumulated stock.

A detailed breakdown of the social capital stock in gross terms is available from the Economic Planning Agency [1986], which can be supplemented by annual data on administrative investment. The latter includes the purchase of land. The consistency between annual flow from the SNA and the EPA social capital data is fairly good, as can be seen from the table (5890 billion yen and 6170 billion yen, respectively). Some of the items of social capital stock, such as sewerage, waste disposal, water supply, and city parks, are relevant for our purpose here, as well as items under land preservation, agriculture and forestry, and national forest.

The next portion of table 12-4 deals with technology. We have already mentioned that the input structure represents production technology in the respective industrial sectors. Here, we are concerned with technology more directly. First, pollution prevention investment is explicitly recognized as well as the stock thereof (which is equated with cumulative investment). Because of the fact that the underlying statistical data were collected starting mid-1960s, this methodology seriously underestimates some items. Sewerage treatment and municipal waste treatment are cases in question. Fortunately, these items are included among the social capital stock and would serve as a good alternative. It should be noted that the pollution prevention investment listed here relates to the relevant machinery portion only and does not include construction costs, whereas social capital figures include the whole facility. For other items such as heavy oil desulphurization, flue gas desulphurization, and industrial water processing, there is no problem of underestimation because the technology was only just being put into practical use around 1970. Flue gas NOX removal had not even started at that time.

The next item under technology is private R&D expenditures related to environment and energy. The table reveals that R&D for environmental protection was of limited magnitude in 1970 as far as the private sector was concerned. It is known that research projects specifically aimed at development of air pollution prevention technology were effective in the 1970s.

The last item shown in table 12-4 is the measure of the quality of life. We have dealt with this topic in chapter 11 in detail. Environmental damage is estimated based on the cost of removing excess emission in order to clarify a certain environmental quality (which is set as a rule equal to the level in 1955). Environmental standards were not used as the criteria because they were introduced at a later stage in the sequence, and their severity varied according to the region.[9] In other words, there is no single, absolute standard to be adhered to. The list includes SO_2, soot and dust, and automobile exhaust concerning air pollution, water pollutants attributable to industry and household, and industrial and municipal wastes. There are numerous other types of pollution (noise, vibration, odor, heat island, acid rain, global climate change, loss of ozone layer, etc.), and there are other consequences of socioeconomic activities on the environment (such as endangered species, soil erosion, loss of scenic beauty, damage to historical sites, etc.). These are not included at this stage of development of SEEDS, and one can argue that the figures provided here are almost certainly underestimating reality. However, even limiting the scope to items such as the ones included here, one notices that the estimated damage amounts to 8.6% of GDP in 1970 and about 10% of the estimated economic benefits including imputation of various capital stock and leisure. In 1990, the estimated environmental damage has been reduced to about 41% of the 1970 level in absolute terms, or to 1.5% of GDP and 1.9% of the total economic benefits.

Tables 12-1 through 12-4, like other tables of the SNA or input-output framework, are calculated in value terms. Conceivably, the table can have a counterpart in physical terms, which can then be matched with the stock of resources on the part of the exporters.

12.5 From accounting to policy: responding to AGENDA 21

One remaining fundamental question is whether the statistical system such as the one presented here, or any indicator derived from it, will by itself suggest the road leading to sustainable development . The question can be asked of the SEEA suggested by the UN as a manual or of the SEEDS which I have been developing in an empirical context. Apparently, the answer is no.

The UN's SEEA in particular is preoccupied with the natural environment and the economy and needs to be strengthened in three respects: first, social aspects such as demographic trends, urbanization, income and consumption, lifestyle, and quality of life. The human aspect is one of the prime concerns of AGENDA 21. One of the important policy goals has to do with quality of life. Our task is to find a scenario for maintaining and even improving the quality of life while removing the negative impact on the global environment. In this regard, unless we realize the importance of the quality of life, as distinct from the levels of income and consumption, the answer seems to be remote.

Second, the technological aspect has to be included in the scope of the analysis. Within the span of the last 20 years, there have been improvements in energy conservation and pollution prevention. At the same time, however, wide differentials exist between the best technological practice and the reality not between North and South but, more importantly, among the members of the North. Such a gap, in a particular segment of the production process or the whole set of the industrial branch or the structure of the economy as a whole, represents room for improvement based on the technology currently available. If we want to narrow the differentials which exist among the current generation and still preserve the global environment for our future generations, technology has to be a part of the answer. The further enhancement of technological potentials through R&D and the diffusion process of new technology must be among the research agenda.

Third, AGENDA 21 pertains to the global community as a whole. In order to respond to the policy issues spelled out in it, we need to strive to establish linkages with the environmental accounting of individual countries. Although such an endeavor presupposes the development of country accounts in the first place, the individual framework should be developed with future international linkages in mind. Otherwise, the whole exercise will be in vain. If we look at a resource-consuming country in the North, such as Japan, it is apparent that one has to look at the resource-producing countries at the same time in order to be able to gauge the whole process. Also, given the fact that industrial relocation is in progress on a world-wide scale, the time has come to look at the global economic structure, which consists of individual economies intertwined through international trade, direct foreign investment, and the global environment.

We have tried in this work to show how these points can be incorporated in the accounting framework which is intended to support policy decisions. This aim, however, is only partially achieved in this study, and we have to await the development of a policy model based on the framework presented here.

One thing which the proposed framework can realize is to provide pictures of "trees" while keeping the entire "forest" in sight. The entries in the reconciliation table are actually composed of numerous matrices which have their own structure. We have tried to demonstrate this point by fitting the tables developed in the preceding chapters in the reconciliation table. We need a time series of such tables in order to describe the dynamism involving the environment, economy, technology, and society.

The most important role the framework can play would be to promote an interaction between in-depth study on the one hand and the comprehensive framework on the other. It is also essential to stimulate discussion between the natural sciences, social sciences, and technology spheres. This is more easily said than done. But this is exactly what needs to be done today.

FOOTNOTES

[1] For SSDS, see United Nations [1975]. See also United Nations [1978] for its relation with social indicators. A list of social indicators coordinated by the OECD appears, for example, in OECD [1982a].

[2] Particularly noteworthy in this regard is the contribution of Prof. Miyohei Shinohara who headed a group of experts in estimating "NNW" [Economic Council of Japan 1973] and Mr. Hisao Kanamori, Chairman of the Japan Center for Economic Research, who attempted a follow up [Kanamori, Takase, Uno 1977, Kanamori 1980].

[3] Sector 35 corresponds to "packing materials" in our classification but has been omitted in recent input-output tables. "Repair, machine" and "repair, auto" have been added to our standard classification scheme in order to reflect their explicit treatment among service activities in recent input-output tables.

[4] In addition, Japanese tables identify consumption outside the household sector which relates to business consumption, the benefit of which is deemed to accrue to household.

[5] An exception is the directly competitive, nonmarket production of goods and services which is included in the accounts through imputations. Examples are production of goods for own use, housing, etc.

[6] The stock is recognized in the SNA. However, empirical data concerning stock variables are typically a weak part of the accounts. In addition, it is generally true that environmental conditions as the stock is not explicitly recognized in the accounts.

[7] For details of the compilation of input-output tables in Japan, see Uno [1989]. This publication also includes 36-sector summary tables (since 1951) and matching industry-occupation matrices covering the same period based on census data.

[8] Uno [1987] provides information on investment motives.

[9] The development of pollution prevention policy measures including the introduction of environmental standards is discussed in Uno [1987]. See also chapter 5 for environmental standards pertinent to automobile exhaust.

References

An asterisk (*) attached to the title indicates translation by the present writer, in which case original title is given in the parenthesis. Otherwise, the English title is from the original publication.

Ahmad, Yusuf J.; Serafy, Salah El; and Lutz, Ernst. 1989. *Environmental Accounting for Sustainable Development.* Washington, D.C.: The World Bank.

Amano, Akihiro, ed. 1992. *Global Warming and Economic Growth—Modeling Experience in Japan.* National Institute for Environmental Studies, Environmental Agency of Japan.

Anderson, Kym and Blackhurst, Richard, eds. 1992. *The Greening of World Trade Issues.* London: Harvester Wheatsheaf.

Anderson, Victor. 1993. *Energy Efficiency Policies.* London: Routledge

Ando, Jumpei. 1991. "Abatement of Ozone and Acid Rain Precursors." International Symposium on Environmentally Sound Energy Technologies and their Transfer to Developing Countries and European Economies in Transition. Milan.

Angerer, Gerhard. 1992. "Innovative Technologies for Sustainable Development." Diets, F.J.; Simonis, U.E.; and van der Straaten, J. eds. *Sustainability and Environmental Policy.* Berlin: Edition Sigma.

Asian Development Bank. 1992. *Integrated Energy-Environment Planning.* Manila.

Atomic Energy Comission of Japan. Various years. *Atomic Power White Paper** (Genshiryoku Hakusho). Tokyo: Ministry of Finance Printing Bureau.

Ayres, Robert U., and IIASA. 1990. "Eco-restructuring: Managing the Transition to an Ecologically Sustainable Economy." Laxenburg, Austria: IIASA.

Balaceaunu, J.C., Bertrand, A., and Lacour, J.J. 1989. "The Greenhouse Effect and Its Connection with the Area of Fossil Fuels: the Concept of CO2 Emission Intensity." *Energy Technologies for Reducing Emissions of Greenhouse Gases.* Paris: OECD.

Banister, David and Burron, Kenneth, eds. 1993. *Transport, the Environment and Sustainable Development.* London: E & FN Spon.

Barrett, Scott. 1989. "Economic Incentives for Environmental Regulation in the UK." Prince Bertil Symposium on Economic Instruments in National and International Environmental Protection Policies. Stockholm.

Bartelmus, Peter; Stahmer, Carsten; and Jan van Tongeren. 1991. "Integrated Environmental and Economic Accounting: Framework for a SNA Satellite System." *Review of Income and Wealth.* Series 37, No.2.

Blavin, Christopher. 1992. "Building a Bridge to Sustainable Energy." Brown, Lester R, et al. *State of the World 1992.* New York: W.W. Norton & Co.

Bleviss, Deborah Lynn. 1992. "Efficiency Improvements and Fuel Substitution in the Transport Sector." International Symposium on Environmentally Sound Energy Technologies and their Transfer to Developing Countries and European Economies in Transition. Milan.

Blok, K.; Farla, J.; Hendriks, C.A.; and Turkenburg, W.C. 1992. "Carbon Dioxide Removal: A Review." International Symposium on Environmentally Sound Energy Technologies and their Transfer to Developing Countries and European Economies in Transition. Milan.

Boland, John J. 1989. "Environmental Control Through Economic Incentive: A Survey of Recent Experience." Prince Bertil Symposium on Economic Instruments in National and International Environmental Protection Policies. Stockholm.

Bowes, Michael D. and Krutilla, John V. 1985. "Multiple Use Management of Public Forestland." Kneese, Allen V. and Sweeney, James L. eds. *Handbook of Natural Resoruce and Energy Economics.* Vol.II. Amsterdam: Elsevier Science Publishers B.V.

Braun, Gerald W. 1992. "Improving Efficiency in Electricity Generation: Advanced Conversion Systems." International Symposium on Environmentally Sound Energy Technologies and their Transfer to Developing Countries and European Economies in Transition. Milan.

Brown, Lester R. et al. Various issues. *State of the World.* New York: W.W. Norton & Co.

Bureau of Statistics. Various issues. *Report on the Survey of Research and Development.* (Kagaku Gijutsu Kenkyu Chosa Hokoku). Tokyo: Nihon Tokei Kyokai.

—. Various issues. *Report on the Survey of Research and Development on Energy taken as a Supplement of the Survey of Research and Development.* (Kagaku Gijutsu Kenkyu Chosani Futaisuru Enerugi Kenkyu Chosa Hokoku).

Tokyo.

—. Various issues. *Annual Report on the Family Income and Expenditure Survey*. Tokyo: Japan Statistical Association.

—. 1991. *A Report on a Survey on the Compilation of Input-Output Tables and Related Statistics in Foreign Countries.** (Shogaikokuni Okeru Sangyo Renkan-hyoto Sakusei Jokyo Chosa Kekka Hokokusho). (mimeographed). Tokyo.

—. and Economic Planning Agency. 1973. *1970 National Wealth Survey of Japan, Vol.6, Household*. Tokyo: Ministry of Finance Printing Bureau.

Chapman, Neil A. and McKinley, Ian G. 1987. *The Geological Disposal of Nuclear Waste*. Chichester: John Wiley & Sons.

Colombo, Umberto. 1992. "Sustainable Energy Development." *Science and Sustainability*. Laxenburg: International Institute for Applied Systems Analysis.

Constanza, Robert, ed. 1991. *Ecological Economics: The Science and Management of Sustainability*. New York: Columbia University Press.

Daneka, Gregory A. and Lawrence, Andy. 1982. "Life-quality Accounting Systems and the Energy transition." Daneke, Gregory A. *Energy, Economics, and the Environment*. Lexington, Mass.: D.C. Heath and Company.

Darmstadter, Joel, ed. 1992. *Global Development and the Environment*. Washington, D.C.: Resources for the Future.

Denison, Edward F. 1971. "Welfare Measurement and the GNP." *Survey of Current Business*. January.

—, and Chung, William K. 1976. *How Japan's Economy Grew So Fast, The Sources of Postwar Expansion*. Washington, D.C.: The Brookings Institution.

—. 1979. *Accounting for United States Economic Growth, 1929-1969*. Washington, D.C.: The Brookings Institution.

Economic and Social Commission for Asia and the Pacific, United Nations. 1991. *Energy Policy Implications of the Climatic Effects of Fossil Fuel Use in the Asia-Pacific Region*. ST/ESCAP/1007. Bangkok.

Economic Council of Japan. 1973. *Measuring Net National Welfare of Japan*. Tokyo: Ministry of Finance Printing Bureau.

Economic Planning Agency. Various issues. *Annual Report on National Accounts*. Tokyo: Ministry of Finance Printing Bureau.

—. Various issues. *Current Consumption Survey* (Shohi Doko Chosa). Tokyo: Ministry of Finance Printing Bureau.

—. 1973. *Project COSMO*. (mimeographed.)

—. 1986. *Social Capital Stock in Japan** (Nihonno Shakai Shihon). Tokyo:

Gyosei.

—. 1988. *Report on National Accounts from 1955 to 1969.* Tokyo: Ministry of Finance Printing Bureau.

—. 1991a. *Report on National Accounts from 1955 to 1989.* Tokyo: Ministry of Finance Printing Bureau.

—. 1991b. *The World and Japan in an Age of Globalization** (Chikyuka Jidaino Sekaito Nihon). Tokyo: Ministry of Finance Printing Bureau.

Elsom, Derek. 1992. *Atmospheric Pollution, A Global Problem.* Second Edition. Oxford: Blackwell Publishers.

Environment Agency. Various issues. *Environment White Paper** (Kankyo Hakusho). Tokyo: Ministry of Finance Printing Bureau.

—. 1977. *A Long Range Plan for Environmental Preservation** (Kankyo Hozen Choki Keikaku). Tokyo: Nihon Kankyo Kyokai.

Food and Agriculture Organization. 1985. *Forest Resources 1980.* Rome.

—. 1993. *FAO Forest Product 1980-1991.* Rome.

Forsund, Finn R. 1985. "Input-Output Models, National Economic Models, and the Environment." Kneeze, Allen V. and Sweeney, James L. eds. *Handbook of National Resources and Energy Economics.* Amsterdam: North-Holland.

Gehrisch, W., Haapalainen, T., Park, Y.M., Stevens, G.H., Tadani, K. 1989. "The Potential Longer Term Contribution of Nuclear Energy in Reducing CO2 Emissions in OECD Countries." *Energy Technologies for Reducing Emissions of Greenhouse Gases.* Proceedings of an Experts' Seminar. Paris: OECD.

Gilbert, A.J. and Braat, L.C. 1991. *Modelling for Population and Sustainable Development.* London: Routledge.

Greyson, Jerome. 1990. *Carbon, Nitrogen, and Sulphur Pollutants and Their Determination in Air and Water.* New York: Marcel Dekker, Inc.

Griffin, James M. and Steele, Henry B. 1896. *Energy Economics and Policy,* Second Edition. Orlando: Academic Press.

Gross, Bertram M. and Michael Springer. 1967. "A New Orientation in American Government." *Annals of the American Academy of Political and Social Science.* May.

— and —. 1967. "New Goals for Social Information." *Annals of the American Academy of Political and Social Science.* Sept.

Harris, DeVerle P. 1993. "Mineral Resource Stocks and Information." Kneese, Allen V. and Sweeney, James L. eds. *Handbook of Natural Resource and Energy Economics.* Vol.III. Amsterdam: Elsevier Science Publishers B.V.

The Institute of Energy Economics. Various issues. *Comprehensive Energy Statistics* (Sogo Enerugi Tokei).

Intergovernmental Panel on Climate Change. 1990. "Policymakers Summary of the Scientific Assessment of Climate Change." Report to IPCC from Working Group 1, prepared by the IPCC Group at the Meteorological Office.

International Energy Agency, Organisation for Economic Co-operation and Development. 1987. *Energy Conservation in IEA Countries.* Paris.

—. 1988. *Emission Controls in Electricity Generation and Industry.* Paris.

—. 1989a. *Energy and the Environment: Policy Overview.* Paris.

—. 1989b. *Energy Technologies for Reducing Emissions of Greenhouse Gases.* Paris.

—. 1989c. *Electricity End-Use Efficiency.* Paris.

—. 1990a. *Energy Statistics of OECD Countries 1987-1988.* Paris.

—. 1990b. *Energy Policies and Programmes of IEA Countries, 1989 Review.* Paris.

—. 1990c. *Substitute Fuels for Road Transport. A Technology Assessment.* Paris.

—. 1990d. *World Energy Statistics and Balances 1985-1988.* Paris.

—. 1991a. *Energy Efficiency and the Environment.* Paris.

—. 1991b. *Greenhouse Gas Emissions: The Energy Dimension.* Paris.

—. 1992a. *The Role of IEA Governments in Energy.* Paris.

—. 1992b. *Energy Policies in IEA Countries. 1991 Review.* Paris.

—. 1993. *World Energy Outlook to the Year 2010.* Paris.

The Japan Association of Industrial Machinery Producers. Various issues. *The Production of Pollution Prevention Equipment** (Kogai Boshi Sochino Seisan Jissekini Tsuite). Tokyo.

Japan Broadcasting Corporation. 1991. *1990 National Time Budget Study** (1990nendo Kokumin Seikatsu Jikan Chosa). Tokyo: Nihon Hoso Shuppan Kyokai.

The Japan Information Center of Science and Technology. 1993. *Corporate R&D in Japan—As Analyzed by JICST Bibliographic Database—.* Tokyo.

Juster, F. Thomas. 1973. ""A Framework for the Measurement of Economic and Social Performance". Milton Moss, ed. *The Measurement of Economic and Social Performance.* National Bureau of Economic Research, Studies in Income and Wealth Volume 38. New York: Columbia University Press.

—; Courant, Paul N.; and Dow, Greg K. 1980. "The Theory and Measurement of Well-being: A Suggested Framework for Accounting and Analysis." Working Paper Series No.8007. Institute of Social Research, The University of Michigan.

—, Courant, P.N., and Dow, G.K. 1981. "A Theoretical Framework for the Measurement of Well-being." *Review of Income and Wealth.* 27, 1-31.

Kagaku Kogyosha. 1973. *Process Flow Diagrams of Pollution Prevention Equipment** (Kogai Boshi Setsubi Koteizushu). Tokyo: Kagaku Kogyosha.

Kanamori, Hisao; Takase, Yuriko; and Uno, Kimio. 1977. *Economic Growth and Welfare—An Estimation of NNW in Japan —** (Keizai Seichoto Fukushi—Nihonno NNWno Suikei). Research Report No.41. Tokyo: The Japan Economic Research Center.

—. 1980. "Japanese Economic Growth and Economic Welfare." Tsuru, Shigeto, ed. *Economic Growth and Resources, Problems Related to Japan.* London: The Macmillan Press.

Kendrick, John W. 1972. "The Treatment of Intangible Resources as Capital." *The Review of Income and Wealth.* March.

Kirschen, E.S. et al. 1964. *Economic Policy in Our Time.* Vol.I General Theory. Amsterdam: North-Holland Publishing Company.

Kneeze, Allen V. and Sweeney, James L. eds. 1985. *Handbook of Natural Resource and Energy Economics.* Amsterdam: North-Holland.

Kosobud, Richard F. 1981. "Civilian Nuclear Energy and Weapons Proliferation: An Analysis of Japanese and U.S. Views." The Japan Seminar.

Land Agency. Various issues. *Land White Paper** (Tochi Hakusho). Tokyo: Ministry of Finance Printing Bureau.

Larson, Eric D. 1992. "Trends in the Consumption of Energy-intensive Basic Materials in the Industrialized Countries and Implications for Developing Regions." International Symposium on Environmentally Sound Energy Technologies and their Transfer to Developing Countries and European Economies in Transition. Milan.

Lashof, Daniel A. and Tirpak, Dennis A., eds. 1990. *Policy Options for Statilizing Global Climate.* New York: Hemisphere Publishing Corporation.

Lecht, Leonard A. 1965. *The Dollar Cost of Our National Goals.* National Planning Association. Washington, D.C.

Lenssen, Nicholas. 1992. "Confronting Nuclear Waste." Brown, Lester R. *State of the World 1992.* New York: W.W. Norton & Company.

Lovins, A.B. 1977. *Soft Energy Paths.* Friends of the Earth Inc.

Malinvaud, Edmond. 1979. "Costs of Economic Growth." Edmond Malinvaud, ed. *Economic Growth and Resources. Vol. 1: The Major Issues.* London and Basingstoke: The Macmillan Press.

Management and Coordination Agency. 1990. *1985 Input Output Tables** (Showa 60nen Sangyo Renkanhyo). Tokyo: Zenkoku Tokei Kyokai Rengokai.

Meadows, D.H. et al. 1972. *The Limits to Growth.* Washington, D.C.: Potomac Associates.

Ministry of Construction. Various issues. *Annual Report on Building Construction Statistics** (Kenchiku Tokei Nenpo). Tokyo: Kensetsu Bukka Chosakai.

Ministry of Health and Welfare. 1993. *Welfare White Paper 1991.* Tokyo: Gyosei.

Ministry of Home Affairs. Various issues. *Administrative Investment Records.* Tokyo: Chiho Zaimu Kyokai.

Ministry of International Trade and Industry. Various years. *Plant and Equipment Investment Plans in Major Industries** (Shuyo Sangyono Setsubi Toshi Keikaku). Tokyo: Ministry of Finance Printing Bureau.

—. Various issues. *Plant and Equipment Investment Plans in Major Industries — Current Status and Issues —** (Shuyo Sangyono Setsubi Toshi Keikaku—Sono Genjoto Kadai). Tokyo: Ministry of Finance Printing Bureau.

—. Various issues. *Census of Manufactures** (Kogyo Tokeihyo). Tokyo: Ministry of Finance Printing Bureau.

—. Various issues. *International Trade White Paper.* Tokyo: Ministry of Finance Printing Bureau.

—. Various issues. *Input-Output Tables* (Extension Tables).* (Sangyo Renkanhyo (Enchohyo). Tokyo: Tsusan Tokei Kyokai.

—. 1991. *Energy New Trends in the Global Age** (Chikyu Jidaino Enerugi Shinchoryu). Tokyo: Denryoku Shinposha.

—. 1992. *Energy '92** (Enerugi '92). Tokyo: Denryoku Shinposha.

Ministry of Transport. Various issues. *Statistical Handbook of Transport Economy* (Unyu Keizai Tokei Yoran). Tokyo: Ministry of Finance Printing Bureau.

Mishan, E.J. 1969. *Growth: The Price We Pay.* London: Staples Press.

Moriguchi, Chikashi, ed. 1987. *Economic Study Concerning Energy** (Enerugini Kansuru Keizaigakuteki Kenkyu). Osaka University Social and Economic Research Institute. (mimeographed).

Moss, Milton, ed. 1973. *The Measurement of Economic and Social Performance.* National Bureau of Economic Research Studies in Income and Wealth, No.38.

New York:Columbia University Press.

Mounfield, Peter R. 1991. *World Nuclear Power.* London: Routledge.

Nilsson, Lars J. 1992. "Energy Efficiency in the Industrial Sector." International Symposium on Environmentally Sound Energy Technologies and their Transfer to Developing Countries and European Economies in Transition. Milan.

Nordhaus, William and Tobin, James. 1972. "Is Growth Obsolete?" National Bureau of Economic Research Fiftieth Anniversary Colloquium Series. *Economic Growth.* New York: Columbia University Press. Also in Moss, M., ed. 1973. *The Measurement of Economic and Social Performance.* National Bureau of Economic Research, Studies in Income and Wealth. Vol.38. New York and London: Columbia University Press.

Nuclear Energy Agency, Organisation for Economic Co-operation and Development. 1985. *The Economics of the Nuclear Fuel Cycle.* Paris.

—. 1986. *Decomissioning of Nuclear Facilities: Feasibility, Needs and Costs.* Paris.

Organisation for Economic Cooperation and Development. 1977. *Environmental Policies in Japan.* Paris.

—. 1982a. *The OECD List of Social Indicators.* Paris.

—. 1982b. *Economic and Ecological Interdependence.* Paris.

—. 1984. *Emission Standards for Major Air Pollutants from Energy Facilities in OECD Member Countries.* Paris.

—. 1985a. *Environmental Policy and Technical Change.* Paris.

—. 1985b. *Environmental Effects of Electricity Generation.* Paris.

—. 1986. *Living Conditions in OECD Countries, A Compendium of Social Indicators.* Paris.

—. 1987. *Energy and Cleaner Air: Costs of Reducing Emissions.* Paris.

—. 1988a. *Environmental Impacts of Renewable Energy.* Paris.

—. 1988b. *Transport and the Environment.* Paris.

—. 1989. *Economic Instruments for Environmental Protection.* Paris.

—. 1991a. *The State of the Environment.* Paris.

—. 1991b. *OECD Environmental Data, Compendium 1991.* Paris.

—. 1992. *Science Responds to Environmental Threats, Synthesis Report.* Paris.

—. 1993. *OECD Environmental Data, Compendium 1993.* Paris.

Pasqualetti, Martin J. 1990. *Nuclear Decommissioning and Society, Public Links to a New Technology.* London: Routledge.

Pearce, David W. and Warford, Jeremy J. 1993. *World without End; Economics,*

Environment, and Sustainable Development. The International Bank for Reconstruction and Development. New York: Oxford University Press.

Peskin, Henry M. and Lutz, Ernst. 1990. *A Survey of Resource and Environmental Accounting in Industrialized Countries.* Environment Working Paper No.37. Washington, D.C.: The World Bank.

Pronk, Jan and Haq, Mahbubul. 1992. *Sustainable Development, From Concept to Action.* Geneva: United Nations Conference on Environment and Development

Repetto, Robert; Magrath, William; Wells, Michael; Beer, Christine; and Rossini, Fabrizio. 1989. *Wasting Assets: Natural Resources in the National Income Accounts.* Washington, D.C.: World Resources Institute.

——, et al. 1991. *Accounts Overdue: Natural Resource Depreciation in Costa Rica.* Washington, D.C.: World Resources Institute.

Resources and Energy Agency. Various issues. *Comprehensive Energy Statistics** (Sogo Enerugi Tokei). Tokyo: Tsusho Sangyo Kenkyusha.

——. 1974. *Collected Materials on Mineral Resources** (Kinzoku Kobutsu Kankei Shiryo). Tokyo.

——. 1974. *Reference Materials on the Overseas Mineral Resources Development** (Kaigai Kobutsu Shigen Kaihatsu Kankei Sanko Shiryo). Tokyo.

——. 1990. *Energy Conservation Handbook** (Sho Enerugi Binran). Tokyo: Energy Conservation Center.

——. 1991. *Earth-Friendly Energy System—All about Petroleum Cogeneration** (Chikyuni Yasashii Enerugi Shisutemu—Sekiyu Kojenereshonno Subete—). Tokyo: Tsusho Shiryo Chosakai.

Ryan, John C. 1992. "Conserving Biological Diversity." Brown, Lester R. et al. *State of the World 1992.* New York: W.W. Norton & Co.

Sametz, A. W. 1968. "Production of Goods and Services." Sheldon, Eleanor Bernert and Moore, Wilbert E., eds. *Indicators of Social Change, Concepts and Measurements.* New York: Russel Sage Foundation.

Sasaki, Nobuhiko. 1989. "Contribution of Energy Conservation and Alternative Energy Technologies to the Problems of Global Environment." *Energy Technologies for Reducing Emissions of Greenhouse Gases.* Paris: International Energy Agency.

Schmidheiny, Stephan. 1992. *Changing Course, A Global Business Perspective on Development and the Environment.* Cambridge, Massachusetts: The MIT Press.

Science and Technology Agency. Various issues. *Science and Technology White Paper** (Kagaku Gijutsu Hakusho). Tokyo: Ministry of Finance Printing Bureau.

—. Various issues. *Indicators of Science and Technology* (Kagaku Gijuts Yoran). Tokyo: Ministry of Finance Printing Bureau.

Simonis, Udo E. 1992. "Sustainable Development - Will the Technology Option Work?" Diets, F.J.; Simonis, U.E.; and van der Straaten, J. eds. *Sustainability and Environmental Policy.* Berlin: Edition Sigma.

Skjoeldebrand, Robert, and Avenhaus, Rudolf. 1973. "Materials Accountability and Pollution Control." *Proceedings of IIASA Planning Conference on Energy Systems.* Laxenburg, Austria: International Instutute for Applied Systems Analysis.

Sprenger, Rolf-Ulrich. 1989. "Economic Incentives in Environmental Policies: The Case of Western Germany." Prince Bertil Symposium on Economic Instruments in National and International Environmental Protection Policies. Stockholm.

Steinmetz, K., Maisonnier, Ch., and Darvas, J. 1989. "Thermonuclear Fusion: A Potential Energy Source for the Next Century." *Energy Technologies for Reducing Emissions of Greenhouse Gases.* Proceedings of an Experts' Seminar. Paris: OECD.

Teitelbaum, Michael S. and Winter, Jay M., eds. 1989. *Population and Resources in Western Intellectual Traditions.* Cambridge: Cambridge University Press.

Tietenberg, T.H. 1989. "Designing Marketable Emission Permit Systems: Lessons from the U.S. Experience." Prince Bertil Symposium on Economic Instruments in National and International Environmental Protection Policies. Stockholm.

Tominaga, Hiroo, ed. 1989. *Handbook on Resources** (Shigen Handbook). Tōkyo: Maruzen.

Turner, Kerry, ed. 1993. *Sustainable Environmental Economics and Management, Principles and Practice.* London: Belhaven Press.

United Nations. Various issues. *Energy Statistics Yearbook.* New York.

—. Various issues. *Statistical Yearbook.* New York.

—. Various issues. *Yearbook of International Trade Statistics.* New York.

—. Various issues. *World Energy Supplies.* New York.

—. 1969. *A System of National Accounts.* ST/STAT/SER.F/2/Rev.3. New York.

——. 1975. *Towards a System of Social and Demographic Statistics.* Series F, No.18. New York.

——. 1977. *The Feasibility of Welfare-Oriented Measures to Supplement the National Accounts and Balances: A Technical Report.* Series F, No.22. New York.

——. 1978. *Social Indicators: Preliminary Guidelines and Illustrative Series.* Series M. No.63. New York.

——. 1979. *Studies in the Integration of Social Statistics: Technical Report.* Series F. No.24. New York.

——. 1980. *Classification of the Functions of Government.* Series M, No.70. New York.

——. 1988. *Concepts and Methods of Environment Statistics: Human Settlements Statistics—A Technical Report.* ST/ESA/STAT/SER.F/51. New York.

——. 1989. *Concepts and Methods of Environmental Statistics: Statistics of the Natural Environment—A Technical Report.* Series F. New York.

——. 1990. *SNA Handbook on Integrated Environmental and Economic Accounting.* Preliminary Draft on the Part 1: General Concept. New York.

——. 1992. *Agenda 21: The United Nations Programme of Action from Rio.* New York.

United Nations Department of Economic and Social Development. 1992. *Handbook of National Accounting, Integrated Environmental and Economic Accounting.* New York.

——. 1993a. *Integrated Environmental and Economic Accounting.* ST/ESA/STAT/SER.F/61. New York.

——. 1993b. *System of National Accounts.* ST/ESA/STAT/SER.F/2/Rev.4. New York.

United Nations University. 1993. *Environmental Pollution Control, the Japanese Experience.* Tokyo.

Uno, Kimio. 1974. "Social Indicators and Macroeconomic Framework (Shakai Shihyo—Hattenno Keifuto Makuro Shakai Shihyo Moderu)." *Journal of Japan Economic Research.* No.3.

——. 1978. "A Social Indicators Approach to Economic Development." Tsuru, Shigeto, ed. *Growth and Resources Problems Related to Japan.* Tokyo: Asahi Evening News. Republished in 1980 in London and Basingstoke: The Macmillan Press.

——. 1985a. "Social, Economic, and Environmental Statistics—Accounting

Framework for Policy Analysis—." Paper prepared for the 19th General Conference of the IARIW, Noordwijkerhout, the Netherlands.

—. 1985b. "Resource Allocation by Social Purpose and the Input-Output Framework." Paper prepared for the International Meeting on Problems of Compilation of IO-Tables, Baden, Austria.

—. 1987. *Japanese Industrial Performance.* Amsterdam: Elsevier Science Publishers.

—. 1989a. *Measurement of Services in an Input-Output Framework.* Amsterdam: Elsevier Science Publishers.

—. 1989b. "Economic Growth and Environmental Change in Japan: Net National Welfare and Beyond." F. Archibugi and P. Nijkamp, eds. *Economy and Ecology: Towards Sustainable Development.* Dordrecht: Kluwer Academic Publishers.

—. 1990a. "National Accounting and Environment." Paper prepared for the United Nations University/WIDER Project on the Environment and the Emerging Development Issues. Helsinki.

—. 1990b. *Demography and Labor Supply Model.* Statistical Data Bank Project Report No.66. Tsukuba: The University of Tsukuba.

—. 1990c. "Annual Input-Output Tables in Japan, 1975-1985." Statistical Data Bank Project Report No.63. Institute of Socio-Economic Planning, the University of Tsukuba.

—. 1991a. "Environmental Quality and Technology: Lifestyle Options." United Nations Economic and Social Commission for Asia and the Pacific. *Energy Policy Implications of the Climatic Effects of Fossil Fuel Use in the Asia-Pacific Region.* Bangkok.

—. 1991b. "Quality-of-Life and Environmental Accounting: Assessment of Pollution Prevention Investment." Paper submitted to Special IARIW (International Association for Research in Income and Wealth) Conference on Environmental Accounting. Baden, Austria.

—. 1991c. "Produce-Consume-and-Recycle: Operationalizing the Concept of Sustainability." Paper submitted to the United Nations Conference on Environment and Development. Geneva.

—. 1991d. "Diffusion of Environment-related New Technology and Policy Implications." Paper submitted to the International Symposium on Environmentally Sound Energy Technologies and Their Transfer to Developing Economies and European Economies in Transition. Milan.

——. 1991e. *Technology, Investment, and Trade.* New York: Elsevier Science Publishing Co.

——. 1993. "Policy Formulation and Information Needs." Paper presented at the International UNU Symposium on Ecorestructuring held at the United Nations University. Tokyo.

——. "Social, Economic, and Environmental Data Set (SEEDS)." 1994. Paper presented at the National Accounts and the Environment Meeting. London.

van Velzen, Daniel, ed. 1991. *Sulphur Dioxide and Nitrogen Oxides in Industrial Waste Gases: Emission, Legislation, and Abatement.* Dordrecht: Kluwer Academic Publishers.

Vesilind, P. Aarne and Peirce, J. Jeffrey. 1983. *Environmental Pollution and Control.* Second Edition. Boston: Butterworths.

Winteringham, F. Peter W., ed. 1992. *Energy Use and the Environment.* Boca Raton: Lewis Publishers.

World Commission on Environment and Development. 1987. *Our Common Future.* Oxford: Oxford University Press.

Wyman, Richard L., ed. 1991. *Global Climate Change and Life on Earth.* New York: Chapman and Hall.

Ygdrassil, Alexis; Gelobter, Michael; Holnicki, Piotr; Anderberg, Stafan; and Schlenzig, Christoph. 1989. "A Critical Review of Energy Projections for the Study of Long-term, Large-scale Interactions between Development and Environment." Toth, Ferenc L.; Hizsnyik, Eva; and Clark, William C., eds. *Scenarios of Socioeconomic Development for Studies of Global Environmental Change: A Critical Review.* RR-89-4. Laxenburg: International Institute for Applied Systems Analysis.

Young, John E. 1992. "Mining the Earth." Brown, Lester R. et al. *State of the World 1992.* New York: W.W. Norton & Co.

Index

Economy & Environment

1. F. Archibugi and P. Nijkamp (eds.): *Economy and Ecology: Towards Sustainable Development.* 1989 ISBN 0-7923-0477-2
2. J. Bojö, K.-G. Mäler and L. Unemo: *Environment and Development: An Economic Approach.* 1990 ISBN 0-7923-0802-6
3. J. B. Opschoor and D. W. Pearce (eds.): *Persistent Pollutants: Economics and Policy.* 1991 ISBN 0-7923-1168-X
4. D.J. Kraan and R. J. in 't Veld (eds.): *Environmental Protection: Public or Private Choice.* 1991 ISBN 0-7923-1333-X
5. J.J. Krabbe and W.J.M. Heijman (eds.): *National Income and Nature: Externalities, Growth and Steady State.* 1992 ISBN 0-7923-1529-4
6. J. Bojö, K.-G. Mäler and L. Unemo: *Environment and Development: An Economic Approach* (revised edition). 1992 ISBN 0-7923-1878-1
7. T. Sterner (ed.): *Economic Policies for Sustainable Development.* 1994 ISBN 0-7923-2680-6
8. L. Bergman and D.M. Pugh (eds.): *Environmental Toxicology, Economics and Institutions.* The Atrazine Case Study. 1994 ISBN 0-7923-2986-4
9. G. Klaassen and F.R. Førsund (eds.): *Economic Instruments for Air Pollution Control.* 1994 ISBN 0-7923-3151-6
10. K. Uno: *Environmental Options: Accounting for Sustainability.* 1995 ISBN 0-7923-3513-9

KLUWER ACADEMIC PUBLISHERS – DORDRECHT / BOSTON / LONDON